— *Illustrated Lecture Notes* —

GENERAL PATHOLOGY

Illustrated Lecture Notes

GENERAL PATHOLOGY

J J Rippey

ILLUSTRATED BY

Elizabeth Rippey

WUP WITWATERSRAND UNIVERSITY PRESS

Witwatersrand University Press
1 Jan Smuts Avenue
Johannesburg
2001 South Africa

Revised edition

First published in 1981 by Juta & Co, Ltd.

Revised edition published by Witwatersrand University Press, 1994
Reprinted 1999, 2006, 2010, 2012

ISBN 978 1 86814 240 8

Cover design by Sarah Palmer of Desktop Impressions

Printed and bound by Creda Communications

FOREWORD

This book first saw the light of day in 1981 after a long gestation period of seven or eight years during which I had been lecturing on the subject to third year medical students. I returned to South Africa in 1993 as a visiting professor to give the course on general pathology, and was gratified to find that the book was still in use by both medical students and postgraduate students who were required to know the elements of general pathology. In the light of recent advances in the knowledge of pathology I therefore took the oppurtunity to produce a revised edition.

As for the purpose of the book, I cannot do better than quote from the original foreword:

'General pathology covers the study of pathological or disease processes in general. The body's reactions to the causes of disease are limited and thus a particular type of change, such as inflammation, has many common features regardless of anatomical location. General pathology provides the basis for the study of medical and surgical pathology of special tissues and organs.

Most modern textbooks give good descriptions of general pathology, but often at considerable length. In some the emphasis seems to be concentrated on the molecular biological features of disease, others emphasize experimental pathology.

I have prepared these lecture notes to cover the essential features of general pathology, placing most weight on those changes which can be seen, either with the naked eye or with the microscope. The emphasis is thus upon morphological pathology.'

As before my wife's line drawings help immeasurably in the understanding of the processes which have been described in words. They crystallise and preserve the underlying concept of simplicity which we have striven for throughout. Thanks are also due to my trusty laptop computer into which I have keyed the whole of the revised text.

It is my earnest hope that the book, now published by the Witwatersrand University Press, will continue to be of use for some years to come to all those who need a basic knowledge of general pathology.

Perth
Western Australia

J J Rippey
December 1993

I append the names of some recent standard textbooks on pathology and general pathology to which I have referred constantly and which are recommended to those who seek more detailed information or references to original material.

Cotran RS, Kumar V, Robbins SL. *Robbin's Pathologic Basis of Disease.* 4th ed. Philadelphia: WB Saunders, 1989.

Kumar V, Cotran RS, Robbins SL. *Basic Pathology.* 5th ed. Philadelphia: WB Saunders, 1992.

MacSween RNM, Whayley K. *Muir's Textbook of Pathology.* 13th ed. London: Edward Arnold, 1992.

O'Gee JO'D, Isaacson PG, Wright NA. *Oxford Textbook of Pathology.* Oxford: Oxford University Press, 1992.

Rubin E, Farber JL. *Pathology.* Philadelphia: Lipincott, 1988.

Walter JB, Israel MS. *General Pathology.* 6th ed. Edinburgh: Churchill Livingstone, 1987.

TABLE OF CONTENTS

Chapter		Page
1	Introduction to pathology and the causation of disease	1
2	The normal cell	7
3	Cell injury	19
4	Cell death and necrosis: Gangrene	27
5	Amyloid	37
6	Calcification	45
7	Pigmentation	51
8	Jaundice	57
9	Fluid and electrolyte balance	65
10	Oedema	75
11	Hyperaemia and congestion	83
12	Haemorrhage	89
13	Shock	93
14	Thrombosis and embolism	103
15	Ischaemia and infarction	115
16	Acute inflammation	125
17	The classification of inflammation	141
18	Chronic inflammation	147
19	Healing and repair	155
20	Infection and some infectious diseases	165
21	Tuberculosis	175
22	Other infectious and granulomatous diseases	185
23	Disorders of growth	195
24	Tumours and their classification	205
25	The characteristics of malignancy	215
26	The spread of malignant tumours	225
27	Carcinogenesis	235
28	Some tumours of epithelial tissue: Skin tumours	247
29	Some tumours of connective tissue	259
30	Some tumours of lymphoid tissue and bone marrow	271
31	Some tumours of childhood	281
32	The effects of tumours	289
33	Tissue response to ionising radiation	295
34	Disorders of carbohydrate metabolism: Diabetes mellitus	303
35	Some inherited and metabolic disorders	313
36	Nutritional disorders	321
37	Transplant rejection and localised autoimmune disease	329
38	Generalised autoimmune disease	337

GENERAL PATHOLOGY - CHAPTER ONE

INTRODUCTION TO PATHOLOGY AND THE CAUSATION OF DISEASE

To understand pathology it is necessary to learn a new language. The words of this new language each have a specific meaning. By learning the definitions of these words it is then possible to use them to describe a pathological process briefly and precisely. To use the language without knowing the exact meaning and definition of the terms used can lead to confusion and cause a breakdown in communication.

Lesson Number One then is to learn the definitions of the terms used in pathology at an early stage. This lesson will continue throughout our study of **general pathology**.

What is the definition of pathology?

Pathology is the scientific study of disease and disease processes

Pathology used in its widest sense includes a number of sub-disciplines which are brought into play in the scientific study of disease. Some of these are:

- **Anatomical Pathology** which includes **morbid anatomy** or post-mortem examination, **surgical pathology** or the study of tissues removed at surgical operation; **cytology** is the study of individual cells obtained from a patient.
- **Chemical Pathology** which is concerned with the study of the chemistry of body fluids and tissues.
- **Haematology** or the study of the blood and its properties.
- **Immunology** which is concerned with the study of immunity and hypersensitivity and the formation of antibodies.
- **Microbiology** which includes **bacteriology**, the study of bacteria which cause disease; **virology**, the study of viruses; **mycology**, the study of fungi; and **parasitology**, the study of parasites.

It is unfortunate that these branches of pathology have become separated from each other, when there are close links between them. It is a mistake to compartmentalise what is learnt in bacteriology, chemical pathology, haematology, or even general pathology, for all may be applied to one and the same patient whose disease state is being studied.

Each branch of pathology should not be isolated, nor should the whole study be separated from other areas of medicine, for pathologists work together with physicians and surgeons and all other branches of the medical profession for the benefit of the patient. Pathology is a bridge between the basic sciences and the clinical subjects.

The pathology covered in this book is more of a basic science, comparable with anatomy and physiology, which provides a background of knowledge of disease processes in general on which a study of special systematised pathology can be built. In fact, the study of pathology underpins the whole of clinical medicine and surgery and a good knowledge of pathology will be most valuable when you undertake clinical studies. Learn pathology well now and it will stand you in good stead in the future.

The basis of this scientific study that is pathology is observation, and most of our knowledge of the changes found in disease is founded first and foremost upon observation. Anatomical Pathology is a descriptive science and many pathological diagnoses are based solely upon visual appearances of one sort or another. Following this, knowledge may be gained about the causes of disease and the alterations in function which accompany the changes that may be seen in the cells, tissues and organs of the body. It is at this stage that the scientific method of formulating a hypothesis and then conducting an experiment to prove or refute that hypothesis can be brought into operation. This is known as **experimental pathology.**

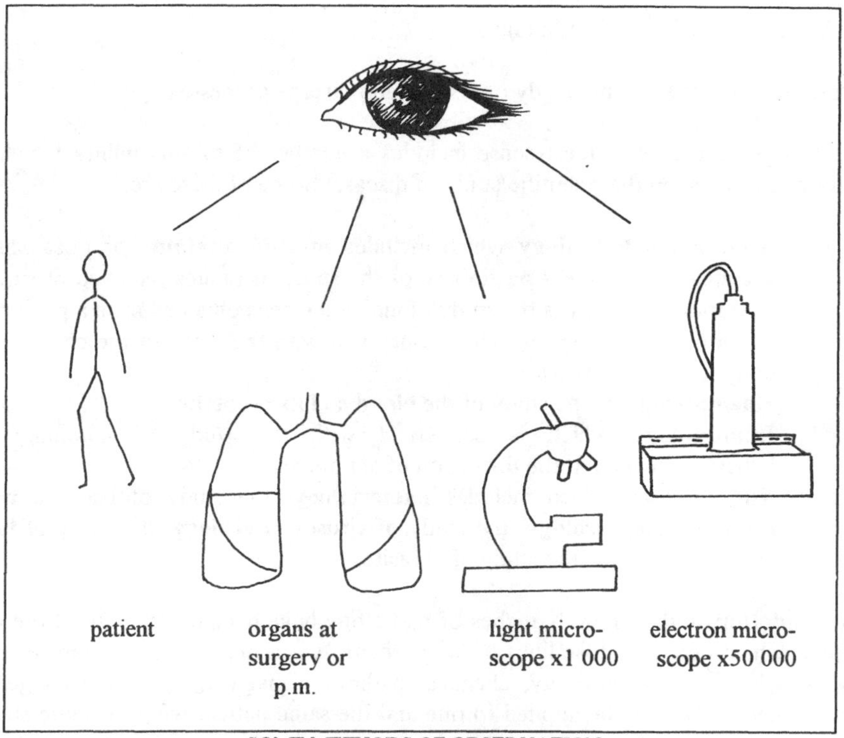

SOME METHODS OF OBSERVATION

The diagram illustrates some of the ways in which observations can be made. It should be noted that our views of the patient are limited in space and time.

For example, a microscopic section, which can be examined at very high magnifications, is in only two dimensions, and represents only a very small portion of tissue. Thus when selecting tissues for examination in this way the possibility of sampling error exists.

A disease is not a static entity but a dynamic, continuously changing series of conditions. For example it cannot be stated that the pathology of any particular disease is such and such. It is necessary to know the changes that occur at every stage of the disease process; from this knowledge a diagnosis can be made and from knowing what has gone before, the future can be predicted and a prognosis given. So when a specimen from a sick patient is examined it should be remembered that what is seen represents only one frame from a moving picture which covers all stages of the disease.

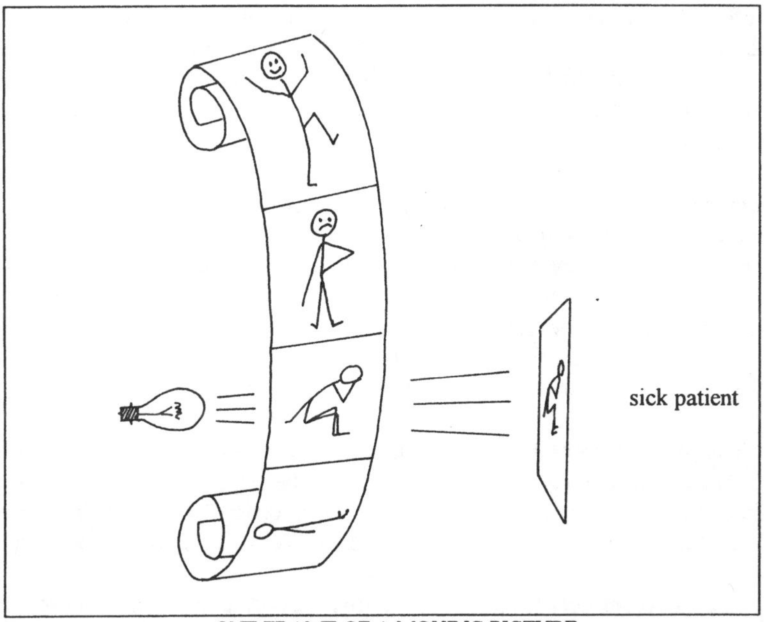

ONE FRAME OF A MOVING PICTURE

Some of these terms can now be defined:

Disease is a condition of the organism which limits its life in either its powers, enjoyment or duration. (World Health Organisation)

This is a very wide definition. For example, medical students may limit their teacher's enjoyment of life, particularly at examination times! Can this be taken to mean that medical students are a disease? I rather think not, although they could perhaps be regarded as an aetiological factor in the development of stress-induced disease. However this all-embracing definition does help us to understand the very wide range of conditions that can fall under the heading of **dis-ease** and which pathologists may be called upon to study in as scientific a manner as possible.

Diagnosis is the identification of a disease process.

Prognosis is a forecast of the course of a disease viewed in the light of the patient's special circumstances.

Two more new terms may be introduced here:

Aetiology is the cause or origin of a disease.

The aetiology involves not only the direct cause but also associated factors which predispose to the development of disease. Thus the tubercle bacillus is the cause of tuberculosis; but tuberculosis occurs most frequently in poor people who are undernourished and who live in crowded conditions. In this day and age immune deficiency also is a predisposing factor and tuberculosis commonly occurs in those who suffer from AIDS. The aetiology also includes the means by which the patient becomes exposed to the causative agent; in the case of tuberculosis the bacillus is often spread by infected persons coughing up sputum which is loaded with organisms. Bacilli survive in the dried state and may be inhaled by other family members or those living in the same house.

Pathogenesis is the way in which a pathological condition is produced, that is it is the mechanism of development of a disease state.

Thus to use the previous example we know that the tubercle bacillus causes tuberculosis. The pathogenesis of tuberculosis describes *how* this happens, how the bacillus enters the body, lodges in the lungs and causes a special type of inflammation which may result in cavity formation at the apex of the lung.

There are three steps in the diagnosis of disease:

- The **physiological** diagnosis of disease, where disorder of function is noticed. The patient complains of symptoms and the physiological stage of diagnosis can be summarised by the answer to the question 'What?'. This is usually the first question the doctor asks the patient, 'What is troubling you?' or 'What is the matter?'.
- The **anatomical** localisation of the abnormality comes next. From a knowledge of anatomy and physiology the doctor should be able to localise the symptoms to find the seat of the disease, that is the answer to the question 'Where?'.
- The **pathological** diagnosis determines the actual nature of the disease and provides the answer to the questions 'How?' and 'Why?'.

THE DIAGNOSIS OF DISEASE

PHYSIOLOGICAL	FUNCTION DISTURBED	PATIENT COMPLAINS OF SYMPTOMS	**WHAT** IS THE TROUBLE?
ANATOMICAL	STRUCTURE ALTERED	DOCTOR FINDS SIGNS	**WHERE** IS THE TROUBLE?
PATHOLOGICAL	DISEASE PROCESS IDENTIFIED	PATHOLOGICAL EXAMINATION	**WHY** AND **HOW**?

At this stage it is worth while considering the classification of disease into various types or groups according to aetiology. The following classification can be applied to various organs and systems.

CLASSIFICATION OF DISEASE

- **Congenital Diseases**. A 'congenital' condition is one which is present at birth. There are two forms, the inherited and those which are acquired during intra-uterine life. In the congenital diseases the inherited ones are transmitted by the genes, and although the tendency to the disease is, of course, present at birth the actual manifestations of the disease may not occur until adult life.
- **Acquired Diseases**. These form a much larger group:-

  - **Injury** is a cause of disease according to the definition. A car accident, acid burn, injury or damage to the tissues by some physical agent such as heat, cold or X-rays are all diseases.
  - **Intoxication** causes disease. The medical use of the term is not restricted to changes caused by excessive alcohol intake but includes the effects of other drugs and chemicals on the body.
  - **Infection** by micro organisms such as bacteria, viruses or fungi

Thus **I**njury, **I**ntoxication and **I**nfection, the three **I**'s, are the first three groups of acquired disease.

  - **Deficiency diseases** form another group. Here it is the absence of an essential substance that is the cause of disease rather than the presence of an identifiable causative agent. The first such disease to be described was scurvy, due to lack of Vitamin C. This disease used to occur in sailors who spent long periods at sea living on preserved foods and totally deprived of fresh fruit and vegetables.
  - **Neoplastic diseases** are characterised by the overgrowth of tissues locally causing tumours. Cancer is a word used to describe a malignant tumour which, if untreated, can spread throughout the body and kill the sufferer. The word neoplasia means new growth.
  - **Degenerative diseases** can affect many parts of the body but particularly the arteries (for example, hardening of the arteries or arteriosclerosis), and the joints (for example, degenerative joint disease or osteoarthritis).
  - **Metabolic diseases** such as diabetes or gout occur when there is a disturbance of the body's normal metabolic processes.
  - **Endocrine diseases** are characterised by the overactivity (hyperfunction) or underactivity (hypofunction) of the endocrine glands of the body. Overactivity of the thyroid gland or thyrotoxicosis is one of the commonest examples of this.
  - **Immune diseases** affect the immune system of the body which responds to foreign substances or antigens and makes protective antibodies.
    This process may go wrong with either an excessive response to antigens as in hypersensitivity or the production of antibodies to the body's own tissues in what is called auto-immune disease.

Immune deficiency may occur as a result of certain types of drug treatment, such as chemotherapy, or in certain disease states such as AIDS.

Psychiatric diseases or mental diseases are not generally accompanied by observable pathological changes and will not be considered further at this stage.

What do pathologists do? Doctors have been carrying out post-mortem examinations since the time of Hippocrates to try and demonstrate anatomical changes in relationship to disease. In modern times the autopsy (post mortem examination or necropsy) is still an important diagnostic tool. However it was only during the last century, with the invention of the microscope, that more detailed observations became possible. Virchow (1821 -1905) has been described as 'the father of cellular pathology' and virtually founded the science of histopathology or tissue pathology as we know it today. At the same time Pasteur was also using the microscope to determine the nature of some infectious diseases and to establish the science of bacteriology.

Examination of the body or tissues with the unaided eye or gross examination is supplemented by microscopic examination. If tissues taken from the body are fixed in formalin or any other fixative, dehydrated and then embedded in paraffin wax, very thin (5μ) sections of tissue can be cut and stained with haematoxylin (which stains nuclei blue) and eosin (which stains cytoplasm red). These sections can then be viewed by the light microscope which permits magnification up to 1,000x or 1,500x. Most pathological diagnoses can be made using this technique, perhaps with the additional use of special stains to demonstrate specific substances or structures.

Nowadays specific or monoclonal antibodies can be labelled with an immunofluorescent dye or with an enzyme. This technique allows for the recognition of special antigens or cell markers in or on cells.

In some cases, particularly in the diagnosis of unusual or poorly differentiated tumours or to aid in the diagnosis of renal biopsies or myocardial biopsies, the electron microscope may be used. This permits magnifications in the order of 50,000x and enables pathologists to identify very small structures within cells and tissues.

GENERAL PATHOLOGY - CHAPTER TWO

THE NORMAL CELL

Before going on to consider cellular pathology and the abnormalities of cellular structure and function related to disease, it may be worth while to review some features of the normal cell.

The cell can be regarded as the basic building unit of the body. Cells are grouped together to form tissues and organs, which are collections of like cells grouped together to fulfil special functions. Organs and the intervening tissues, together with the extracellular matrix they have made, form the substance of the body, which contains about 10^{13} cells.

DIFFERENTIATION OF CELLS

There are many different types of cells of all degrees of specialisation or differentiation, but they all have certain features in common. Before describing the typical cell it is necessary to describe some of the major categories which are of importance in the study of pathology. The following definitions may not be strictly anatomically correct, but they are in constant and everyday use by pathologists.

- **Epithelial cells** have two main functions. They cover surfaces and they form secretions.

  - The skin, respiratory tract and alimentary tract, for example, are covered or lined by various types of epithelial cells which form a protective layer between the internal structures of the body and the outside.

    In some instances, for example the alimentary tract, these covering cells include a number of secretory cells.
  - Secretory types of epithelial cells may be arranged to form glands of varying degrees of complexity; these glands may be exocrine or endocrine. In addition, major organs which may have secretory functions, such as the liver and the kidney, are composed of epithelial cells.

  A feature that epithelial cells have in common is that they tend to be very closely packed together. In fact, it has been calculated that there are only fifteeen to twenty nanometres between them (1 metre = 10^9 nm), that is just about twice the thickness of a cell membrane.

  Epithelial cells are often arranged in a cohesive sheet with an underlying basement membrane

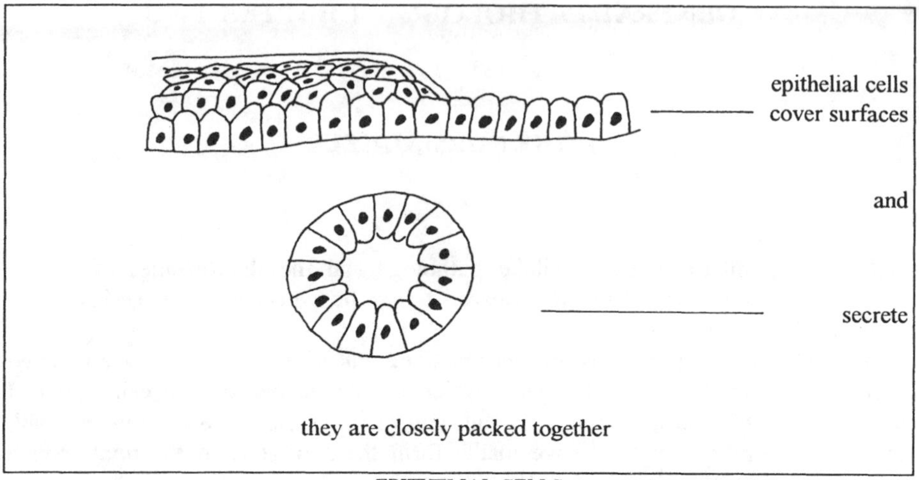

EPITHELIAL CELLS

- **Connective tissue cells** form another major group and include many different types of cells. The connective tissue cells differ in that they are set in a matrix of ground substance containing collagen and they are not closely packed together. Types of connective tissue cells vary enormously according to the extracellular matrix that they produce. Tendons, bone and cartilage are included in the connective tissues. Also sometimes included in the overall group are cells with a more specialised function such as the muscle fibres (which are more closely packed together and adhere to each other). In these cells the cytoplasm is differentiated to permit contraction. In nerve cells the cytoplasm is differentiated to conduct impulses.

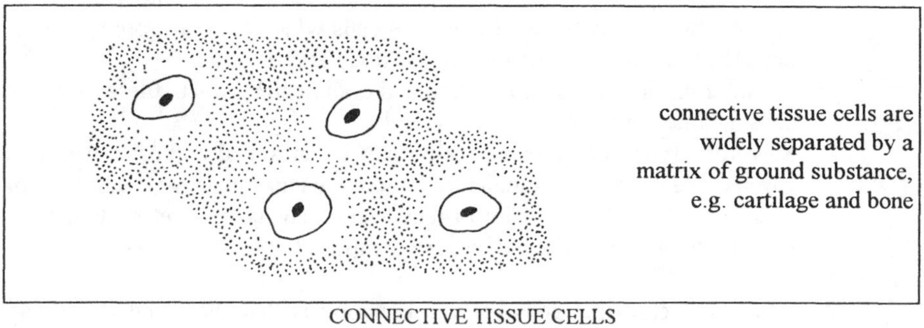

CONNECTIVE TISSUE CELLS

There are also:

- **Endothelial cells** which line the blood and lymphatic vessels.
- **Mesothelial cells** which line the serous cavities of the body (the pleura, pericardium and peritoneum).
- **Other cells,** of the blood and lymphatic system and of the central nervous system for example, may be categorised separately.

The differences between the types of cells, particularly epithelial cells, are important in the study of tumours which may vary according to their tissue of origin.

STRUCTURE OF CELLS

The diagram shows the light microscopic appearance of typical epithelial cells.

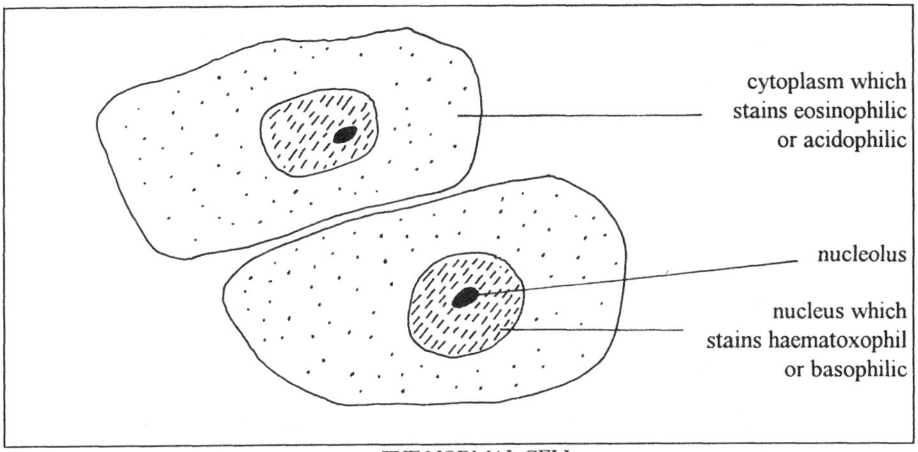

THE NORMAL CELL

Most cells have a similar basic architecture consisting of a cell membrane which surrounds the cytoplasm. Contained within the cytoplasm there is a nucleus also surrounded by a membrane. Within the cytoplasm there is a variety of small structures or organelles, which are best seen and appreciated under the electron microscope (EM).

THE NUCLEUS

This is the central data store of the cell. It is enclosed by a double membrane perforated by holes or nuclear pores. The outer part of the membrane bears ribosomes on the surface and is continuous with the endoplasmic reticulum. The inner portion is supported by filamentous polypeptides (lamins) on the inner surface forming the nuclear lamina.

The nucleus contains 46 chromosomes, consisting of 22 pairs of autosomes and one pair of sex chromosomes. The chromosomes, only seen when the cell divides by mitosis or meiosis, are specifically shaped condensations of deoxyribonucleic acid (DNA) in the form of intensely coiled chains of nucleosomes. Each chromosome is one molecule of DNA about 5μm in length. They are closely packed in association with histones to form nucleosomes which are the basis of the electron dense, deeply basophilic staining heterochromatin. Where the DNA is less closely packed it is less electron dense and more lightly staining with haematoxylin and is known as euchromatin.

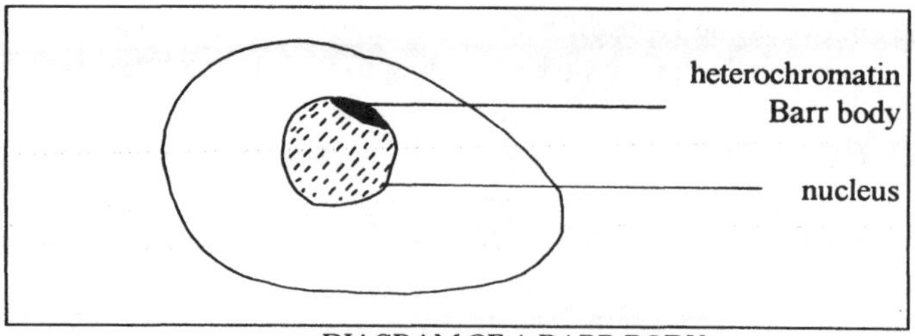

DIAGRAM OF A BARR BODY

The inactivated X chromosome in female cells which rests against the inner surface of the nuclear membrane is formed of condensed DNA or heterochromatin. This is called a Barr Body.

The double helix of the DNA molecule is well known. The chains which form the helix are composed of sugar-phosphate complexes arranged in linear fashion and joined in pairs by purine (adenine and guanine) and pyrimidine (thymine and cytosine) bases. The order of the bases in the DNA forms the genetic code. A sequence of three bases, a base triplet or codon, codes for one amino acid. The actual manufacture of amino acids, polypeptides and proteins takes place in the ribosomes in the cytoplasm. The code sequence is passed from the nuclear DNA to the ribosomes by messenger ribonucleic acid (mRNA), which has transcribed the code from a portion of the DNA molecule known as a gene. Ribosomes pass along the mRNA adding amino acids as coded to form a polypeptide. Not all of the nuclear DNA actually codes mRNA. That part of the gene which does is known as the exon.

It is worth noting that reverse transcription can occur, for example an RNA tumour virus has DNA polymerase activity that can bring about the formation of specific nuclear DNA.

In any individual cell only a few genes seem to be working at any one time. It is postulated that regulator genes form repressor substances which do not allow the operator gene, which can code for polypeptides, to function.

DNA also has the property of being able to replicate itself by copying the genetic code from the template and manufacturing another molecule, (a process termed semi-conservative replication). DNA synthesis and replication is a necessary precursor to nuclear and cell division.

The nucleolus within the nucleus is the site of formation of ribosomal RNA. On electron microscopy it can be seen to have both granular and fibrillar areas. Pale fibrillary areas (which are argyrophilic, that is stain positively with a silver stain) are known as nucleolar organising regions (NORs) and are prominent in dividing cells and in the cells of malignant tumours.

THE CELL MEMBRANE

The cell membrane or outer covering of the cell forms the interface between cell and cell or between cell and the tissue fluids. It appears on electron microscopy as a laminated structure with two dark electron dense layers surrounding a lighter central electron lucent area.

It is 7.5nm in thickness. It is composed of lipid material to which and through which protein molecules are fixed. Carbohydrate polysaccharide chains attached to protein molecules on the outer side of the cell membrane form the glycocalyx.

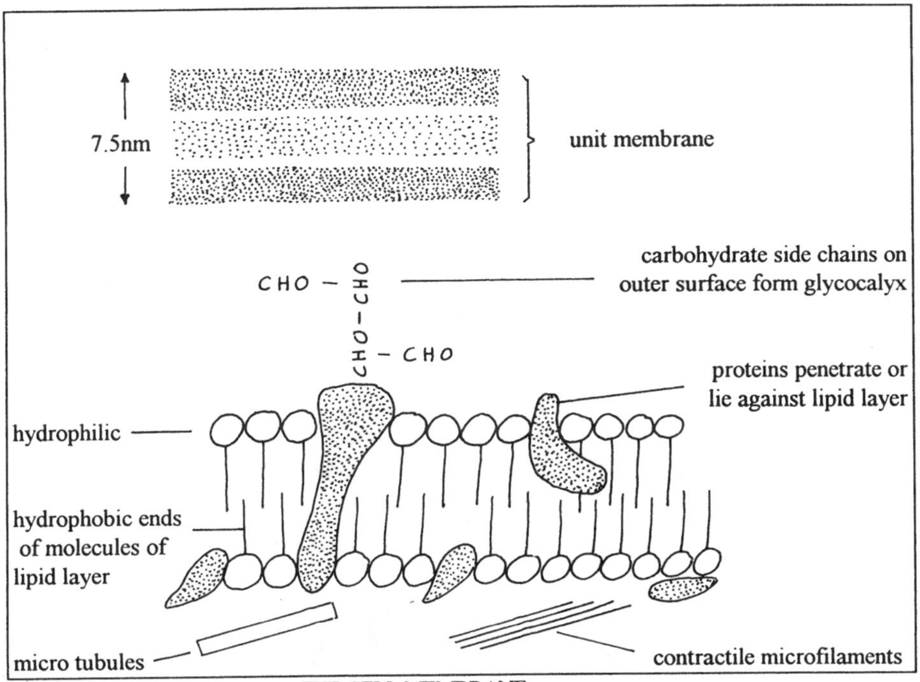

THE CELL MEMBRANE

The cell membrane may vary in shape, with microvilli being formed at absorptive surfaces, for example, in renal tubular cells or small intestinal cells.

The outer part of the cell membrane plays a part in cell recognition and cell adhesion as well as being the site of surface receptors. The cytoskeleton of microtubules and contractile microfilaments is related to the internal surface of the cell membrane both anatomically and functionally in respect of membrane movements..

Functions of the cell membrane

- **Transport**: The cell membrane maintains the cell's integrity and individual chemical composition by allowing material to enter and leave the cell either by diffusion as through a semi-permeable membrane or by active transport against a physico-chemical gradient, for example, the sodium pump mechanism.

  However there are also the mechanisms of exocytosis and endocytosis for the transport of larger molecules or particles across the cell membrane.

  In **exocytosis** a membrane bound granule or vacuole within the cell fuses with the cell membrane allowing its contents to be discharged to the exterior of the cell.

In **endocytosis**, which includes **phagocytosis** (incorporation or swallowing of solid material) and **pinocytosis** (incorporation of liquid material into the cell), the cell surface surrounds the material to be internalised and ingests the material contained within a portion of the cell membrane. The membrane bound vesicle thus formed may then fuse with a lysosome which will break down the contents.

Endocytosis may be receptor mediated where specific substances become attached to surface receptors and are then internalised. The molecules which attach to the receptors in this way are called **ligands**. Often there is a layer of protein, called **clathrin,** adherent to the internal surface of the cell membrane at the site of such receptors and the resultant ingested vesicle is coated with clathrin. Following ingestion, the vesicle will fuse with a lysosome which digests the contents and releases the receptors. The receptors then migrate back to the cell surface. For example, low density lipoproteins (LDL) which are rich in cholesterol, are taken up by fibroblasts in this fashion. The absence of the receptors for LDL can lead to familial hypercholesterolaemia.

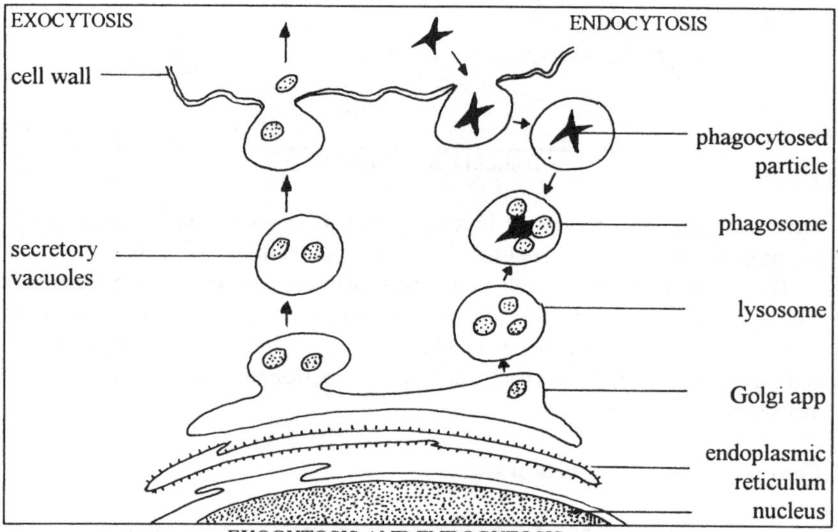

EXOCYTOSIS AND ENDOCYTOSIS

- **Sites of receptors and antigens**: The cell membrane may carry receptors for organisms, drugs or hormones. Some hormones, such as adrenaline and ACTH, act through receptors which stimulate adenyl cyclase activity which converts adenosine triphosphate ATP to adenosine 3,5 cyclic phosphate or cyclic AMP which acts as a 'second messenger'. Other receptors act through affecting the binding of calcium ions to a protein, **calmodulin.**

  Antigens such as histocompatibility antigens and blood group antigens are attached to the cell membrane.
- **Cell movement**: Movement is mediated by the flow of cytoplasm accompanying changes in the shape of the cell and cell membrane brought about through the microtubules and microfilaments of the cytoskeleton.

- **Inter-cell relationship**: The cell membrane is joined to other cells by various junctions which provide for intercellular adhesion and inter cellular communication. The cell surface may also control the growth and replication of cells, for example by contact inhibition.

THE CYTOPLASM

The cytoplasm is 90 per cent water. It contains electrolytes such as K, Mg and PO_4 in relatively high concentrations as opposed to Na, Cl and HCO_3 which predominate in the extracellular fluid. The osmotic pressure within the cells is equal to that of the extracellular fluid. The cytoplasm contains stored materials and formed organelles.

The major membrane systems are:

- **The endoplasmic reticulum and the Golgi apparatus**: These membranes surround spaces or cisternae which communicate with the perinuclear cisternae. Membrane bound products such as secretory granules, lysosomes and microbodies which are formed from the endoplasmic reticulum are also a part of the system

  The endoplasmic reticulum is a system of interconnected membrane bound spaces. The rough or granular endoplasmic reticulum has ribosomes attached to the outer surface. These measure 15nm in diameter, are constituted of RNA and are associated with protein synthesis, most of which takes place in the rough endoplasmic reticulum. Some ribosomes appear to lie free in the cytoplasm. Smooth endoplasmic reticulum has no granules on the outer surface and is not involved in protein synthesis but has metabolic functions concerning, for example, lipid metabolism, drug detoxification and steroid hormone synthesis.

  The Golgi apparatus is formed of stacks of cisternae placed near the free surface of the cell. It is thought to sort and package secretory products, made by the rough endoplasmic reticulum, which are placed in bound vacuoles ready for discharge by exocytosis.
- **Lysosomes and microbodies**: Lysosomes are membrane bound vesicles containing hydrolytic enzymes which are active at acid pH. They are formed by the rough endoplasmic reticulum and released from the Golgi complex as primary lysosomes. They fuse with ingested material or with worn out cell components to form phagosomes or secondary lysosomes. Indigestible material within lysosomes may persist within the cell as residual bodies which can be seen under the light microscope as brown lipofuscin pigment. The absence of a lysosomal enzyme will result in a form of storage disease.

  Microbodies are formed by budding off from the rough endoplasmic reticulum. They contain enzymes involved in the breakdown of hydrogen peroxide and other oxidative metabolic processes.

Mitochondria are sausage shaped structures with an inner and an outer membrane. The inner membrane is folded into shelf like projections. They are responsible for the oxidative metabolism of the cell and make adenosine triphosphate (ATP) which is the source of energy for many functions of the cell. Mitochondria contain a small amount of DNA in the inner space (like a bacterial chromosome).

They have been observed to divide and have a limited capacity to manufacture protein. Mitochondria resemble bacteria and it has been suggested that they are derived from the combination of nucleated cells and anuclear oxidative bacteria in an early phase of cellular evolution.

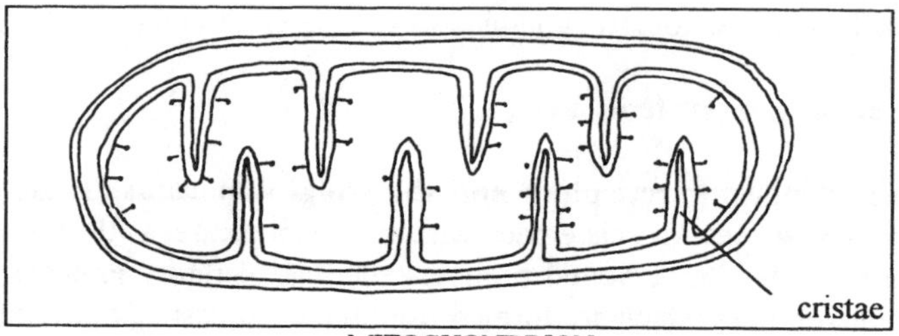

MITOCHONDRION

The cytoskeleton has three major components:

- **Microtubules** are 23nm in diameter and of variable length. They are built up from subunits (heterodimers) of a cytoplasmic protein, **tubulin.**
  - Cytoplasmic microtubules play an important role in stabilising the position of the cellular organelles and membrane systems.
  - Spindle microtubules are essential for cell division by mitosis or meiosis. Division can be arrested in the metaphase if the spindle is disrupted by the drug, colchicine.
  - Axonemal microtubules are an essential part of the ciliary motor system.
  - The centrioles, which form the poles of the mitotic spindles, divide to give rise to the basal bodies of the cilia which have a similar appearance under electron microscopy. They are composed of a ring of nine triplets of microtubules. Cilia have a central axoneme in which nine doublets of microtubules are arranged around a pair of centrally placed microtubules.

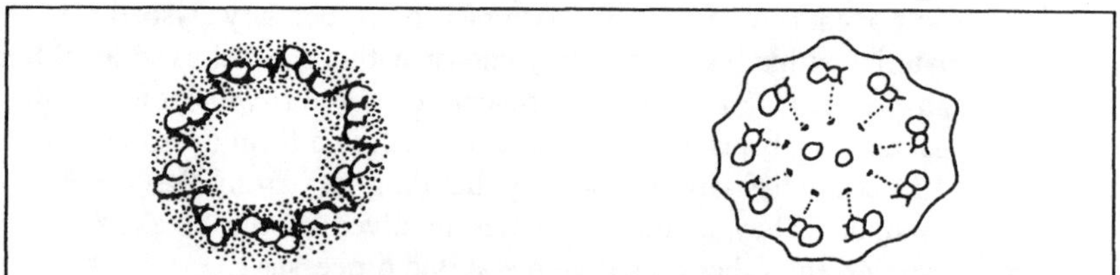

MICROTUBULAR STRUCTURES

- **Actin Filaments** or microfilaments are 7nm in diameter and are present in all cells where they form 8-14 per cent of total protein. They appear to have a stiffening function, as they are present in the cores of microvilli.

  Actin filaments may also be associated with the cell membrane, and may play a part in cell movement.
- **Intermediate filaments** have a diameter of 10nm intermediate between thin actin filaments and thick myosin filaments.

  They are much less labile than microtubules or actin filaments and appear to be associated with the mechanical and supportive function of the cytoskeleton. The major classes of intermediate filaments include **cytokeratin, vimentin, desmin, neurofilaments** and **glial** filaments. The nature of the intermediate filaments in a cell can be used to diagnose the nature of malignant tumours. For example, carcinomas contain cytokeratins and sarcomas contain vimentin.

Cytoplasmic granules can take a variety of forms and distinctive granules may be found in a number of cells. There may be secretory granules in exocrine and endocrine cells as well as melanosomes in melanocytes, Weibel Palade granules in endothelial cells and Birbeck granules in Langerhans cells.

CELL INTERRELATIONSHIPS

This topic has already been mentioned briefly as one of the functions of the cell membrane. The contact of one cell with another is important in a wide range of cell activities including communications, movement, growth habit and growth control and differentiation. Cell to cell contact is obtained at specialised cell junctions which are morphologically recognisable as well as by surface glycoproteins called cell adhesion molecules (CAMs).

Specialised cell junctions include:

- **Tight junctions (zonula occludens)** in which there is apparent fusion of the outer lamina of adjacent cell membranes. The junction encircles the cell towards the apical or luminal margin. This junction is impermeable.
- **Adherens junction (zonula adherens)** may also encircle the cell like a belt. Actin filaments are attached to the interior of the cell membrane at the site of the junction.
- **Desmosomes** are disc like structures formed on adjacent cell membranes (usually epithelial cells). Intermediate filaments (tonofilaments) from each cell appear to terminate in the desmosomes. Hemidesmosomes occur where cells are attached to a basement membrane.

These three junctions are often found in proximity to one another and are known as a junctional complex.

- **Gap junctions** are specialised junctions in which fine tubules seem to cross from one cell to the next.

Cylinders with a fine central channel extend from each cell to meet in the intercellular space. These are called **connexons**. Ions and small molecules may pass through them.

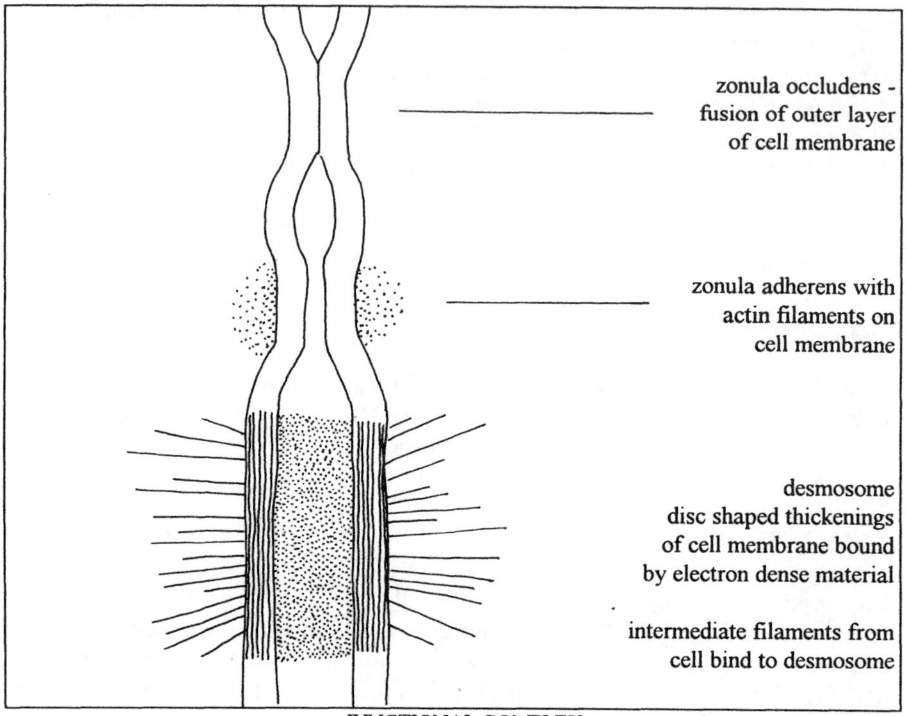

JUNCTIONAL COMPLEX

Finally, not all cells fit into the picture just described of a cell with a single nucleus. Some such as red cells and platelets have no nuclei. There are also multinucleate giant cells which can occur in a number of conditions.

MULTINUCLEATE GIANT CELLS

- **Physiological** giant cells include osteoclasts, megakaryocytes and placental syncytiotrophoblastic cells.
- **Damaged striated muscle fibres** may have multiple nuclei, for example, regenerating sarcolemmal cells in damaged voluntary muscle
- **Macrophages**: macrophages are phagocytic cells which, when faced with insoluble or large pieces of material, may fuse together. Examples include:

  - Langhans giant cells that are found in infections, especially tuberculosis
  - Foreign body giant cells, found, for example, around sutures and around keratin in tissues.
  - Touton giant cells that contain lipid material
  - Aschoff giant cells that are formed from myocardial macrophages in rheumatic fever.

- **Induced by viruses.** Examples include multinucleate epithelial cells of mouth or genitalia in herpes virus infection, and multinucleate macrophage-like cells such as Warthin Finkeldey giant cells in lymphoid tissue in measles.

- **Tumour giant cells.** Any tumour may contain giant cells which can be formed when the rate of nuclear division exceeds the rate at which the cytoplasm divides. They occur particularly:

  - In tumours from tissues in which giant cells are normally found, such as osteoclastoma (giant cell tumour of bone) and choriocarcinoma (malignant tumour of placenta).
  - In Hodgkin's disease the bi-nucleate mirror image giant cell, the Reed-Sternberg cell, is diagnostic.
  - In any tumour, for example, giant cell variants of lung and thyroid tumours may occur.

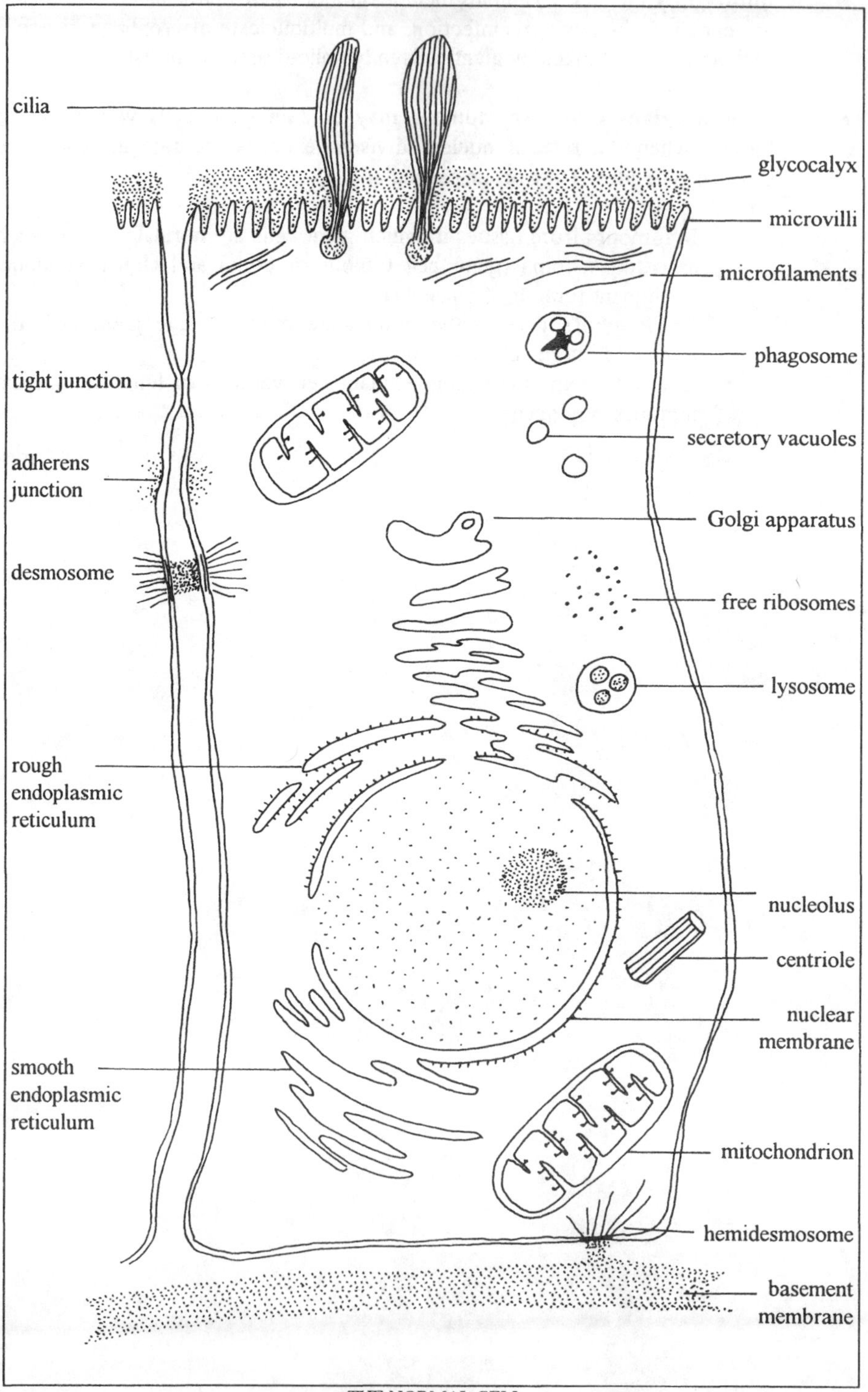

THE NORMAL CELL

GENERAL PATHOLOGY - CHAPTER THREE

CELL INJURY

The manner in which injurious agents act upon the cell is one of the basic studies of pathology. The changes that appear after cell injury may be classified according to the microscopic appearances in descriptive fashion. As more is learned about cellular and molecular biology, the inadequacies of this sort of classification become apparent, inasmuch as the changes seen under the light microscope do not necessarily reflect the electron microscopic changes, much less the actual nature of the injury at molecular level.

The original damage to the cell in most instances is a biochemical one. It may be eight or twelve hours before the changes due to this damage become observable by the ordinary light microscope, even although irreversible injury may take place within twenty to sixty minutes.

The causes of injury to the cell reflect the causes of disease generally (which have been listed in Chapter 1) and include:

- anoxia or ischaemia
- physical injury
- chemical injury
- injury due to biological agents
- (such as infection)
- injury due to immune mechanism
- genetic defects causing injury
- nutritional injury
- aging

The functional effects of such injury on the cell may be to depress or stop vital functions such as:

- Production of energy needed for the cell's normal metabolic processes by damage to **mitochondria**.
- Synthesis of proteins, both enzymic and structural by damage to the **endoplasmic reticulum**.
- Maintenance of the ionic and osmotic steady state (homeostasis) of the cell by damage to the **cell membrane**.
- Preventing reproduction by harmful effect on the **nucleus**.

Mild injury or stimulus may result in the cell adapting and continuing to function normally. More severe injury will result in damage to the cell, but recovery may still be possible; however, even more severe injury will result in irreversible changes in the cell, eventually resulting in cell death.

In simplified form:

INJURY→ NORMAL CELL→ ADAPTATION→ SICK CELL→ DEAD CELL

SICK CELL: reversible damage	DEAD CELL: irreversible damage

The structural and morphological changes occurring in cells in response to non-fatal injuries may be referred to as **degenerations**. These changes include abnormal accumulations of substances within the cell, which may be referred to as **infiltrations**.

If the injury to the cells is more severe there may be irreversible damage with death of the cells followed by **necrosis.**

Necrosis may be defined as the morphological changes which occur following cell death within the living body.

All these changes of degeneration, infiltration or necrosis take some time to develop. Cells have a somewhat limited ability to respond to injury but the changes that do occur depend upon:

- The site of the injury. For example, liver cells may respond differently from kidney cells.
- The nature of the injury. For example, alcoholic injury to cells may cause fatty change in liver cells, whereas diabetic injury may cause the accumulation of glycogen within liver cells.

Ischaemia is a condition of inadequate blood supply to an area of the body resulting in a reduction in oxygen supply or **hypoxia**, that is insufficient oxygen, or **anoxia,** that is no oxygen.

Hypoxic or Ischaemic Cell Injury results in a failure of the cell's aerobic respiratory function, that is, oxidative phosphorylation with the formation of energy rich ATP, which takes place in the mitochondria. The cell ceases to function (for example, heart muscle cell). Anaerobic respiration with a breakdown of glycogen results in the accumulation of lactic acid and phosphates, with a resultant reduction of pH (increase of acidity of the cell).

Failure of the sodium pump because of insufficient ATP results in the accumulation of sodium, loss of potassium, and net gain of water by the cell, which enlarges.

The endoplasmic reticulum becomes swollen. Ribosomes become detached from the rough endoplasmic reticulum and protein synthesis is diminished. The cell membrane may lose microvilli, develop blebs and show increased permeability.

The entire cell is thus enlarged and waterlogged. At this stage the injury is reversible. Indeed it is thought that actually much of the damage we see is due to the restoration of blood supply and re-oxygenation of previously ischaemic tissues. It is during this phase that damage occurs and this is thought to be due to the action of free radicals.

Free Radicals are chemicals that have an odd number of electrons with a single unpaired electron in the outer orbit. They may be looked on as possessing half a bond and are highly reactive. If they react with non-radical substances another free radical is formed, thus potentiating the reaction. Free radicals may be the agents for cell injury in a number of different circumstances. Some important examples include

- $O_2^{-\cdot}$ (superoxide)
- H_2O_2 (hydrogen peroxide)
- $OH\cdot$ (hydroxyl ion)

These are known as toxic oxygen species. Free radicals may be formed within cells by the action of UV light or X-rays, by intracellular oxidative reactions, or by the metabolism of chemicals or drugs. They act on the cell membrane, and on lipids and on proteins and DNA nucleotides.

Free radicals may be the direct cause of cell damage in:

- **hypoxia or ischaemia**
- **chemical injury**
- **inflammation** (leucocyte mediated)
- **bacterial destruction** (leucocyte mediated)
- **radiation damage**
- **aging**

ACCUMULATION OF WATER IN THE CELL

This may take place in stages:

- **Cloudy swelling** is an almost universal morphological expression of reversible injury. The cells swell due to the accumulation of water within them. This may occur in hypoxic or other forms of injury. The cytoplasm takes on a granular appearance (due to swelling of mitochondria and endoplasmic reticulum). The changes are mostly seen in the parenchymatous cells of organs, particularly in the liver and the cells of the tubules of the kidneys.

  An affected organ is swollen, of soft consistency, pale in colour (because some blood may have been squeezed out) and shows bulging of the cut surface.

  If the accumulation of water continues the cell may swell further and vacuoles rather than granules may be seen in the cell cytoplasm.
- **Hydropic degeneration** occurs when there is considerable waterlogging of the cell, and the vacuoles have either coalesced or burst. The cells are swollen and rounded and the cytoplasm appears pale and homogeneous. Although even at this stage the change may be reversible, eventually the whole cell may rupture and die.

  Hydropic change as an extreme form of water accumulation may be seen in renal tubular cells in potassium deficiency, in epidermal cells in virus infections and in liver cells following the action of poisons such as chloroform or carbon tetrachloride.

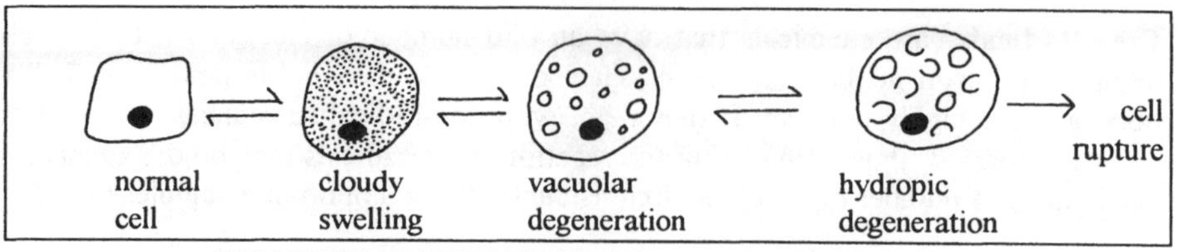

THE CELLULAR DEGENERATIONS - WATER ACCUMULATION

ACCUMULATION OF FAT WITHIN THE CELL

This is best referred to as **fatty change** rather than by the older terminology of fatty degeneration or infiltration.

Fat collects in cells at first in small membrane bound inclusions (liposomes) close to the endoplasmic reticulum. Fat vacuoles grow in size to become visible. The size of the fat vacuoles reflects the amount of fat in the cell.

Fatty change is a common type of response to cell injury. It may be preceded by cloudy swelling and, although it is still reversible, can indicate more severe cell damage and lead to cell death. In the liver from a subject with carbon tetrachloride poisoning, for example, the changes of cloudy swelling, hydropic degeneration, fatty change and necrosis may all be seen in a single microscopic field.

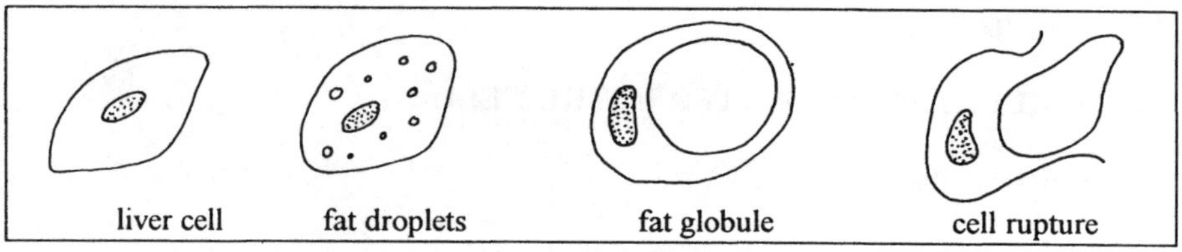

THE CELLULAR DEGENERATIONS - FATTY CHANGE

How may fatty change be recognised? The fat occurs in the cells in vacuoles - but water or glycogen can also form cytoplasmic vacuoles. However, in normally processed histological sections the tissues have passed through fat solvents and hence we do not actually see the vacuoles, but rather the spaces they once occupied. Whilst the site and nature of the vacuoles may provide an indication that they were occupied by fat, for confirmation a frozen section must be cut and stained with special stains for fat such as the Sudan stains or Oil Red O. True fat droplets or vacuoles will then stain red.

Fatty change is most commonly encountered in the liver, heart and kidneys because the cells in these organs are either involved in fat metabolism or depend upon lipids for their energy sources.

The liver grossly may appear enlarged and flabby, yellowish in colour and have a greasy cut surface.

The most severe fatty change may be seen in alcoholics and fatty change may be followed by scarring and cirrhosis of the liver. In malnutrition, especially kwashiorkor, marked fatty change may be noted.

enlarged fatty liver

Other conditions such as diabetes mellitus, obesity, congestive heart failure with congestion and anoxia of the liver, and the action of poisons such as CCl_4, $CHCl_3$, phosphorus, anaesthetics such as halothane may also result in fatty change.
Some drugs, such as tetracycline and warfarin cause fatty change. Other causes include acute fatty liver of pregnancy, hepatitis C, intestinal bypass operations, Reye's syndrome and certain inherited metabolic disorders.

In the liver cells the fat first accumulates as liposomes, but as the amount of fat increases the membranes rupture, forming vacuoles. These often coalesce into one large globule, which may occupy most of the cytoplasm of the cell and displace the nucleus to one side. Sometimes cells may grow so large that they rupture and the fat globules fuse to form fat cysts within the liver.

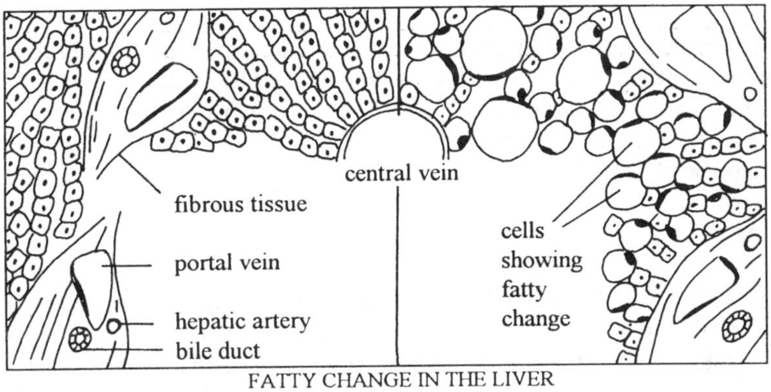

FATTY CHANGE IN THE LIVER

The heart may exhibit fatty change in the myocardium. This may take two forms:

- In moderate to severe hypoxia such as may be produced by severe anaemia (for example, Vit B_{12} deficiency anaemia) there may be yellowish streaking or a mottled appearance sub-endocardially. This is the so-called 'thrush breast' or 'tabby cat' appearance. The areas of myocardium furthest from the blood vessels become most short of oxygen and show the most marked changes, thus accounting for the irregular distribution of fatty change.
- A more severe diffuse form of fatty change may result from very severe hypoxia or toxic effect, for example, in diphtheritic myocarditis there may be diffuse yellowing and flabbiness of the myocardium.

Microscopically in both forms the fat is dispersed throughout the cytoplasm of the myocardial fibres in the form of very fine droplets which may only be demonstrated by the use of special stains.

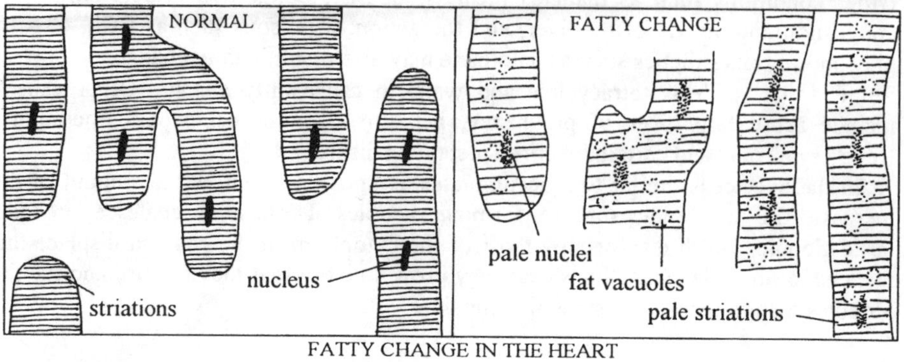

FATTY CHANGE IN THE HEART

The kidneys may show little gross change, although there may be slight pallor and enlargement. Fatty change occurs most often in the cells of the proximal convoluted tubules. It is usually due to severe hypoxia, but some poisons and renal diseases such as the nephrotic syndrome may also cause tubular fatty change. The fat is usually in the form of small droplets.

THE PATHOGENESIS OF FATTY CHANGE.

Most work has been done on fatty change in the liver. Fat may reach the liver either from fat stores or from the intake of dietary fat. - usually in the form of free fatty acids. In the liver cell these then become triglycerides, that is true fat. The fat is bound to protein and leaves the cell as lipoprotein. Thus fat can accumulate in liver cells:

- If too much fatty acid is transported to the liver, as may occur in starvation when fat stores are mobilised
- If there is increased triglyceride formation from fatty acids, as may occur in alcohol toxicity
- If fat cannot leave the cell because of lack of conversion factor to lipoprotein, as in CCl_4 poisoning or protein malnutrition.

The clinical consequences of fatty change are mainly those of the condition causing it and depend on how severe this is.

The liver has a large functional reserve, but very marked fat accumulation may affect function, even causing liver failure (as in acute fatty liver of pregnancy). Fatty change may be a precursor to cirrhosis of the liver.

The heart is not generally affected, but problems may arise such as impaired function, arrhythmias or low blood pressure.

The kidneys are not usually affected enough to interfere with tubular function.

STROMAL FATTY INFILTRATION

This condition should not be confused with fatty change. There is growth of true fatty tissue as connective tissue into organs such as heart or voluntary muscle. It may be found in the heart in obese patients, in the muscles of patients with muscular dystrophy or as a part of the aging process in the thymus, parathyroids, pancreas and lymph nodes.

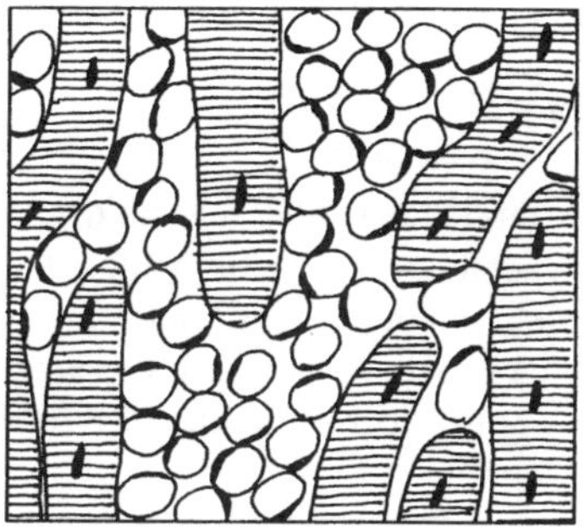

STROMAL FATTY INFILTRATION
OF HEART MUSCLE

ACCUMULATION OF OTHER SUBSTANCES WITHIN THE CELL

ACCUMULATION OF PROTEIN

This used to be called hyaline degeneration because of the appearance of homogeneous glassy eosinophilic or hyaline material in the cytoplasm of the cell. Note that a number of substances (such as amyloid and collagen) can have a hyaline appearance outside the cell. Intracellular hyaline degeneration may denote serious damage to the cell and be a precursor of cell death.

Examples include:

- **Hyaline droplets** in the kidney tubule cells form when there is heavy proteinuria. They consist of resorbed protein fused with a lysosome to form a phagolysosome.
- **Russell bodies** are large eosinophilic hyaline inclusions in plasma cells which consist of synthesised immunoglobulin (within dilated endoplasmic reticulum).
- **Mallory's hyaline** may be found in the cytoplasm of liver cells in alcoholic hepatitis and cirrhosis. It forms an irregular network in the cell. It consists of an accumulation of intermediate filaments in damaged cells

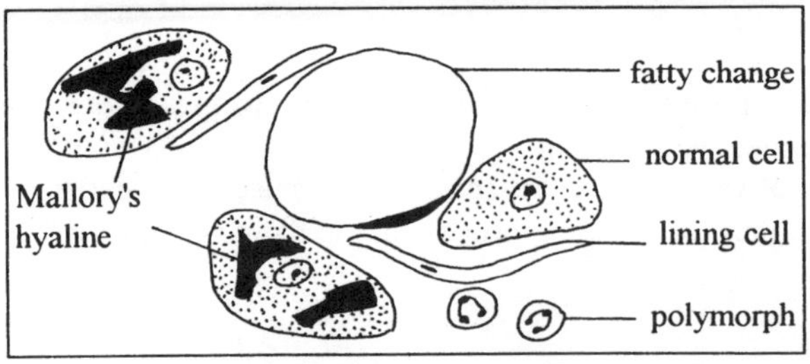

MALLORY'S HYALINE IN LIVER CELLS

GLYCOGEN INFILTRATION

This is a regressive, non-fatal, reversible alteration in cells. It is not usually a reflection of serious injury and is not accompanied by impaired cellular function. Glycogen is normally present in most cells, particularly in muscle and liver cells. Excessive accumulations can occur, producing vacuoles in the cytoplasm and nucleus. The cells do not appear to be functionally damaged. Glycogen infiltration does not lead to cell death.

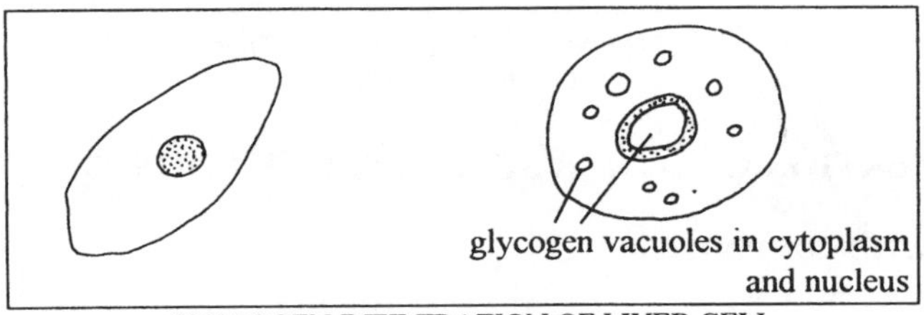

GLYCOGEN INFILTRATION OF LIVER CELL

Excess glucose may lead to increased glycogenesis - as happens in diabetes mellitus

Reduced glycogenolysis may result in a build-up of glycogen - as happens in the glycogen storage diseases, where there are inborn errors of metabolism in which there is a deficiency or absence of glycogenolytic enzymes.

Generally glycogen accumulates in the liver, voluntary or striated muscle and in the kidneys.

- **The liver** normally contains much glycogen. Excess accumulation may be suspected when, in addition to foaminess of the cytoplasm there is apparent vacuolation of the nucleus.
- **The kidney.** In the kidney sites of glycogen deposition are the sites of glucose resorption, the terminal straight portion of the proximal convoluted tubule and the loop of Henle. The cytoplasm is cleared and the nucleus pushed basally. Other accumulations, for example, hydropic or vacuolar degeneration, must be distinguished and specials stains may be needed.
- **Muscle.** Only in severe glycogen storage disease can vacuolation of muscle be seen.

As glycogen is water soluble, tissues are best fixed in alcohol if it is wished to demonstrate glycogen. Special stains include the periodic acid Schiff (PAS) reaction and Best's carmine stain, both of which stain glycogen red.

GENERAL PATHOLOGY - CHAPTER FOUR

CELL DEATH AND NECROSIS: GANGRENE

Necrosis has already been defined as the morphological changes which occur (in cells) subsequent to cell death within the living body.

Point of death: We cannot actually tell with any exactitude when a cell dies, just as we cannot tell the exact moment when a human being dies. For example, the heart may go on beating following the cessation of respiration. Incidentally, recent legislation defines death of a person as occurring when there is no persistent recognisable electrical activity in the brain.

So in the cell, synthetic activity (for example, protein making) may go on after the cellular respiratory mechanisms are destroyed, although not for long. Thus it is difficult to pinpoint with any certainty the point of no-return. However, the structural changes of necrosis follow on quite considerably later; only hours or even days after biochemical death of a cell within the body can these changes be recognised by microscopical examination.

Incidentally, in a patient who dies suddenly, for example, of a myocardial infarct ('heart attack') our present routine techniques cannot show up any structural change in the heart muscle. The patient must survive for at least twenty four hours before obvious necrosis of the myocardial fibres can be demonstrated histologically. Estimation of enzymes released by dead myocardial fibres, such as AST, LDH and CK*, and special staining techniques allow somewhat earlier clinical and pathological recognition.

THE DEVELOPMENT OF NECROSIS

Two major mechanisms are responsible for the development of necrosis. These include:

- Irreversible damage to mitochondria and failure of ATP generation. As a result of failure of oxidative phosphorylation, anaerobic respiration continues with the production of lactic acid which accumulates causing the intracellular pH to drop, i.e. the cell becomes more acid. The lowered pH activates lysosomal enzymes and proteolytic digestion of the cell follows. *Influx of water.*
- Damage to the cell membrane occurs with loss of phospholipid and further damage due to lipid breakdown products; the cytoskeleton becomes detached from the cell membrane and there is also injury caused by free radicals.

*AST = **AS**partate **T**ransaminase (formerly SGOT or **S**erum **G**lutamic **O**xaloacetic **T**ransaminase) LDH = **L**actic **D**e**H**ydrogenase CK = **C**reatine **K**inase

The membrane then allows enzymes, for example AST, to escape into the extracellular fluid. It also allows the influx of calcium ions into the cell after reperfusion. The effect of the calcium is to cause permanent damage to mitochondria, to inhibit cellular enzyme action and to denature proteins, causing the characteristic cellular changes that will be described under the heading of coagulative necrosis.

The first indications are most obvious in cell nuclei:

- **Pyknosis** is the shrinking and condensation of the nucleus which becomes more basophilic, that is, it stains deeper blue with haematoxylin stain.
  The nuclear membrane becomes wrinkled with the shrinkage, and the chromatin, which is usually finely divided or dispersed, clumps and becomes more homogeneous and darker staining. Normal cell nuclei take up basic stain because of the acidity of nucleic acids. As nucleoprotein is hydrolysed or dissolved, more nucleic acid is made available to combine with the basic dye - hence the deeper staining as pyknosis develops.
- **Karyorrhexis** occurs when the nuclear membrane ruptures and the nucleus fragments with dispersion of the particles as 'nuclear dust'.
- **Karyolysis** occurs where the nuclear membrane does not rupture but the basophilia gradually fades until the nucleus can hardly be made out. In this case the hydrolytic process has continued and broken down the nucleic acids further.

Eventually after two or three days the nucleus will disappear entirely from the cell.

the structural changes of necrosis occur first in the nuclei

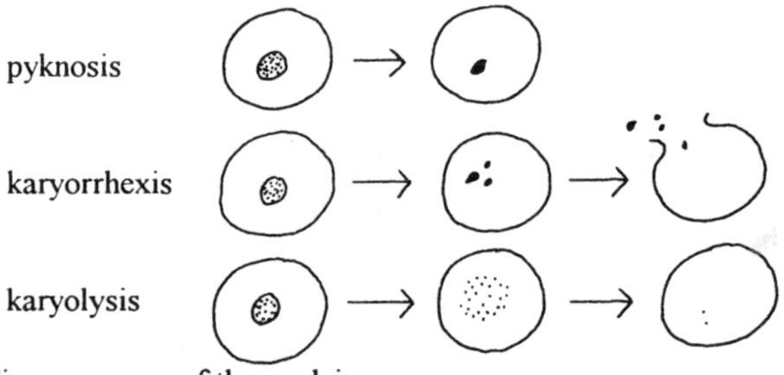

culminating in disappearance of the nuclei

NUCLEAR CHANGES OF NECROSIS

Meanwhile in the cytoplasm cloudy swelling or fatty degeneration has progressed, and the cytoplasm of the cell may be replaced by large clumps of phospholipids in the form of myelin figures. These have the appearance of diffuse hyaline coagulative acidophilic material replacing the cell cytoplasm.

At this point, the cell membrane may rupture and adjacent cells fuse into an indistinct mass. Then:

- Necrotic cells are dissolved by enzymes (intracellular and extracellular) and absorbed
- Fused cells are fragmented, phagocytosed by leucocytes and removed.

Timing: The changes take some time to develop. In one experiment no histological changes could be seen in tissues killed by cutting off the blood supply until eight hours later, even although the cells would not survive on transplantation after one hour.

Under the electron microscope changes may be seen in the mitochondria in fifteen minutes and in the nucleus in thirty minutes.

It would appear that lysosomes may be intact after two hours, so they do not initiate necrosis.

TYPES OF NECROSIS

Special patterns of necrosis are described and recognition of these may be of value when trying to determine the aetiology of necrosis.

- **Coagulative necrosis** is the type commonly due to cutting off of the blood supply - when the lesion is called an **infarct**.

  An **infarct** is a localised area of ischaemic necrosis.

  There is coagulation of the cell to an eosinophilic opaque mass, usually with loss of the nucleus, but with some preservation of the basic cell shape and the architecture of the tissue preserving a 'shadow' or 'ghost' outline, despite the loss of intracellular detail. This is sometimes referred to as structural necrosis.

  Incidentally, areas of necrosis act as irritants to the adjacent living tissue which shows an inflammatory response, and hence is surrounded by dilated vessels and inflammatory cells. This is called the **vital reaction** to necrosis and helps to distinguish necrosis during life from post-mortem change, also known as autolysis or self digestion.

  Eventually the shadow or ghost outline will disappear after days or even weeks, with further changes leading to the removal of dead cells, as can be seen in infarcts of kidney, prostate, heart, spleen and lung.
- **Colliquative necrosis** or liquefactive necrosis occurs when there is rapid and total enzymatic dissolution of tissue and complete destruction of cells. This is most often seen in the brain for example in brain infarcts. A brain infarct is one cause of a stroke.
- **Suppurative necrosis** occurs in bacterial infection that lead to the formation of pus, and is a form of colliquative necrosis.
- **Caseous necrosis** is a particular type of necrosis most commonly associated with tuberculosis (TB). The cells are converted into a granular eosinophilic mass of amorphous fat and protein with a total loss of cell detail and outline.

It is called caseous necrosis because of its supposed gross resemblance to cheese, a sort of soft friable cream cheese.

It is thought to occur as a result of the action of toxic substances in the waxy capsule of TB bacilli.

A similar change can occur in other bacterial and fungal infections, or in the centre of rapidly growing tumours that have outgrown their blood supply; however the term caseous necrosis tends to be used mainly to describe TB.

- **Gummatous necrosis** occurs in late syphilis (tertiary syphilis) and is rather similar to caseous necrosis, although nuclei tend to persist in the centre of necrosis and polymorphonuclear leucocytes may be present. Characteristic changes at the periphery of the necrosis (such as plasma cell infiltration and vessel thickening) aid in the diagnosis. In both caseous necrosis and gummatous necrosis it may be difficult to demonstrate the causative organisms.
- **Fat necrosis** can occur and is of two types:

  - **Enzymatic fat necrosis**: In acute pancreatitis enzymes escape from the pancreas into surrounding tissues. Lipases in the pancreatic and peri-pancreatic fat hydrolyse true fat into glycerol and fatty acids. These fatty acids saponify or form soaps which appear as white opaque masses or plaques in the fat. Calcification may occur later. The appearances are virtually diagnostic of acute pancreatitis; surgeons must be able to diagnose this condition on sight because acute pancreatitis often presents as an acute abdominal emergency and may be investigated surgically (by laparotomy). Acute pancreatitis is not amenable to surgical treatment and early diagnosis by the surgeon is important.
  - **Traumatic fat necrosis** is focal necrosis of fatty tissue which may occur anywhere but is particularly likely to occur in the fat of the female breast. There is not always a history of noticeable trauma.

    There is no enzymatic breakdown of fat, but rupture of the cell membrane will release neutral fat into the tissues where it is phagocytosed by macrophages which take on a foamy appearance. There is a chronic inflammatory reaction with subsequent fibrosis and scarring.

- **'Fibrinoid' necrosis** is worth mentioning at this stage although it is not truly necrosis. It occurs in the walls of small arteries in hypertension and in vasculitis of immune origin (immune complex disease) and in collagen in the connective tissue disorders such as rheumatoid arthritis. It is called fibrinoid necrosis because there is deposition of eosinophilic material which stains like fibrin but also may contain complement and immunoglobulins
- **Gangrenous necrosis** is necrosis with superadded putrefactive infection and will be discussed in more detail shortly.

The causes of necrosis or cell injury were listed at the beginning of the chapter on cell injury.

Apoptosis is a term used to describe the death of single cells in an organ or tissue, and is thought to be a process by which cells that have reached the term of their natural life are shed from the body. It is sometimes referred to as programmed cell death. Individual cells shrink and show marked eosinophilic staining of the cytoplasm. Then the cell breaks into pieces; some fragments may be shed from the surface whilst others are phagocytosed. Unlike necrosis, single cells rather than groups of cells are affected, and the process of apoptosis does not excite an inflammatory or vital reaction. It is thought to be a form of physiological, as opposed to pathological, cell death which occurs, for example when foetal organs shrink following birth and in endometrial cells at the time of menstruation. It can be found accompanying pathological processes such as atrophy.

POST MORTEM CHANGES

So far we have been talking about the death of groups of cells within the living body, that is necrosis. However cell death can occur in parts removed from the body at operation, or in all cells following on death of the person. This is known as **somatic death** and is recognised when the heart stops, breathing ceases and there is no demonstrable electrical activity of the brain. Individual cells and tissues may remain alive for a period of time after death, but not for very long. As aerobic respiration ceases anaerobic glycogenolyis can continue, metabolic products will accumulate, the cell pH will drop and intracellular lysosomes may start to digest the cell contents, just as described in necrosis. An important difference is that there is no surrounding inflammatory reaction or **vital reaction** in the surrounding living tissues. The self-digestion of cells in the dead body or in a part removed from the body is known as **autolysis**. Tissues autolyse at different rates: the mucosa of the stomach, gall bladder and intestine autolyse rapidly. The brain and renal tubular cells also show fairly rapid autolysis, whereas connective tissues autolyse much more slowly.

In addition to autolysis the following changes may occur after death:

- Cooling of the body
- Development of muscular rigidity
- Gravitation of blood to dependent parts
- Clotting and haemolysis of blood
- Putrefaction

These changes are important in the practice of forensic pathology, when for example, a pathologist is performing post mortem examination on a murder victim.

- **Cooling of the body** (*algor mortis*). This depends on the outside temperature, but equilibrium is usually reached in thirty six to forty hours. The rate of cooling depends on such factors as the environmental temperature, clothing and size of the body, but averages 1.5 or $2^{\circ}C$ per hour for the first few hours, gradually decreasing to $0.5^{\circ}C$ per hour until the environmental temperature is reached.

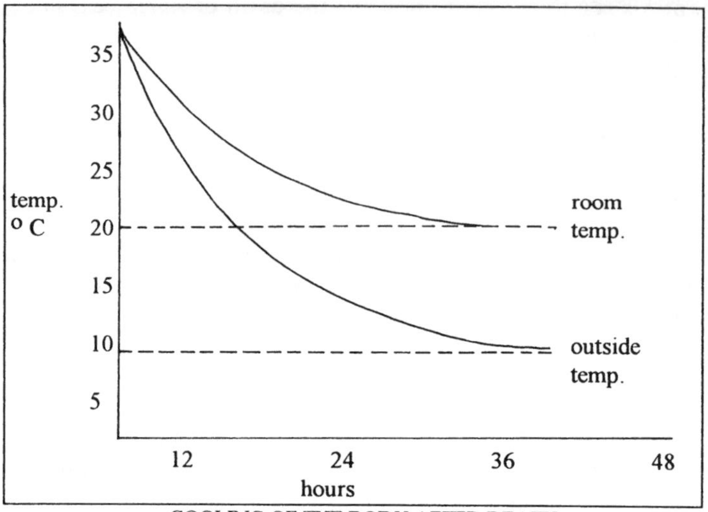

COOLING OF THE BODY AFTER DEATH

- **Muscular rigidity** (*rigor mortis*) develops soon after death. This is due to chemical changes in the body and skeletal muscles with the coagulation of protein. The process begins four to ten hours after death and passes off in three to four days. It starts with the jaw, then the head and neck and then proceeds to the lower extremities and the whole body. It can be broken and will not recur. It is not a very reliable indication of the time of death because it is variable.

  Violent exercise, high fever, exhaustion and high outside temperature before death bring it on more quickly, whilst wasting disease or low outside temperature retard the onset of *rigor*.
- **Post mortem lividity** (*livor mortis*). Blood sinks by gravity to the dependent areas of the body which become reddish purple. The vessels dilate because of loss of muscular tone in the wall. Internal organs such as the lungs may be affected as well as the skin. The process can take up to ten to twelve hours and depends on the position of the body. Because of changes in the blood, lividity persists in the same place if the body is moved after it has occurred.

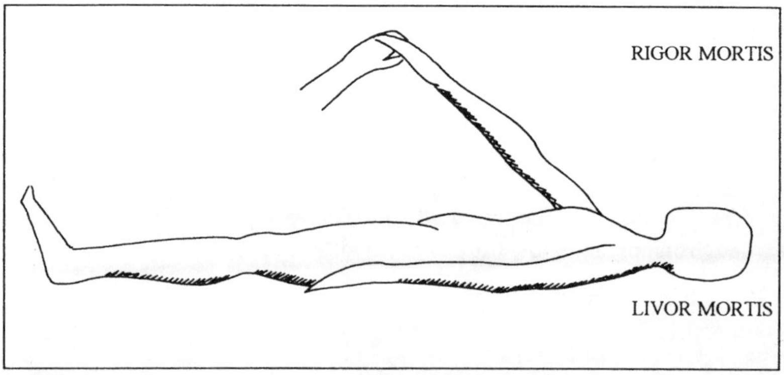

- **Changes in the Blood**

  - **Clotting of the blood** occurs early after death or even during the process of dying. (Something that occurs during the process of dying is termed agonal).

    Post mortem clots must be distinguished from **ante mortem thrombi.** Generally post mortem clots are smooth, with a shiny surface, soft and floppy. Because clotting occurs slowly the red cells may settle out so that the part of the clot containing settled red cells is red (resembling 'red currant jelly') and the other upper part is pale (resembling 'chicken fat'). These clots conform to the shape of the vessels in which they are formed and are not adherent to the wall.

    Ante mortem thrombi are rough, friable granular and firm. They show layering and lamination of fibrin with blood elements trapped between. They may not necessarily have the same shape as the vessel in which they are found, since they are often embolic, that is formed in another part of the circulatory system. They may be adherent to the vessel wall where they are formed.
  - **Haemolysis of blood** occurs after death and will stain the lining of blood vessel walls and surrounding tissues pink with haemoglobin.

- **Putrefaction** is due to multiplication of saprophytic organisms in the body. The organisms are usually from the intestine and gain access to the body tissues after death. There is greenish-black discolouration of tissues and they may be foamy or spongy due to the production of hydrogen sulphide gas.

Early and effective refrigeration may prevent some of these changes and preserve body structure and histology for some considerable time after death.

GANGRENE

Gangrene may be defined as necrosis of tissue with superadded putrefaction; that is, there is an invasion of the necrotic tissue by saprophytic putrefactive organisms from the gut or from the soil. Clinically the word gangrene is an old descriptive term applied to any black foul-smelling area which is in continuity with living tissue.

Gangrene can only occur in living tissue (similar changes post mortem are referred to as putrefaction) to which free living saprophytic organisms can gain access, for example in skin, mouth, bowel and lungs. The causes are the same as those of necrosis, but in addition there must be an opportunity for the dead tissue to become infected. As putrefactive organisms are commonly found in the gut, gut necrosis is almost invariably followed by gangrene. This type of gangrene is called **moist gangrene** because the affected tissues are swollen with fluid (oedema).

Examples:

- In the gut: The cutting off of the blood supply to a loop of gut which becomes twisted (volvulus) or strangled in a hernial sac leads to ischaemic necrosis which is rapidly followed by gangrene.
  In addition, a secondary putrefactive infection may arise in an already infected and inflamed part - for example, acute appendicitis may progress to gangrenous appendicitis.
- In the lung, inhalation of a foreign body (for example a piece of tooth during dental extraction) may be followed by abscess formation and gangrene (from organisms in the mouth carried down with the foreign body).
- Gangrene of the female genital tract may follow damage to tissue during childbirth (puerperal gangrene).
- In severely malnourished or debilitated patients following infection or surgical operation gangrene may occur in the oral region or around the genitalia.

In summary, dead tissue from any cause or at any site can develop gangrene if infected with saprophytic organisms. These generally belong to the *Clostridia* but *Streptococci*, anaerobic *Streptococci* and *Bacteroides* species may also be responsible. These are anaerobic organisms which do not grow in the presence of oxygen and can only multiply in dead, unoxygenated tissue. Whenever there are large quantities of dead tissue, there is a risk of gangrene occurring and so one of the basic principles of surgery is to clean wounds and remove dead tissue as soon as possible.

The local effects of gangrene include:

- Swelling, pain and ulceration
- Foul smell (from gases produced by putrefactive bacteria)
- Discolouration (from sulphides and breakdown products of haemoglobin)
- Loss of function of the affected part
- Perforation of gangrenous gut is a common complication and gives rise to a generalised peritonitis

In general there is absorption of bacterial toxins and the products of tissue breakdown resulting in severe 'toxaemia'. Untreated gangrene is fatal.

Dry Gangrene is, in fact, a misnomer which is applied to the changes taking place in an ischaemic limb or extremity. In reality it is a form of ischaemic necrosis or infarction and not true gangrene at all. Slow narrowing of an artery, for example by degenerative arterial disease or atherosclerosis, can lead to so-called dry gangrene of the extremities usually commencing at the tips of the digits. The tissue becomes dry and blackened (due to haemoglobin and myoglobin breakdown products); there is no oedema, foul smell or gas formation. The affected area is sharply demarcated and very painful. The only treatment is by amputation and if this is not performed surgically it can happen spontaneously when the dried, blackened, rigid digit can just break off.

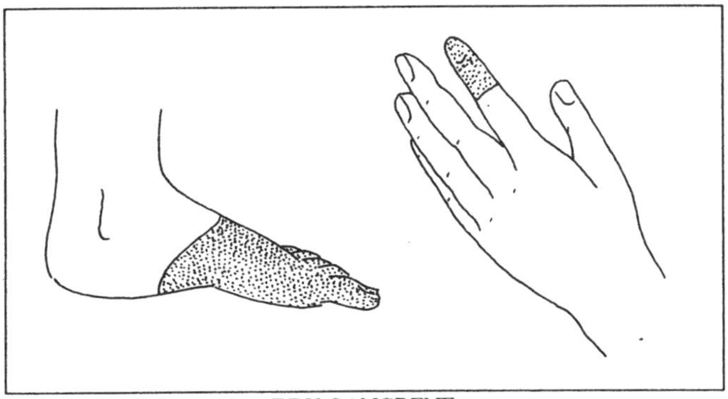

DRY GANGRENE

There are a number of causes of dry gangrene, the most common being atherosclerotic narrowing of the arteries to the limb. This is particularly likely to occur in diabetic patients and in elderly patients. Other causes include Raynaud's disease in which there is spasm of arteries; ergot poisoning which causes intense vasoconstriction; thrombo-angiitis obliterans which is a disease with marked narrowing of blood vessels found in smokers; many other causes of vessel blockage such as thrombosis and embolism will be discussed later. A similar sort of change can take place in severe frostbite.

Gas gangrene is a condition in which Clostridial organisms invade dead or damaged tissue to cause a rapidly spreading infection with widespread death of connective tissue and muscle.

Clostridia are anaerobic, spore-bearing Gram positive bacilli and the most common organisms to be involved include *Clostridium perfringens, C. oedematiens* and *C. septicum*.

As has been mentioned there must be dead tissue present for the organisms to gain a foothold in the body. Generally war wounds and crush wounds following motor vehicle accidents are likely precursors, especially if there has been contamination by soil.

The organisms produce a large variety of enzymes and toxins which can break down and destroy tissue. Infection takes several days to produce its effects which include swelling, inflammation and widespread necrosis of skeletal muscle. The affected area is painful. The wound appears black and has a foul smell. Clear blood fluid may be exuded from the wound. The organisms form gas from the fermentation of sugars and there are gas bubbles within the tissues which give a crackling feeling when palpated. There is severe toxaemia and there is also generalised haemolysis of blood. Left untreated the disease is fatal.

Treatment with antitoxins and antibiotics can cure the disease. Prevention by the thorough cleaning of contaminated wounds and the excision of dead tissues, together with prophylactic antibiotic therapy is the best course.

GENERAL PATHOLOGY - CHAPTER FIVE

AMYLOID

Amyloid consists of abnormal protein material which may be deposited extracellularly in various organs and tissues, often in the walls of small blood vessels in the condition of **amyloidosis**. Amyloid is inert and does not excite an inflammatory reaction.

In fact amyloid is not one protein but a number of chemically different proteins which have very much the same appearance under both light and electron microscopy and which share similar staining reactions. Amyloid is a fibrillar protein arranged in β-pleated sheets and it is thought that it is this physico-chemical characteristic which accounts for the similar properties of the various proteins.

THE RECOGNITION OF AMYLOID

Grossly organs which are affected by amyloid may show little change. If severely affected they may be enlarged, pale and firm with a somewhat waxy appearance. Extracellular amyloid deposition may increase the rigidity of tissues.

Both grossly and microscopically amyloid may stain a mahogany brown colour when treated with iodine. If 1% sulphuric acid is then added, the amyloid turns dark blue. This stain can be used on whole tissues such as liver or kidney slices. In the kidneys particularly, affected glomeruli may stand out as blue dots on the cut surface of the stained organ.

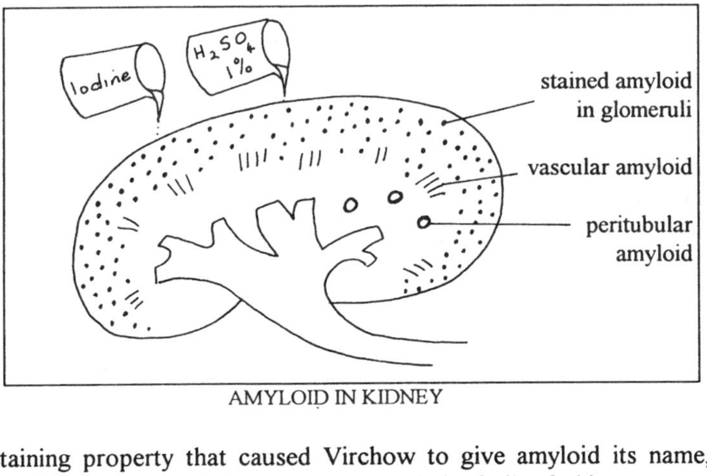

AMYLOID IN KIDNEY

It is this staining property that caused Virchow to give amyloid its name, which means starch-like, because cellulose and starch stain in similar fashion.

Seen under the microscope in sections stained routinely with haematoxylin and eosin amyloid appears as hyaline eosinophilic amorphous extracellular material, often within or close to the wall of a blood vessel. To demonstrate amyloid more specifically special stains must be used. The most specific stain in general use is Congo Red which stains amyloid a salmon-pink colour; when viewed under polarised light amyloid stained with Congo Red exhibits an apple green birefringence. Amyloid will also take up a fluorescent dye, Thioflavine T and thus stained will show fluorescence when viewed by ultra-violet light. Immunohistochemical staining methods are now available which will distinguish between various chemical types of amyloid.

STAINING METHODS FOR AMYLOID

Iodine: amyloid stains mahogany brown + 1%H_2SO_4----blue
this can be used grossly on fresh tissues or on cut surfaces

Methyl violet: stains amyloid rose pink; other tissues violet. This is known as a metachromatic reaction

Congo Red: stains amyloid salmon pink; green birefringence under polarised light

Thioflavine T: fluoresecent dye with affinity for amyloid

Immunohistochemical stains: using immunofluorescence or immunoperoxidase methods will differentiate between AA and AL.

A reliable and sensitive way of identifying amyloid is by electron microsocopy which shows that amyloid is composed of non-branching fibrils arranged in bundles or a meshwork; smaller doughnut shaped structures are also present between the fibres. Fibrils constitute 90 per cent of the amyloid and it is these which show the β-pleating on X-ray diffraction. The fibrils are 7.5-10nm in diameter and of indefinite length. The doughnut shaped structures are pentagonal and arranged in stacks. They measure some 10nm in diameter. These structures form the P component and make up ten per cent of amyloid. Serum amyloid P is present in the circulation and is related to C-reactive protein. P substance is found in basement membranes and in elastic tissue.

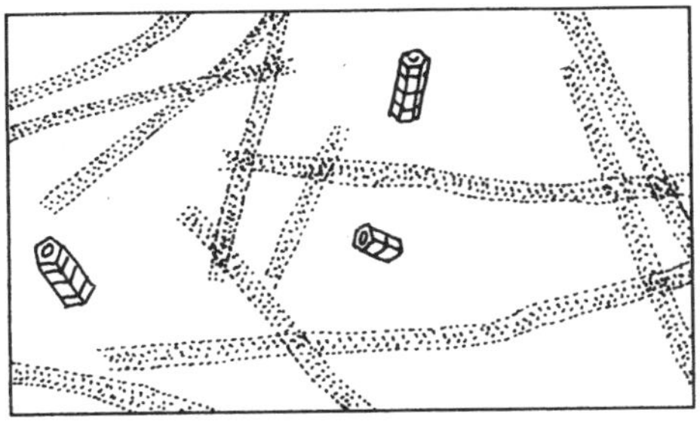

ELECTRON MICROSCOPIC APPEARANCE OF AMYLOID X 100 000

TYPES OF AMYLOID

TYPE OF AMYLOID	ASSOCIATED DISEASES	AMYLOID PROTEIN & PRECURSOR	FAVOURED SITES OF DEPOSITION
Generalised			
Immunocytic dyscrasias (primary amyloidosis)	multiple myeloma and other B-cell neoplasm with monoclonal gammopathy	**AL** (Amyloid Light chain) derived from immunoglobulin light chain mainly λ	(kidney), heart, tongue, g.i.tract, skin, nerve, joint
Chronic inflammatory disease (secondary amyloidosis)	TB, leprosy, bronchiectasis, osteomyelitis rheumatoid arthritis, Crohn's and ulcerative colitis Ca kidney, Hodgkin's disease	**AA** (Amyloid Associated protein) derived from Serum Amyloid Associated protein (**SAA**)	kidney, liver, spleen, intestine, lymph nodes, adrenals
Hereditary syndromes Familial Mediterranean Fever Neuropathic forms		**AA** with **SAA** precursor **Prealbumin** (transthyretin)	as for secondary amyloidosis peripheral and autonomic nerves
Haemodialysis associated	chronic renal failure	**β_2-microglobulin**	synovium, joints and tendon sheaths; carpal tunnel syndrome
Localised			
Senile cardiac		**Prealbumin** (transthyretin)	heart
Senile cerebral	Alzheimer's disease	**A4** (β-protein) (no P component)	brain
Endocrine	e.g. medullary Ca thyroid e.g. diabetes mellitus	**procalcitonin** from calcitonin from **insulin**	in thyroid tumour islets of Langerhans
Isolated deposits	localised amyloid of lung, skin, genito-urinary tract etc.	**AL** from immunoglobulin light chains	lung, larynx, skin, g-u. tract, tongue etc.

TYPES OF AMYLOID

Amyloid can be classified in a number of ways:

- According to the cause or associated disease
- Whether generalised throughout the body or localised to one site
- According to the chemical composition of the amyloid material deposited
- According to the sites of deposition of the amyloid

All of these factors are considered in the above table.

IMMUNOCYTIC DYSCRASIAS WITH AMYLOID

In these conditions there is proliferation of a clone of immunoglobulin producing cells (B-lymphocytes or plasma cells) and a resultant monoclonal gammopathy. The most common of these conditions causing amyloidosis is multiple myeloma, a cancer formed by malignant new growth of plasma cells in the bone marrow, characterised clinically by holes in the bones.

The plasma cells produce a specific immunoglobulin, often with an excess of light chains, and it is from these that AL is derived. However not every patient with myeloma develops amyloid, even if there is an excess of light chain production. Systemic amyloidosis occurs in five to fifteen per cent of patients with multiple myeloma and this disease accounts for about twenty per cent of cases of AL amyloid. Other conditions which can give rise to this sort of amyloidosis include Waldenstrom's macroglobulinaemia, solitary plasmacytoma and B-cell lymphoma. However most patients with AL amyloidosis do not have a neoplastic condition of this sort. These are the cases that used to be known as 'primary amyloidosis', without any preceding causative condition. These patients tend to have increased immunoglobulins and often an increase of plasma cells in the bone marrow (plasma cells manufacture immunoglobulin). Sometimes these patients will develop multiple myeloma or one of the other conditions mentioned above but the prognosis of the amyloidosis itself is poor and most patients die within a year of diagnosis.

Amyloid tends to be deposited in the muscle of the heart and tongue and in the gastro-intestinal tract in this type of amyloid. However any organ can be affected and there is much overlap between various types of amyloid as regards distribution. The heart is affected in ninety per cent of cases of AL amyloid and death is due to cardiac dysfunction (restrictive cardiomyopathy or arrhythmia due to interference with the conducting system) in about half the cases. The gut may be affected with malabsorption, haemorrhage, perforation or obstruction. The kidneys may also be affected leading to renal failure. Amyloid in nerves and skin is less likely to have serious clinical consequences.

SYSTEMIC AMYLOIDOSIS ASSOCIATED WITH CHRONIC INFLAMMATORY DISEASE

In days gone by this was associated with chronic infections such as tuberculosis, bronchiectasis and osteomyelitis but these are much less common in the antibiotic era. It is worth noting however that amyloid of AA type is a frequent complication of lepromatous leprosy, which occurs in Africa. However in this day and age AA amyloidosis is most frequently seen in association with chronic inflammatory diseases such as rheumatoid arthritis (in 5-20 per cent of cases) and inflammatory bowel diseases such as ulcerative colitis (in 1-8 per cent of cases). It also occurs in conjunction with some tumours, especially renal carcinoma and Hodgkin's disease. Heroin users who inject subcutaneously may develop multiple recurrent skin infections and amyloidosis.

The major organs tend to be affected, particularly kidney, liver, spleen and intestine, but any organ can be affected. Patients often present with proteinuria and chronic renal failure is the most common cause of death (in 40-60 per cent of cases). The heart is less severely affected than in AL amyloid.

The prognosis is that of the associated disease and if this can be treated successfully then amyloid deposition will stop and perhaps may even regress.

FAMILIAL AMYLOIDOSIS

- **Familial Mediterranean Fever** is an autosomal recessive disease occurring in those of Mediterranean stock, especially Jews and Armenians. It is characterised by periodic attacks of fever and inflammation of serous membranes.
  AA amyloid may be deposited, presumably because of the long standing chronic inflammation. Death may occur from renal failure due to amyloidosis affecting the kidney.
- **Neuropathic forms** have been described from a number of areas including Portugal, Japan, Sweden and the USA. Amyloid is laid down in the nerves, both peripheral and autonomic. The protein precursor is prealbumin, which is found in the serum where it transports thyroxine and retinol, accounting for the alternative name transthyretin. Mutant variants of this protein constitute the amyloid in familial cases and also in amyloid associated with aging.

HAEMODIALYSIS ASSOCIATED AMYLOID

This condition is due to deposition of β_2-microglobulin, a blood protein which is not cleared by the haemodialysis process. Deposition occurs in the synovium, joints and tendon sheaths. The carpal tunnel syndrome is a common complication. Amyloid may be found in up to 70 per cent of patients undergoing haemodialysis.

LOCALISED AMYLOIDOSIS

This may be found in aging, in association with certain conditions of endocrine glands, and as localised deposits in various organs, particularly lung, larynx, skin, urinary bladder, tongue and the region around the eye. certainly in some of the cases of localised nodular deposition of amyloid the material is of AL type, suggesting that there may be a local immune reaction.

PATHOLOGICAL APPEARANCES

Amyloid substance always has a similar histological appearance whatever the cause or wherever it may be situated.

The spleen is affected in most cases of AA amyloid, giving rise to a firm enlarged spleen.

The common type of distribution is nodular. Amyloid is laid down in the white pulp of the follicular arteries and nodules of amyloid destroy the white pulp. Grossly the overall effect is that of an organ containing firm grey nodules resembling boiled sago grains, hence the descriptive term 'sago' spleen.

Less commonly, the spleen is diffusely affected. It is enlarged and has a waxy appearance ('lardaceous' spleen). Here the amyloid is laid down diffusely in the walls of the sinuses of the spleen.

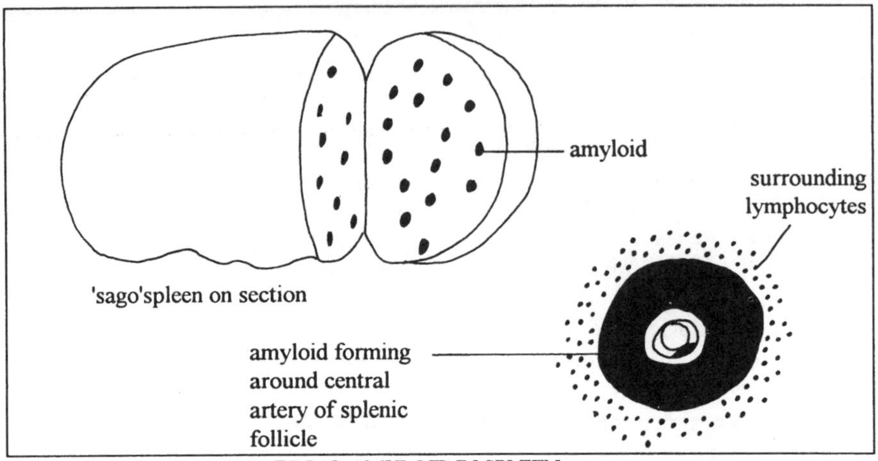

FOCAL AMYLOID IN SPLEEN

In the **kidney**, which may be slightly enlarged and pale, amyloid is laid down in relationship to the glomerular basement membrane. Deposition first occurs between the basement membrane and the endothelial cell, before spreading to involve the epithelial surface of the basement membrane and the mesangium. Eventually glomerular capillaries become obliterated. Amyloid is also laid down in association with tubular basement membranes and may extend into the peritubular connective tissue. There is also thickening of the walls of arteries and arterioles due to amyloid infiltration and there may be consequent ischaemic changes in the kidney.

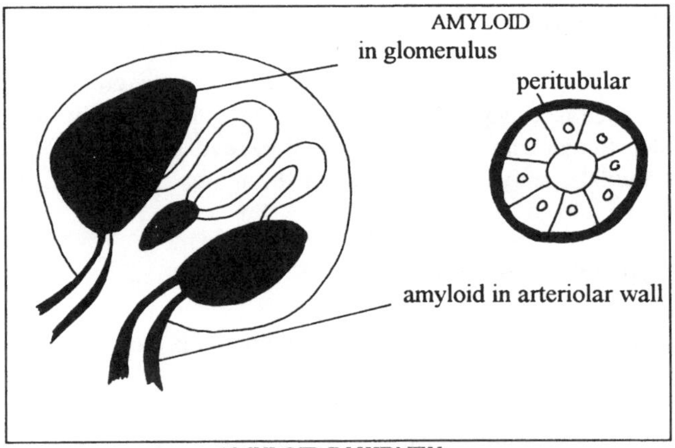

AMYLOID IN KIDNEY

The **liver** may be enlarged pale and firm. Amyloid is deposited between the walls of the sinuses and the liver cells in the space of Disse starting in the mid-zonal region of the liver lobule. Vessels in the portal tracts may also be affected. Even though there may be marked atrophy of liver cells, liver function is rarely affected.

The **gastro-intestinal tract** is usually involved within or around small vessels in the submucosa.

It may be involved throughout its length from the tongue to the anus. Amyloid can extend from the vessels in the form of plaques in the submucosa, muscularis or subserosa.

THE DIAGNOSIS OF AMYLOID

The definitive diagnosis of amyloid is by histopathological means and is based almost entirely on the staining characteristics of amyloid, predominantly the Congo Red reaction, viewed under transmitted light and polarised light when the apple green birefringence is diagnostic. AA protein will not stain after pre-treatment with potassium permanganate, whereas other types of amyloid will.

Sites of biopsy when amyloid suspected:

- Renal biopsy (if proteinuria)
- Rectal biopsy
- Gingival biopsy
- Fat aspiration

Liver biopsy is best avoided because of the danger of bleeding post-biopsy.

MECHANISMS OF DISEASE IN AMYLOIDOSIS

- Pressure effect on parenchymal cells leading to atrophy and destruction
- Narrowing of blood vessels leading to ischaemia
- Restricted movement of cardiac muscle
- Increased glomerular permeability leading to proteinuria

However despite all this it is surprising how massive amyloid deposits must be before clinical effects become obvious.

GENERAL PATHOLOGY - CHAPTER SIX

CALCIFICATION

Heterotopic calcification is the term used to describe the deposition of calcium salts in tissues other than those in which they are normally found, that is bones and teeth. Calcium shows up on X-ray examination and is thus clinically obvious, a fact which is made use of in radiographic examination of the breast or mammography, where focal calcification in early carcinoma of the breast may be used to define and diagnose the lesion. Macroscopically calcification is obvious as hardness or grittiness in tissues which resists the surgeon's of pathologist's knife. Microscopically calcium appears as dark blue basophilic granules resembling bacteria but showing more variation in size and shape; calcium may also appear as larger plaques and sometimes in some tumours laminated calcified bodies known as psammoma bodies may be found. Calcium salts may also be laid down on the surface of foreign parasites such as schistosome ova and on elastic fibres in arteries and in the lungs.

Calcium may be stained specifically by Alizarin Red S which stains it red. The von Kossa stain is a silver stain which reacts with phosphates and carbonates. However when these are precipitated they are almost invariably associated with calcium and hence the von Kossa stain is used as a stain for calcium.

There are two main types of heterotopic calcification:

DYSTROPHIC CALCIFICATION

This is the term used to describe local deposition of calcium salts in dead or degenerate tissue when there is no abnormality of calcium metabolism.

In dead tissue it occurs:

- in areas of caseous necrosis of tuberculosis.
- around dead parasites such as schistosome ova, hydatid cysts and around the worm *Trichinella spiralis* in muscle.
- in thrombi in veins, especially pelvic veins, forming **phleboliths.**
- in areas of fat necrosis, especially pancreatic, where calcium combines with released fatty acids to form calcium soap.
- rarely in a retained dead foetus calcification will occur with the formation of a **lithopaedion** (stone child).

In degenerate tissues it occurs:

- In scar tissue. This is especially important in the heart where calcification of heart valves damaged by rheumatic fever leads to more marked dysfunction (stenosis or incompetence). Calcification following on tuberculous pericarditis can constrict the heart.
- In tumours. The focal calcification occurring in carcinoma of the breast has already been mentioned. Laminated calcified psammoma bodies occur in papillary tumours of the ovary and thyroid and in benign meningioma of the coverings of the brain.
- In arteries where there has been atherosclerotic degeneration calcification is a common complication, for example in the aorta or in the coronary arteries. It can also occur in the hyalinised media of arteries in the condition known as Monckeberg's medial sclerosis, where rings of calcification form, especially in the limb arteries.

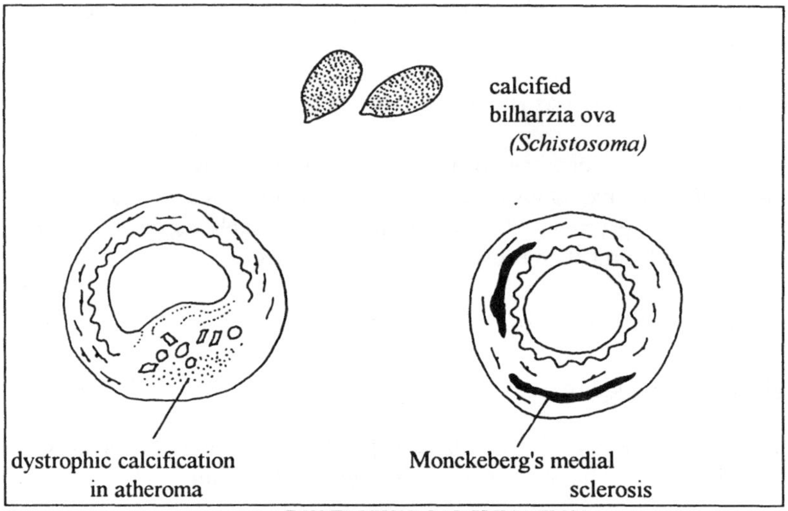

DYSTROPHIC CALCIFICATION

Dystrophic calcification is usually irreversible. Heterotopic bone formation can occur in areas of calcification with transformation of fibroblasts to osteoblasts and laying down of true bone.

METASTATIC CALCIFICATION

This occurs when there is a disturbance of calcium metabolism usually associated with a rise in the circulating blood levels of calcium or hypercalcaemia. There are certain anatomical sites where metastatic calcification tends to occur:

- The kidney is the most important site. Calcium salts are deposited in relation to peritubular basement membranes giving rise to nephrocalcinosis, which may interfere with renal function. In addition, stones are likely to form in the pelvis of the kidney or in the ureter, and renal failure may occur.

- In the lung calcium salts are laid down on elastic fibres in the alveolar walls.
- In the stomach calcification occurs around fundal glands and in the wall of the stomach.
- In blood vessel walls calcification may occur on the internal elastic lamina of arteries or in the walls of small blood vessels, especially in the brain
- In the cornea of the eye.
- In the joints, calcium crystal deposition arthritis may develop caused by the formation of calcium pyrophosphate crystals (pseudogout or chondrocalcinosis) or calcium phosphate (as hydroxyapatite) crystals in the joint space. This may occur as an isolated event, or as a familial condition or as a result of trauma without there necessarily being hypercalcaemia.

An explanation which has been given for the occurrence of metastatic calcification in kidney, lung and stomach is that in these sites excretion of acid products takes place and this renders the tissue relatively alkaline and favours the precipitation of calcium.

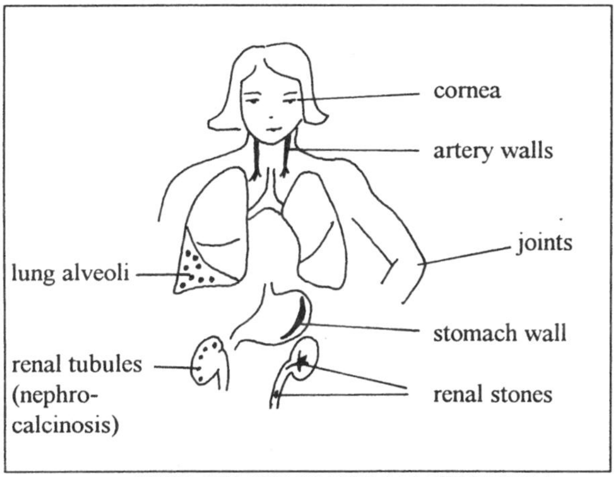

SITES OF METASTATIC CALCIFICATION

The normal total plasma calcium is 2,3-2,7mmol per litre (9-11mg/dl) of which about one half (1,1mmol or 4,5mg/dl) is ionised calcium in diffusible form; it is the ionised fraction which is regulated by parathyroid hormone. If the plasma proteins are normal the total plasma calcium is a good guide to the level of ionised calcium.

The causes of metastatic calcification are associated with disorders of calcium metabolism and are mostly related to hypercalcaemia.

The causes of hypercalcaemia include:

- **Primary hyperparathyroidism**: This is caused by overactivity of the parathyroids due to tumours (adenoma 80 per cent and carcinoma 2-3 per cent) and hyperplasia (15 per cent).
  Clinically this is known as 'bones, stones and groans' disease because there are changes in the bones, formation of kidney stones and intense pain due to these lesions.

In the bones there is excess osteoclastic resorption of bone with giant cell accumulation, cyst formation and fibrous replacement, which is known as osteitis fibrosa cystica. Pathological fractures may occur as may more generalised osteoporosis. The hypercalcaemia causes clinical tiredness, weakness and nausea. Haemorrhage into affected bone may give rise to swellings known as brown tumours.

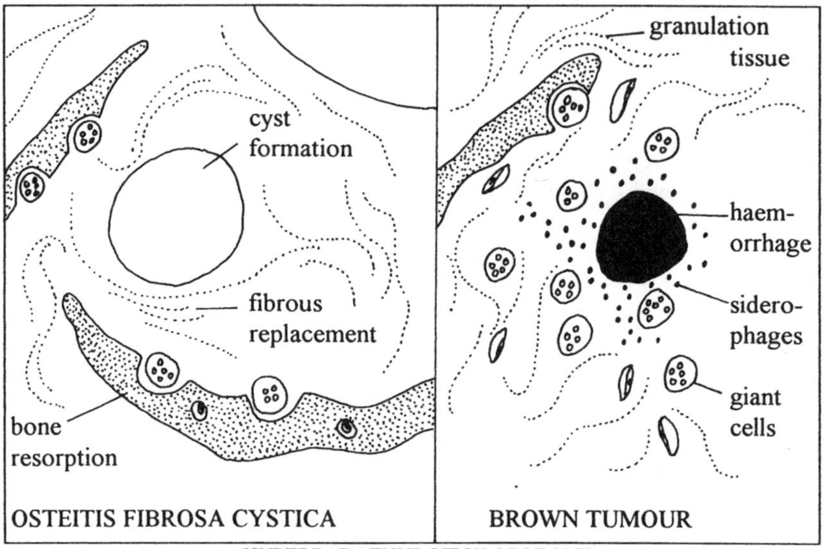

HYPERPARATHYROIDISM IN BONE

Metastatic calcification occurs in the kidney around tubules; there is also calculus or stone formation within the renal pelvis or urinary tract. This used to occur in 60 per cent of cases of primary hyperparathyroidism, but now with earlier diagnosis occurs in less than 25 per cent of cases. Metastatic calcification may take place elsewhere and in a number of cases there may be calcium pyrophosphate deposition in joints.

Biochemically there is high plasma calcium and low plasma phosphorus, with increased excretion of calcium in the urine and a raised serum alkaline phosphatase.

- Excessive absorption of calcium from the bowel may be caused by:
  - Excess Vitamin D following overenthusiastic therapeutic administration.
  - Vitamin D hypersensitivity, for example in sarcoidosis and idiopathic hypercalcaemia of infants.
  - Milk (alkali) syndrome. Increased milk intake increases the amount of calcium taken in and absorbed. If there is also ingestion of alkalis as may occur in the treatment of peptic ulceration, this may make the nephrocalcinosis worse.
- Release of calcium from the bones may occur in:
  - Osteoporosis following prolonged immobilisation.

- Malignant infiltration and destruction of bone as may be found in metastatic carcinoma, myeloma, Hodgkin's disease and the lymphomas.

- Inappropriate secretion of parathyroid hormone-like substance by tumours such as certain carcinomas of the lung.
- Other causes include hyperthyroidism where there is increased metabolic activity in bone; diuretic therapy with thiazides which reduces the urinary output of calcium; and rarely, Addison's disease where there is hypofunction of the adrenal glands.

Causes of metastatic calcification without hypercalcaemia include:

- **Secondary hyperparathyroidism** in which parathyroid activity is stimulated by hypocalcaemia and calcium is taken from the bones to restore blood levels. This usually occurs in renal failure where phosphate retention leads to a low serum calcium. Skeletal lesions with mixed osteomalacia/osteitis fibrosa cystica may be seen. There also may be metastatic calcification.

  Biochemically there is a low or normal plasma calcium. Calcium seems to be withdrawn from the bones and deposited elsewhere. Serum phosphorus is increased.
- **Hypoparathyroidism** Paradoxically in this condition in which there is diminished activity of the parathyroids and a low plasma calcium, soft tissue calcification, particularly in the brain and in the eye may occur.
- **Renal tubular acidosis** is an inherited condition in which the kidney cannot excrete hydrogen ions. In this condition there is metastatic calcification restricted to the kidney.

CALCULI

Calculi or stones are concretions formed in hollow organs or ducts, particularly in the urinary tract in the pelvis, ureter or bladder, in the gall bladder or in ducts of the salivary glands or pancreas and in the prostate. They mostly contain calcium in some combination and are radio opaque but some stones, for example pure cholesterol stones of the gall bladder are radiolucent and need to be demonstrated by ultra sound rather than by X-ray. Hypercalcaemia contributes to the formation of urinary stones.

CALCINOSIS

Calcinosis is a disorder or series of disorders in which calcium salts are deposited in the skin or subcutaneous tissues for no apparent reason. It may be localised or more diffuse.

Tumoral calcinosis is a condition in which large subcutaneous swellings containing plaques of calcium salts occur in relationship to large joints such as the shoulders, elbows, hips and knees. Tumoral calcinosis is common in Black Africans. The cause is unknown but it may be related to trauma and hence a form of dystrophic calcification.

THE MECHANISM OF CALCIFICATION

Two phases are described:

- **Initiation** takes place extracellularly in membrane bound vesicles which have budded off from degenerating or aging cells. Calcium is laid down as hydroxyapatite crystals. A similar process can take place intracellularly in the mitochondria of damaged cells which allow calcium ions in. Mitochondrial calcification is in fact an indication of cell death.
- **Propagation** is the phase in which crystal growth takes place.

GENERAL PATHOLOGY - CHAPTER SEVEN

PIGMENTATION

Exogenous pigments are those which are introduced from outside the body,for example by injection, implantation, inhalation or ingestion.

Endogenous pigments are produced within the body and include melanin, lipofuscins and pigments formed from blood breakdown, that is from the breakdown of haemoglobin from which haemosiderin, haematin, bilirubin and the porphyrins are derived.

EXOGENOUS PIGMENTS

- **Dust**: Inhaled carbon particles from coal dust or smoke accumulate in the macrophages of the lung, regional lymphatics and lymph nodes in most town dwellers. The blackening of the lung is known as anthracosis and has little clinical importance. In coal miners other dusts such as silica may be inhaled and there will be fibrosis of the lung - coal miner's pneumoconiosis.
- **Heavy metals** are usually ingested. Some such as lead and bismuth may be recognised by the deposition of a blue or black line of pigmented, heavy metal compound at the gum margin.
- **Miscellaneous pigments.** Coloured soluble substances derived from food, from drugs such as mepacrine or from sun tan tablets may cause, for example, hypercarotinaemia.

  Various pigments may be introduced into the skin by needle, as in tattoos. They remain for long periods within macrophages or fibrous tissue in the dermis.

ENDOGENOUS PIGMENTS

MELANIN

This is a brown pigment produced by melanocytes and normally found in the skin, the choroid of the eye, the hair and to a lesser extent in mucous membranes and meninges. It also occurs in the cortex of the adrenals and the substantia nigra of the brain. It is an important pigment, chiefly because of its protective effects against sunlight - without it extensive sunburn, blistering and the development of skin cancers can occur. It is not unusual for teenage albinos, who have no melanin in their skin, to present at hospital with squamous cell carcinoma of the face or sun-exposed areas.

Melanin is found in melanocytes, chiefly in the basal layer of the skin. They are thought to be derived from migrated neuro-ectoderm from the neural crest. Melanin is formed from the metabolism of the amino acid tyrosine.

TYROSINE→^{tyrosinase}→DIHYDROXYPHENYLALANINE→MELANIN
(DOPA)

This reaction forms the basis of the 'dopa' test which is used to distinguish melanin *producing* cells (melanocytes) from melanin *containing* cells (melanophages). A histological section is treated with dopa, which is oxidised to melanin by the melanocytes. The melanocytes become blackened and hence can be identified by their dopa positivity.

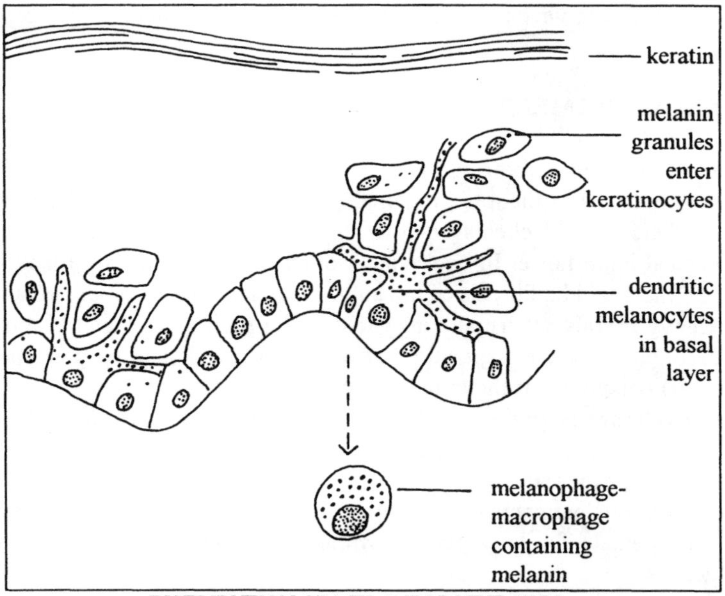

DISTRIBUTION OF MELANIN IN THE SKIN

Melanin will reduce silver salts such as ammoniacal silver and can be bleached by hydrogen peroxide.

Melanin synthesis is brought about by the pituitary melanocyte stimulating hormone (MSH), which is closely related to ACTH which is also produced by the pituitary.

DISORDERS OF MELANIN PIGMENTATION

Increased melanin pigmentation may be seen in:

- Addison's disease which is hypofunction of the adrenal gland. Excess ACTH production occurs. There is pigmentation in exposed and irritated parts and in the buccal mucosa.

- Pregnancy. There may be blotchy pigmentation of the face together with darkening of the nipples and genitalia.
- Haemochromatosis in which there is an accumulation of iron within the tissues of the body. This may be accompanied by excess melanin pigmentation in the skin.
- Patchy or localised melanosis may occur in freckles or *cafe au lait* spots where there is a focal increase in melanin pigment of the skin without proliferation of melanocytes, for example in neurofibromatosis or in Peutz Jeghers syndrome, where patchy pigmentation of the lips and mouth is seen together with polyposis of the small bowel. In lentigines there is an increase in the number of melanocytes as well as in the amount of pigment. There is pigmentation in skin tumours of melanocytes such as moles and malignant melanomas.

Decreased melanin pigmentation may be seen in:

- Albinism. In the absence of the enzyme tyrosinase, melanin cannot be formed and sufferers have white skin pale, blond hair and pink, non-pigmented eyes. Some other less well defined enzyme defects may also be involved. Sufferers are liable to marked sun damage to the skin and skin cancer.
- Localised depigmentation can occur in vitiligo (focal loss of melanocytes within the epidermis) or in association with other disease processes such as leprosy or areas of scarring.

Pigments related to melanin:

- Ochronosis is a condition in which there is an inborn error of metabolism resulting in the accumulation of homogentisic acid. A melanin-like pigment is deposited for example in the cartilages of the ear, nose, larynx, trachea, joints, tendons, ligaments and fibrous tissue.
- Melanosis coli is brown or black discolouration of the mucosa of the large bowel due to the presence of melanin-like pigment in macrophages in the lamina propria of the colon. The origin of this pigment is uncertain.

LIPOFUSCIN

Lipofuscin pigments are golden brown. They are thought to be 'wear and tear' pigments resulting from cell injury or death in a tissue.

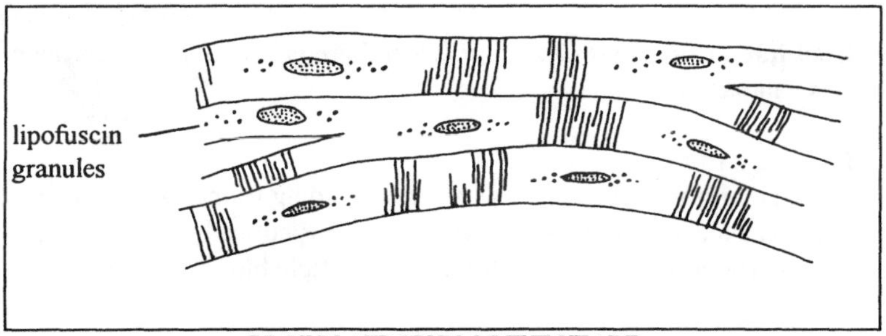

BROWN ATROPHY OF THE HEART

It is thought that free radical injury and lipid peroxidation of cell membranes results in indigestible substances accumulating within lysosomal autophagic vacuoles. Pigment accumulates in cells within the heart (brown atrophy) and liver particularly. Lipofuscins may give positive results with fat stains and with acid fast stains.

Lipofuscins may be autofluorescent when viewed under ultraviolet light

PIGMENTS DERIVED FROM HAEMOGLOBIN BREAKDOWN

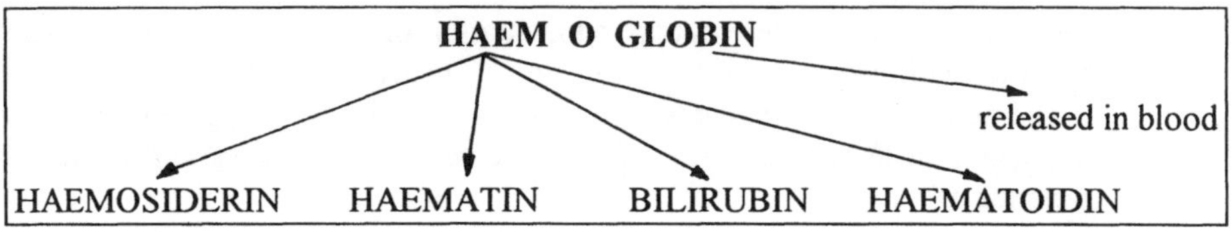

HAEMOSIDERIN

Haemosiderin is the form in which excess iron is stored in the cells and appears as a golden brown pigment (which is a somewhat lighter shade than melanin).

Iron is normally stored within cells in combination with a protein, apoferritin, as ferritin. When there is an excess of iron, ferritin aggregates to form insoluble golden brown haemosiderin granules.

The Prussian blue reaction (also known as Perl's stain) is obtained by treating haemosiderin with hydrochloric acid and potassium ferrocyanide forming blue-green ferric ferrocyanide complex.

Localised haemosiderin deposition is generally seen around areas of haemorrhage such as in tissue injury or haemorrhagic infarcts.

Most of the pigmented breakdown products of haemoglobin can be seen in a bruise. At first this is red or reddish blue due to the escape of red cells into tissues. This becomes greenish with the formation of biliverdin, more yellow with the formation of bilirubin and eventually brownish black as these are absorbed and haemosiderin is left.

Haemosiderin may often be seen in the lung. Small haemorrhages may occur in congestive conditions of the lung such as mitral stenosis and left-sided heart failure. At first haemosiderin granules may be seen within the cytoplasm of macrophages within the alveolar spaces. Later it is also seen within macrophages in the alveolar wall. If haemosiderin escapes from within cells it can give rise to fibrosis and encrustation (ferruginisation) of elastic fibres within the lung tissue.

Generalised haemosiderosis can occur when there is iron overload of the body as a whole. This can happen if:

- Too much iron is absorbed
- There is increased formation of red cells leading to increased iron absorption
- There is excess iron in the diet or given therapeutically
- Excess iron has accumulated following multiple blood transfusions.

Initially excess iron is stored as haemosiderin in the phagocytic cells of the reticulo-endothelial system, for example, in the liver, bone marrow and spleen. Only later, when the iron overload is even greater, does haemosiderin enter parenchymal cells of liver, pancreas, heart and endocrine organs.

Generalised haemosiderosis occurs in:

- Haemochromatosis, which is an inherited abnormality probably at the level of the intestinal mucosa, which results in the absorption of excessive amounts of iron and widespread deposition of haemosiderin in the liver, pancreas and skin. The condition is known as 'bronze diabetes' because of pigmentation of the skin with both haemosiderin and melanin, and because damage to the pancreas results in diabetes. There is also fibrosis and cirrhosis of the liver and hepatocellular carcinoma may develop subsequent to the cirrhosis.

  Pigment deposited in the parenchymal cells of the liver brings about their destruction and fibrous replacement.

  The disease occurs in males more commonly than females, who may be protected by monthly losses of iron. The body normally contains only 5g of iron. In full blown haemochromatosis this may be increased to more than 50g which, according to Robbins, is enough to set off the metal detector alarm at an airport.
- African siderosis results from a high dietary intake of iron, particularly if the diet is inadequate in other respects. The consumption of beer brewed in iron containers predisposes to excess iron absorption as alcohol aids absorption. Childhood anaemia may contribute by conditioning excess absorption for life. Iron deposition is at first reticulo-endothelial and then the parenchymal cells become involved. For unknown reasons, possibly dose-related, the haemosiderin seems to cause less damage than in haemochromatosis.
- Transfusion and haemolytic siderosis may occur after multiple blood transfusions but also due to increased absorption of iron in haemolytic states. Although considerable amounts of haemosiderin may accumulate, fibrosis is less likely than in true haemochromatosis.

Thus haemochromatosis is a specific disease. However, overloading from any cause if of sufficient degree, can give rise to a similar clinical and pathological picture.

HAEMATIN

This term is used to describe a substance or substances found in a number of conditions. It is believed to be derived from haem, probably by the oxidation of the ferrous ion to the ferric form. It is a bluish black pigment which may be found in macrophages in a number of organs, particularly the liver and the spleen, in malaria and in schistosomaisis. In **malaria** it is commonly found in the Kupffer cells in the liver. In **schistosomiasis** it may also be found in Kupffer cells, but is also likely to be found within portal tracts. The pigment may be a metabolic product of haem formed by the parasites. It does not give a Prussian blue reaction, but is birefringent (sometimes weakly) when viewed under polarised light. This feature allows it to be distinguished from haemosiderin.

A similar pigment may be noted in congested organs which have been fixed in formalin for a long time; this is more strongly birefringent.

These pigments may be removed from sections by treatment with a saturated alcoholic solution of picric acid.

HAEMATOIDIN AND BILIRUBIN

Haematoidin is probably the same as bilirubin chemically, but it is a bright yellow crystalline form which may be found locally in tissues around areas of haemorrhage. Bilirubin is the pigment (non-iron containing) found in the bile. Excess bilirubin gives rise to jaundice.

GENERAL PATHOLOGY - CHAPTER EIGHT

JAUNDICE

Bilirubin is a yellow pigment formed from the 'haem' component of haemoglobin after the removal of iron. Most bilirubin (about 80 per cent) is derived from haemoglobin released from worn out red cells which are broken down in the spleen, liver and bone marrow. Much of the rest comes from other 'haem' containing proteins such as myoglobin, catalase and the cytochromes. A small amount is derived from the premature breakdown of newly formed red cells in the bone marrow (ineffective erythropoiesis).

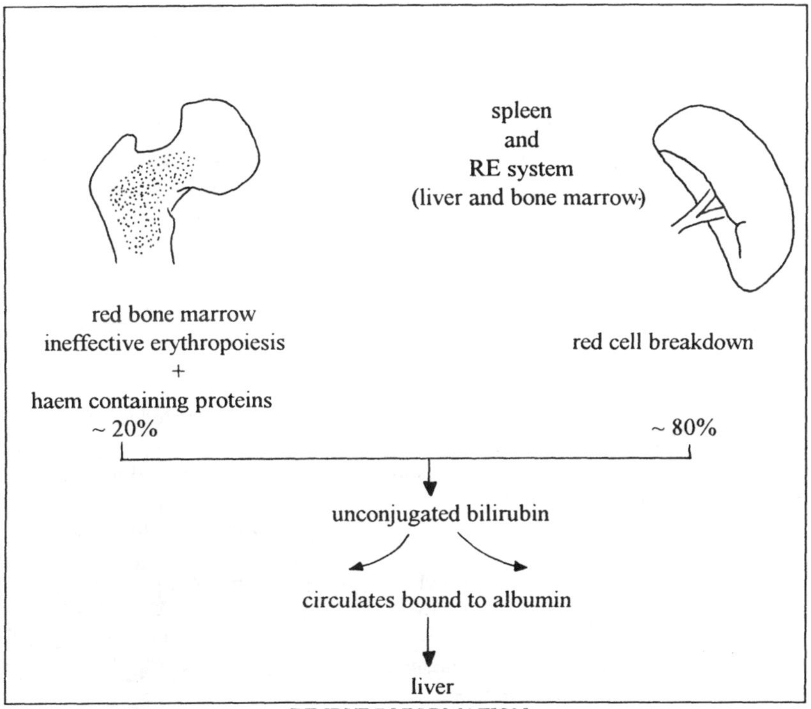

BILIRUBIN FORMATION

Newly formed bilirubin is insoluble in water and is transported through the circulation bound to albumin. In the liver the bilirubin-albumin complex binds to the cell membrane, albumin is released and the bilirubin is taken up by the cell. In the cytoplasm it is bound to another protein (ligandin) and transported to the endoplasmic reticulum where it is conjugated into bilirubin diglucuronide by the action of the enzyme uridine diphosphate (UDP) glucuronyl transferase.

A little monoglucuronide is formed, and there is also a small quantity of unconjugated bilirubin in the cell.

Unconjugated bilirubin differs from conjugated bilirubin in a number of ways:

Unconjugated bilirubin	Conjugated bilirubin
Binds with albumin	Conjugated as glucuronide
Is insoluble in water	Water soluble
Is fat soluble	Not soluble in fat
Not excreted in urine	Can be excreted in urine
Toxic	Non-toxic
*Indirect van den Bergh	Direct van den Bergh

*Note that bilirubin is measured by using the diazo reagent (sulphanilic acid treated with a mixture of sodium nitrite and hydrochloric acid). This reagent forms a red compound with conjugated bilirubin (direct van den Bergh reaction). There is no reaction with unconjugated bilirubin until alcohol is added to dissolve the bilirubin, when the red colour appears immediately (indirect van den Bergh reaction).

Following on conjugation, the bilirubin diglucuronide is transported across the cell membrane into the bile canaliculi by an active process, along with bile acids and other constituents of bile (cholesterol, phospholipids and electrolytes).

Conjugated bilirubin then flows along the biliary tract to the gall bladder and eventually to the intestine, where some is excreted and the rest is converted into urobilinogen by bacterial action. Much of this urobilinogen is excreted directly in the faeces (as stercobilinogen), but some is absorbed in the intestine and passes to the liver to be re-excreted in the bile as bilirubin. A little seems to pass through the liver into the systemic circulation to be excreted by the kidney as urobilinogen.

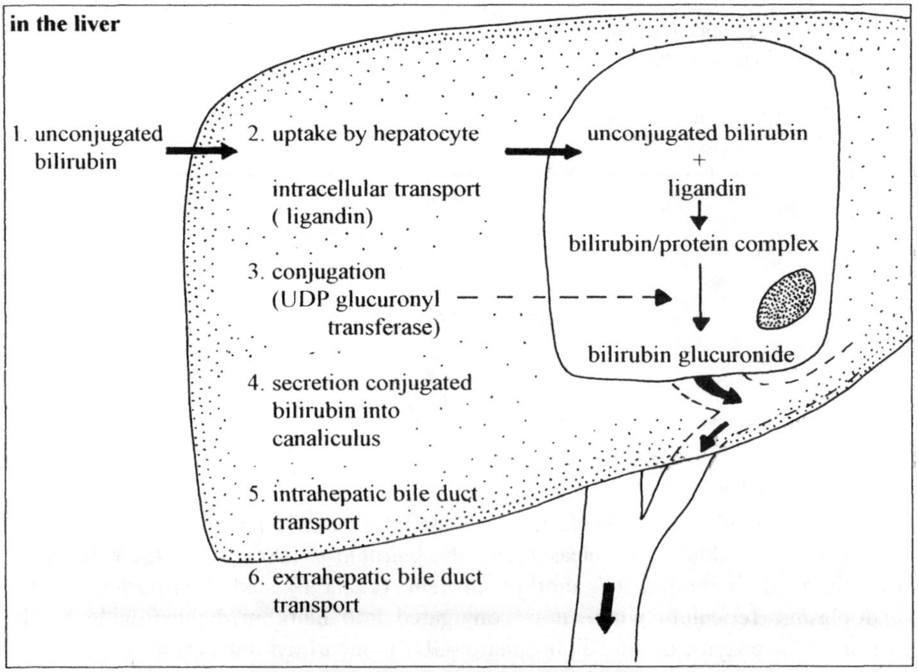

BILIRUBIN METABOLISM IN THE LIVER

in the gut

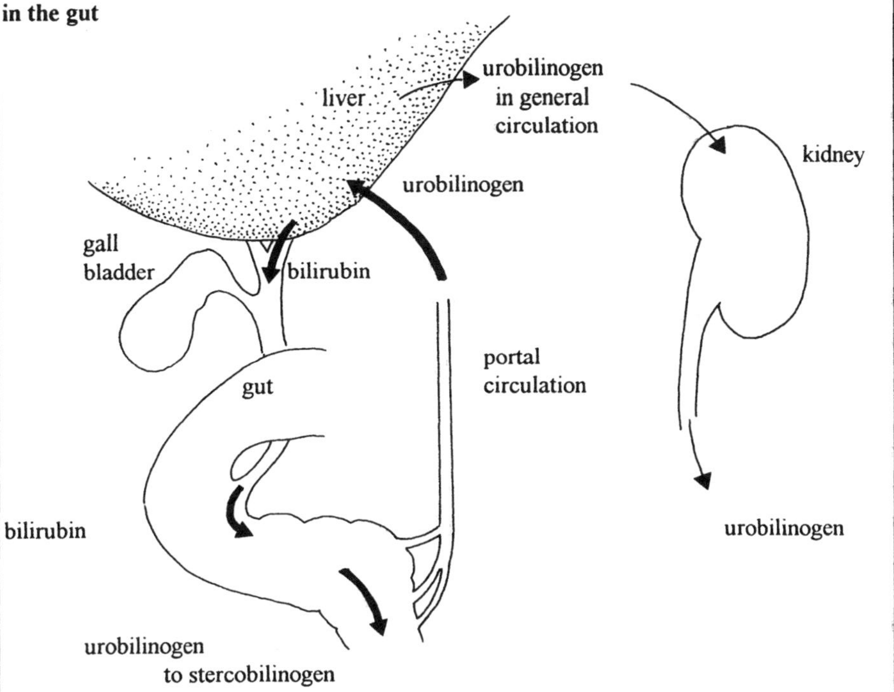

BILIRUBIN METABOLISM IN THE GUT

JAUNDICE

Jaundice occurs when bilirubin accumulates in the tissues and interstitial fluid of the body. It can first be noticed as a yellow discolouration of the sclera of the eyes followed by similar discolouration of the skin, when bilirubin levels reach 35-50μ mol/l (normal serum or plasma bilirubin is 5-17μmol/l).

Jaundice due to the accumulation of unconjugated bilirubin:

- **Overproduction of bilirubin,** for example, due to excessive breakdown of red blood cells.
- **Decreased hepatic uptake of bilirubin,** for example, when hepatic cells are damaged as in viral hepatitis or due to the action of drugs (rifampicin, probenecid).
- **Decreased bilirubin conjugation,** for example, hereditary lack of glucuronyl transferase or interference with its action by drugs (chloramphenicol) or diffuse disease such as hepatitis or cirrhosis.

Jaundice due to the accumulation of conjugated bilirubin

- **Decreased cellular secretion of conjugated bilirubin,** for example, hereditary conditions, drugs such as oestrogens and recurrent jaundice of pregnancy

- **Impaired canalicular bile flow** in some cases of viral hepatitis, alcoholic liver disease or due to drugs such as chlorpromazine or methyl testosterone.
- **Extrahepatic biliary obstruction** due, for example, to gallstones, carcinoma of the head of the pancreas, enlarged lymph nodes in the porta hepatis, stricture or tumour of the bile ducts, extrahepatic congenital biliary atresia.

To enlarge upon someof these causes of jaundice:

Overproduction of bilirubin

This is almost always caused by increased destruction of red blood cells or haemolysis.

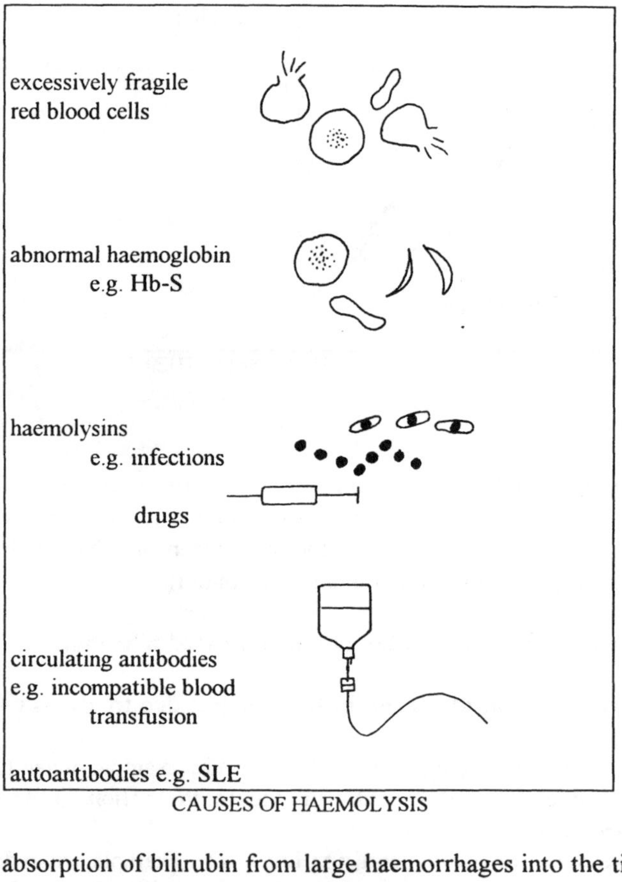

CAUSES OF HAEMOLYSIS

There may be absorption of bilirubin from large haemorrhages into the tissues or the gut.

In foetal-maternal blood group incompatibility, maternal antibodies cause haemolysis of foetal blood cells and accumulation of unconjugated bilirubin. The newborn baby cannot metabolise this because there is usually a deficiency of glucuronyl transferase in liver cells in the first few days of life and there may also be a deficiency of ligandin.

When the unconjugated bilirubin has exhausted the albumin binding capacity of the body, excess unconjugated bilirubin can cross the blood-brain barrier and will dissolve in lipid rich areas of the brain. This causes damage to neurones and stains yellow certain areas of the brain, such as the basal ganglia and hypothalamic nuclei. This is known as **kernicterus.**

Levels of bilirubin of about 340μmol/l (20mg/dl) are needed before there is kernicterus but lower levels can be dangerous if drugs which bind with albumin (sulphonamides and salicylates) are given or if there is a metabolic disturbance which will increase the permeability of the blood brain barrier. The blood-brain barrier to unconjugated bilirubin is not well developed in the newborn. In kernicterus damage to the neurones is irreparable. Clinically the child develops involuntary movements, spasticity and deafness.

Decreased bilirubin conjugation

This, as has been mentioned, can occur in the newborn and there may be jaundice of the newborn without foetal-maternal blood group incompatibility or pathological haemolysis. This is so called 'physiological' jaundice due to low levels of glucuronyl transferase in the first few days of life. Unconjugated bilirubin levels rarely rise above 204μmol/l (12mg/dl). Unconjugated bilirubin is susceptible to photo-oxidation and produces non-toxic metabolites, so one method of treatment is to expose the baby to bright blue or white fluorescent light. Treatment with phenobarbitone will also increase glucuronyl transferase levels.

When there is an increase in the level of unconjugated bilirubin no bilirubin will appear in the urine but if hepatic cells can conjugate some bilirubin there may be an increase in stercobilinogen in the faeces and urobilinogen in the urine.

Decreased cellular secretion of conjugated bilirubin

This occurs in certain hereditary conditions (to be mentioned later) and may be drug induced.

- Chlorpromazine and certain other drugs such as para-aminosalicylic acid seem to cause intrahepatic cholestasis regardless of the dose given, probably on an allergic basis.
- Halothane can cause hepatic damage due to hypersensitivity.
- Certain steroid hormones such as methyl testosterone can also cause cholestatic drug jaundice and will do so in most cases if a large enough dose is given.
- Benign familial recurrent cholestasis is an obscure disease with no known cause.
- Recurrent jaundice of pregnancy give a picture of obstructive jaundice in the last few months of pregnancy and may be due to oestrogen sensitivity.

CHOLESTASIS

Cholestasis is the term given to conditions in which there is an obstruction to the outflow of bile.

When it is due to intrinsic liver disease it is known as **intrahepatic cholestasis** and when it is due to obstruction of the large bile ducts it is known as **extrahepatic cholestasis**.

- In intrahepatic cholestasis greenish-brown plugs of bile can be seen in bile canaliculi in the centrilobular area. There may also be drops of bile within actual liver parenchymal cells and in Kupffer cells. The liver cells show feathery degeneration (due to intracellular accumulation of water).
- In extrahepatic cholestasis there is a build up of bile which gradually extends into the intrahepatic biliary system in bile ducts in the portal tracts and eventually into canaliculi. These can then rupture with release of bile which forms bile lakes surrounded by necrotic or degenerate liver cells.

  This change is generally characteristic of extra hepatic biliary obstruction. If there is ascending infection there may be cholangitis and neutrophils within portal tracts.

  Within portal tracts there may be proliferation of small bile ducts and some increase in portal fibrosis.

In cholestasis of both intrahepatic and extrahepatic types there is an increase in conjugated bilirubin in the blood and conjugated bilirubin and bile salts in the urine. Bilirubin does not reach the gut and stools are pale and fatty because fat is not well absorbed in the absence of bile salts. There is no urobilinogen in the urine. Increased bile salts in the circulation may give rise to a distressing pruritus (itching). There may be hypercholesterolaemia and hyperlipidaemia. The alkaline phosphatase level is raised. Impaired fat absorption may result in deficiency of the fat soluble Vitamin K and hypoprothrombinaemia and a bleeding tendency.

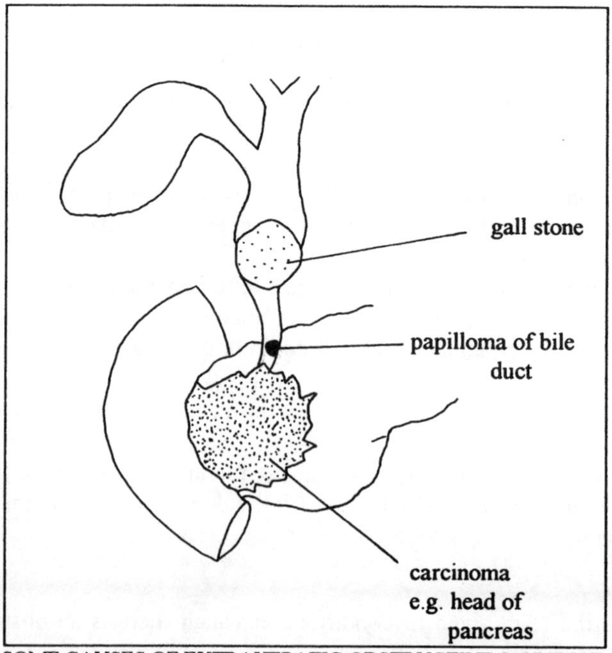

SOME CAUSES OF EXTRAHEPATIC OBSTRUCTIVE JAUNDICE

Hepatocellular damage may cause jaundice for a number of reasons and there may be an accumulation of both unconjugated and conjugated bilirubin.

Liver cell necrosis may be patchy, diffuse, or zonal.

Patchy and/or diffuse necrosis occurs in infections of the liver, in viral hepatitis, or in cirrhosis due to some drugs and poisons.

Zonal necrosis may be:

- Centrilobular - as in chloroform or carbon tetrachloride poisoning or overdose of acetoaminophen; or due to ischaemic injury or marked venous congestion.
- Midzonal necrosis is an unusual pattern seen in yellow fever.
- Peripheral necrosis may be seen in phosphorus poisoning and eclampsia.

Massive or sub-massive necrosis may result from fulminant hepatitis or drug or chemical poisoning.

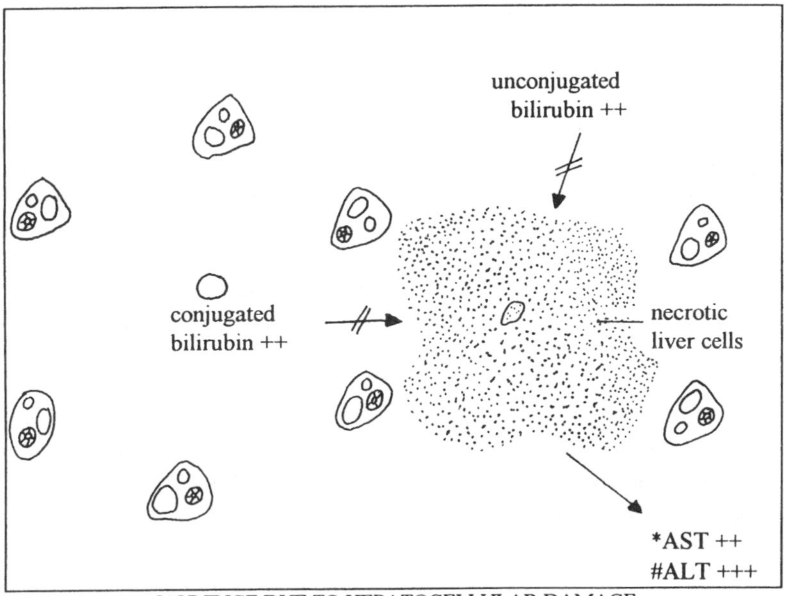

JAUNDICE DUE TO HEPATOCELLULAR DAMAGE

Note: *AST = aspartate transaminase; formerly SGOT or serum glutamic oxaloacetic transaminase
#ALT = alanine transaminase; formerly SGPT or serum glutamic pyruvic transaminase.

Liver cell necrosis must be severe before jaundice occurs. Histology will reveal this necrosis and damage to the liver cells.

Initially the damaged liver cells are not able to conjugate the bilirubin reaching them. Hence there is an accumulation of unconjugated bilirubin. However, in various forms of hepatitis, toxic hepatic damage and cirrhosis there is also swelling of liver cells and blockage of the canaliculi so that whatever bilirubin is conjugated cannot be excreted and hence an obstructive element is added. This means that there may be a mixed picture with both types of unconjugated and conjugated bilirubin in the blood; the changes in the urine and faeces are also variable.

The diagnosis of liver cell damage may be made biochemically because damaged liver cells release alanine transaminase (ALT; SGPT) and aspartate transaminase (AST; SGOT) into the blood. These enzymes are released following damage to cytoplasmic and/or mitochondrial cell membranes. There may also be changes in the serum proteins in prolonged and severe liver disease with diminished prothrombin and albumin synthesis.

Genetic metabolic abnormalities are conditions in which there are disturbances of bilirubin transport mechanisms or enzyme deficiencies leading to hyperbilirubinaemia. Apart from Crigler-Najjar disease they are not very severe causing jaundice without other serious clinical effects.

- **Crigler-Najjar disease** is an hereditary disorder in which there is reduced or absent glucuronyl transferase activity, and hence lack of conjugation of bilirubin. In the more severe form, which is transmitted in autosomal recessive form, unconjugated bilirubin accumulates in early infancy leading to death from kernicterus. In the less severe form (which is transmitted as an autosomal dominant characteristic of variable penetrance) there is some glucuronyl transferase activity and the disease responds to treatment with phenobarbitone.
- **Gilbert's disease** is a milder form of unconjugated hyperbilirubinaemia which is quite common, affecting up to seven per cent of the population. There seems to be a combination of defective mechanisms involved, including deficient glucuronyl transferase activity with decreased conjugation of bilirubin and possibly also decreased hepatic uptake of bilirubin.
- **Dubin-Johnson syndrome** is transmitted as an autosomal recessive. There is an accumulation of conjugated bilirubin due to a defect in the transport of conjugated bilirubin out of the cell and into the canalicular system. There is also accumulation of a melanin-like pigment in the lysosomes of liver cells in this condition, which gives the liver a dark grey appearance macroscopically.
- **Rotor syndrome** is similar to the Dubin-Johnson syndrome except that pigment does not accumulate in the liver cells.

GENERAL PATHOLOGY - CHAPTER NINE

FLUID AND ELECTROLYTE BALANCE

In this chapter the fluid and electrolyte content of the body will be reviewed before considering the circulation and some of its disorders. Water is the principle constituent of the body forming some 60 per cent of the body mass in the normal adult male and a rather smaller proportion (55 per cent) in the female. Fat contains virtually no water - hence the sex difference. Electrolytes are contained within body fluids and all fluids must be maintained within a narrow pH range.

DISTRIBUTION OF BODY WATER

In a normal 70kg adult there are about 42 litres of total body water of which two-thirds are in the cells and one-third is in the extracellular compartment. Of the extracellular fluid about 3 litres is contained within the circulatory system.

intracellular	**extracellular**
28 litres	interstitial 10 litres
	plasma 4 litres
TOTAL	**42 litres**

DISTRIBUTION OF BODY WATER

Mechanisms for the maintenance of body water include:

- Thirst when there is free access to water
- Loss via urine, breathing, skin and gut

Within the body a number of mechanisms exist to maintain the amount of water or sodium (with which water is linked) in the body.

- **Antidiuretic hormone** (ADH) or **arginine vasopressin** (AVP) as it is now called. If water is lost the osmolality of the extracellular fluid increases and osmoreceptors in the hypothalamus are stimulated to produce ADH which increases reabsorption of water by the kidney tubules. If osmolality is low ADH is not secreted and water resorption by the distal tubules is reduced.

If sodium is retained, water is retained with it providing that there is access to water and the above mechanism is active.

Sodium is controlled by:

- **Glomerular filtration rate:** If this is reduced, urinary excretion of sodium is lessened and sodium is retained. This mechanism comes into effect if there is diminished cardiac output, for example, in blood loss or shock.
- **The renin-angiotensin-aldosterone mechanism:** Renin is secreted by the juxtaglomerular apparatus if there is a fall in renal arteriolar pressure, or a drop in sodium concentration in the distal tubules. When renin is released it acts on the angiotensin system to bring about aldosterone secretion by the adrenal cortex. Aldosterone brings about reabsorption of sodium in the distal renal tubules.
- **Atrial natriuretic peptide** (ANP) is released in response to atrial dilatation (due to the expansion of the intravascular compartment) and acts to inhibit the mechanisms outlined above for sodium and water retention.

Intracellular fluid (ICF) remains relatively constant, varying only in extreme conditions. Within the cell a different set of ions, potassium, magnesium, phosphate and protein are in relatively high concentration as opposed to the sodium, chloride and bicarbonate which are found in higher concentration in the extracellular fluid. Normally the cell maintains homeostasis by the active transport of sodium out of the cell against a physico-chemical gradient, but the fluid medium is more or less isotonic with the surrounding extracellular fluid. In damage to the cell, cloudy swelling with water accumulation can occur.

Extracellular fluid (ECF) There is transport of fluid from the intravascular compartment of the interstitial compartment via the endothelium of the vessel wall which acts as a semi-permeable membrane. Protein is retained within the vessels and exerts a colloid osmotic pressure (oncotic pressure) which counterbalances the hydrostatic pressures.

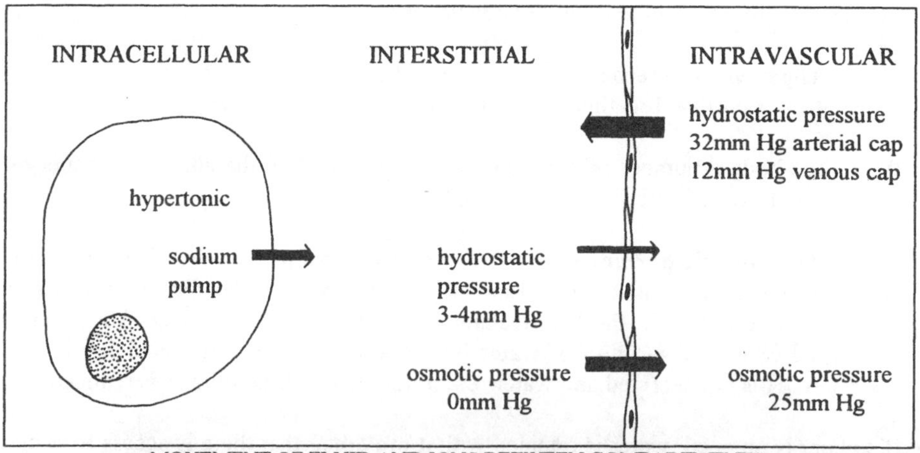

MOVEMENT OF FLUID AND IONS BETWEEN COMPARTMENTS

Damage to vessel walls allows loss of protein from the vascular space, hence loss of colloid osmotic pressure from vessels and extracellular fluid will accumulate.

WATER SODIUM IMBALANCES

These are extremely important and common conditions.

PREDOMINANT WATER DEFICIENCY (Hypo-osmotic fluid loss)

Cause: This occurs following water deprivation or in diabetes insipidus

Effect: ECF concentration rises, there is a transfer of cellular water and hence intracellular dehydration.

Clinically: There is thirst, oliguria with concentrated urine and a concentration of plasma with the rise in Na^+ and protein.

Treatment: Water by mouth or glucose solution by transfusion.

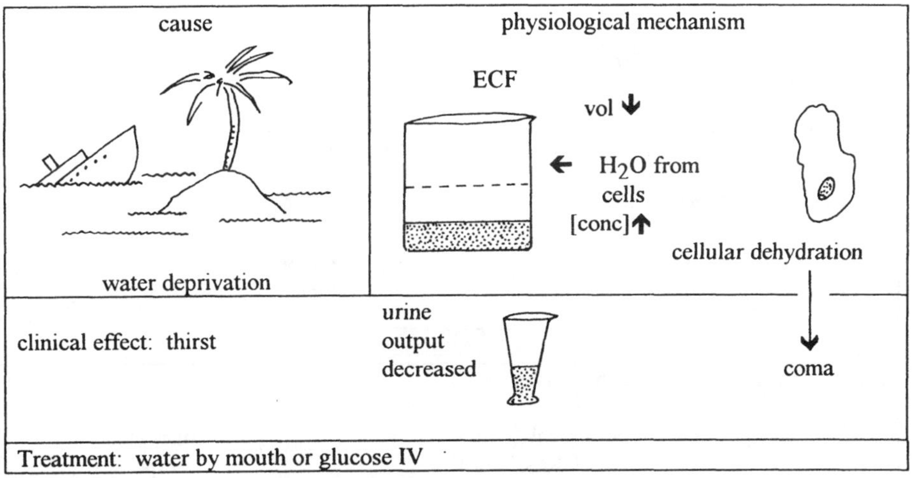

WATER DEPRIVATION

COMBINED SODIUM/WATER DEFICIENCY (Iso-osmotic fluid loss)

Causes: Losses from the GI tract - vomiting and diarrhoea
Excess sweating
Urine - Diuretics in heart failure
- Addison's disease
- Salt-losing nephritis
- Diabetes mellitus

Effect: ECF loses sodium and water, but mainly sodium. Therefore, there is loss of osmotic pressure, and water enters the cells. This water loss leads to diminished ECF volume and shock.

Clinically: There is dehydration. The urine contains virtually no sodium and chloride and there is reduced plasma volume.

Treatment: Saline, usually by transfusion

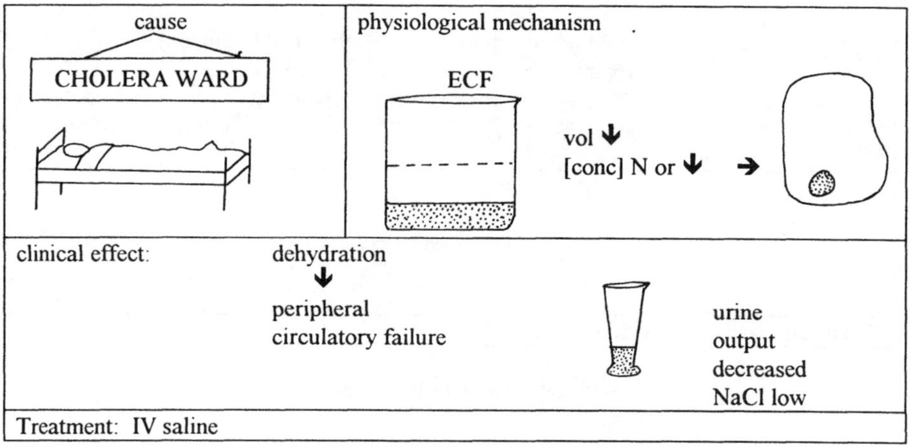

SODIUM AND WATER LOSS

PREDOMINANT WATER EXCESS (Hypo-osmotic fluid retention)

Causes: Following the administration of post operation fluids, glucose and the like,
Water to sodium depleted patients,
Water in renal failure,
ADH excess.

Effect: ECF is distended and dilute - water enters the cells.

Clinically: There is nausea, vomiting, weakness and muscle cramps.
Cerebral manifestations are headache, drowsiness, convulsions and coma.
Plasma shows a reduced concentration of electrolytes. Death can occur due to cellular damage.

Treatment: Hypertonic saline.

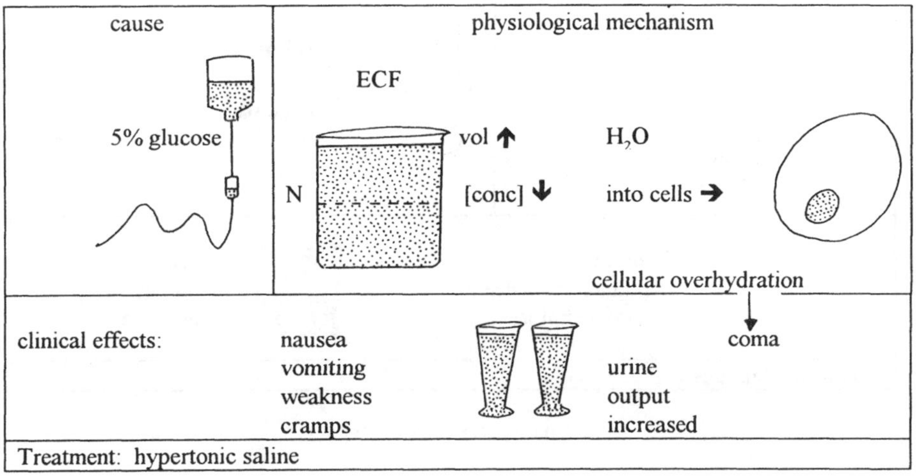

WATER EXCESS

COMBINED SODIUM AND WATER EXCESS (Iso-osmotic fluid loss)

Cause: Excess administration of saline (especially when there is defective renal function).

Effect: There is an even distribution of saline through the extracellular compartment, with increased venous pressure.

Clinically: Fluid overload and increased venous pressure leads to generalised oedema. Pulmonary oedema may also be noted; congestive cardiac failure may develop, and this will aggravate the oedema.

Treatment: Diuretic therapy

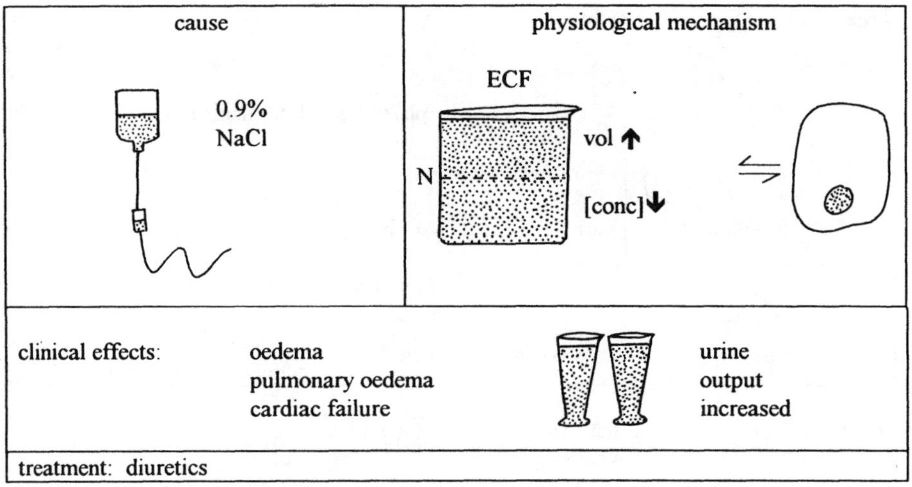

SODIUM AND WATER EXCESS

POTASSIUM is the main intracellular cation; only two per cent of body potassium is in the extracellular compartment. There is a relatively low concentration in plasma normally. Potassium may be lost from the cells in water deprivation.

Potassium is driven into the cells by glucose and insulin. Intracellular potassium is not reflected by its plasma concentration.

Potassium Deficiency

Causes: Loss from the GI tract: Vomiting
Diarrhoea, for example cholera, pancreatic islet cell tumour, potassium losing villous adenoma
Urine: Osmotic diuresis of diabetes
Diuretics
Hyperaldosteronism
Renal disease
Insulin treatment of diabetic coma, driving potassium into the cells and causing hypokalaemia

Effects: There is muscle weakness leading to paralysis and respiratory failure
mental confusion
Renal vacuolation of proximal tubular cells, and
Cardiac irregularities and flattening of T Waves in ECG.

Treatment: Infuse potassium

Potassium Excess

Causes: Renal failure, Addison's disease and excess administration.
Extensive tissue destruction.

Effects: There is confusion
Muscle weakness sometimes
Cardiac irregularities and high T Wave in ECG
Ventricular fibrillation may result and prove fatal.

Treatment: Give glucose and insulin to drive potassium into the cells.

ACID-BASE BALANCE

The normal pH of the body is 7.4. Virtually all the body's metabolic activities produce acid (that is H^+ ions). There are mechanisms which exist, firstly to buffer the effect of added H^+ ions and secondly to remove them from the body.

Plasma buffers. These consist of a strong base and a weak acid formed from it. They include:

• Bicarbonate	HCO_3^-	$+ H^+ \rightleftharpoons H_2CO_3 \rightleftharpoons H_2O + CO_2$
• Monohydrogen phosphate	HPO_4^{--}	$+ H^+ \rightleftharpoons H_2PO_4^-$
• Protein (including haemoglobin)	Protein	$+ H^+ \rightleftharpoons$ H Protein

Mechanisms for H^+ Removal:

- Lungs: In the lungs CO_2 is blown off and this means that one bicarbonate ion can remove one hydrogen ion according to the equation above.
- Kidney: In the urine hydrogen ions are excreted, but bicarbonate ions are retained by the water being excreted but CO_2 in the tubule being resorbed and reformed into bicarbonate.
  Phosphate buffers in the urine carry the excreted hydrogen ion.
  Ammonia in the urine accepts hydrogen ion to form an ammonium ion.

Acidosis and Alkalosis

Acidosis results from a decrease in the ratio $(HCO_3^-) : PCO_2$
Alkalosis results from an increase in the ratio $(HCO_3^-) : PCO_2$
In compensated acidosis or alkalosis the ratio may be normal although values for (HCO_3^-) or PCO_2 may be abnormal.

Respiratory acidosis occurs when there is underventilation (and respiratory failure) with CO_2 retention.

Respiratory alkalosis results from overbreathing, due to hysteria or lung disease, or overventilation, for example during anaesthesia. Too much CO_2 is lost.

Metabolic acidosis occurs with accumulation of H^+ ions and the bicarbonate being used up in buffering. This can occur:

- In hypoxic conditions such as shock, when there is anaerobic metabolism of cells with lactic acid production
- In starvation or uncontrolled diabetes mellitus when body fat stores are being mobilised to an excessive extent and ketone bodies (acid) accumulate
- In renal disease when the kidney is unable to excrete H^+ ions
- Severe diarrhoea
- Uretero-colic anastomosis

Metabolic alkalosis is due to the accumulation of HCO_3^- as in excess ingestion of alkalis or in loss of acid such as may occur with severe vomiting. It may also occur in severe hypokalaemia as in primary hyperaldosteronism or Cushing's syndrome.

		Arterial blood				
Cause		pH	PCO_2	HCO_3^-	K^+	Urine
Respiratory Acidosis	Under ventilation	↓	↑	N or ↑	↑	NH_3^+ and H^+ions
Respiratory Alkalosis	Overbreathing; anaesthesia; high altitude;salicylate OD	↑	↓	N or ↓	↓	HCO_3^+
Metabolic Acidosis	hypoxia; diabetes; renal failure	↓	N	↓	↑	NH_3^+ and H^+ions
Metabolic Alkalosis	alkalis; loss acid eg vomiting severe hypokalaemia	↑	N	↑	↓	HCO_3^+

The table shows some of the changes which can occur in the chemistry of the blood and urine in acidosis and alkalosis before compensatory changes take place.

The body's compensatory mechanisms are then brought into play, thus deep sighing respirations are a feature of metabolic acidosis - attempting to rid the body of excess hydrogen ions through the lungs. Vasodilation may accompany respiratory acidosis as a direct effect of a raised PCO_2. Tetany may accompany alkalosis due to accompanying decrease in ionised calcium.

STATE	PURE WATER LOSS	PURE WATER EXCESS	SODIUM & WATER LOSS	SODIUM & WATER EXCESS
AETIOLOGY	**Decreased intake-** coma, oesophageal obstruction, shipwreck etc. **Increased loss-** diabetes insipidus	**Increased intake-** p.r. fluids non-saline i.v.fluids **Decreased loss-** renal failure excess ADH secretion	1. **GIT**: diarrhoea and vomiting 2. **Skin**: excess perspiration without replacement 3. **Renal**: diabetes mellitus, osmotic diuresis, Addison's disease excess diuresis with CCF 4. **Blood loss**: haemorrhage or plasma loss e.g. burns	i.v. saline over transfusion
Events that occur with compensation	1. ECF increased OP 2. Therefore ICF→ECF 3. To compensate: increased secretion of ADH and aldosterone resulting in oliguria	1. ECF decreased OP 2. Therefore ECF→ICF 3. To compensate:decreased secretion of ADH and aldosterone 4. Increased urine initially 5. With renal damage, oliguria	1. ECF decreased volume 2. OP between ICF and ECF unchanged. Therefore no marked fluid shift 3. To compensate: increased aldosterone and to a lesser extent ADH	1. ECF increased volume 2. OP between ICF and ECF unchanged. Therefore no marked fluid shift 3. To compensate: decreased aldosterone and to a lesser extent ADH
Compartment ultimately affected	ICC - dehydration which is intracellular	ICC - intracellular overhydration	ECC - decreased volume Hypovolaemia	ECC - increased volume. Fluid overload
CLINICAL	Thirst proceeding to coma Late - fall in b.p. with death	Nausea, vomiting, weakness, muscle cramps. Mental changes leading to convulsion, coma and death	Decreased tissue turgor + eyeball pressure, skin veins collapsed, apathetic, tachycardia	Congestive cardiac failure Pulmonary oedema
BIOCHEMISTRY				
Electrolytes	increased concentration	decreased concentration	unchanged	unchanged
Urea	increased concentration	decreased concentration	increased concentration	decreased concentration
Plasma proteins	increased concentration	decreased concentration	increased concentration	decreased concentration
PCV and Hb	unchanged - rbc's also lose fluid	decreased concentration	increased concentration	decreased concentration
Urine	diminished output	increased flow followed by diminished	diminished output	increased output
TREATMENT	Non-saline fluids	Hypertonic saline	Saline fluids	Diuretics

GENERAL PATHOLOGY - CHAPTER TEN

OEDEMA

Oedema is an excessive accumulation of extracellular fluid in the intercellular (interstitial) tissue spaces or body cavities. It may be localised or generalised.

Severe generalised oedema is called **anasarca.**

Oedema in the peritoneal space is called **ascites.**

Oedema in the pleural space is called **hydrothorax** or **pleural effusion**

Oedema in the pericardium is called **hydropericardium** or **pericardial effusion**

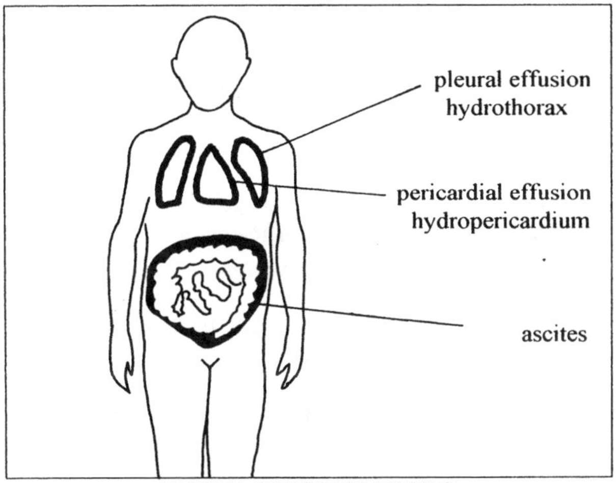

OEDEMA OF BODY CAVITIES

It has already been mentioned that the normal forces governing the distribution of fluid between the vascular compartment and the interstitial fluid are the relative hydrostatic and osmotic pressure.

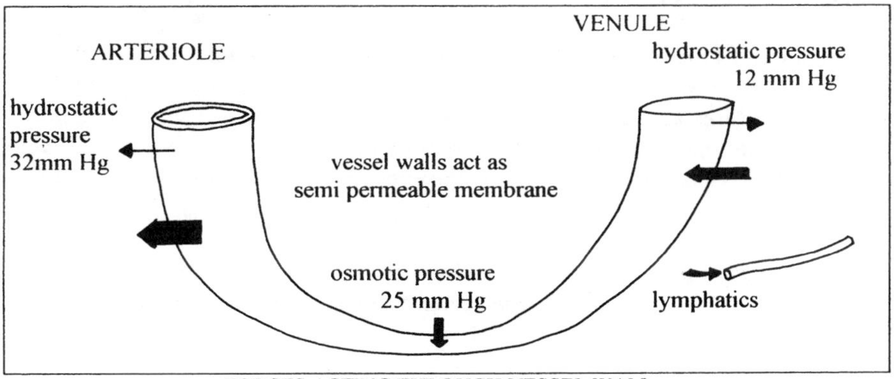

FORCES ACTING THROUGH VESSEL WALL

Disturbances of the balance between fluid in the circulation and the tissues may occur and the following five patho-physiological mechanisms may play a part in the causation of oedema:

- An increase in the hydrostatic pressure within the vessels.
- A decrease in the colloid osmotic pressure within the vessels.
- Changes in fluid balance (with increased extracellular fluid and plasma volume).
- Increased vascular permeability (usually only local).
- Lymphatic obstruction (usually only local).

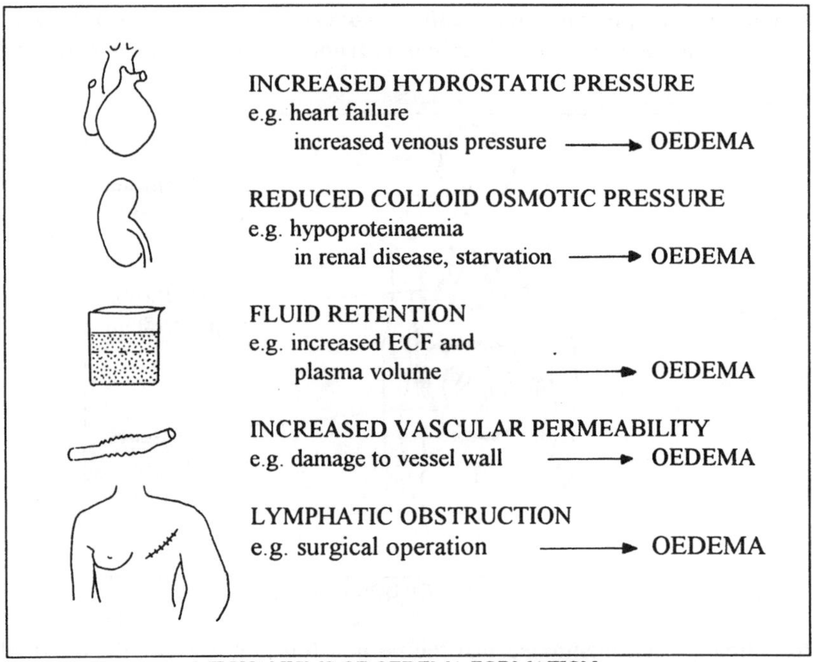

MECHANISMS OF OEDEMA FORMATION

TYPES OF OEDEMA FLUID

There are two types of oedema fluid, **exudate** and **transudate:**

- **Exudates** are due to increased vascular permeability. The proteins pass out of the vessel, colloid osmotic pressure of the vascular system is reduced and there is only tissue tension to stop the outward flow of plasma. Exudates are usually localised and associated with inflammatory changes affecting vessels.

- **Transudates** are due to an imbalance of hydrostatic pressures with normal vascular permeability. These account for the majority of cases of generalised oedema.

	Exudate oedema	Transudate oedema
Protein	Approximately that of plasma (7g/100 ml)	1g/100 ml
Distribution	As in plasma (Alb 5g/100 ml Glob 2g/100 ml)	Nearly all albumin
Fibrinogen	Present, giving spontaneous clotting	Absent, so no clotting
Specific gravity	High, 1018 or more	About 1012
Cells	Numerous inflammatory cells e.g. polymorphs and lymphocytes	Few, mainly mesothelial

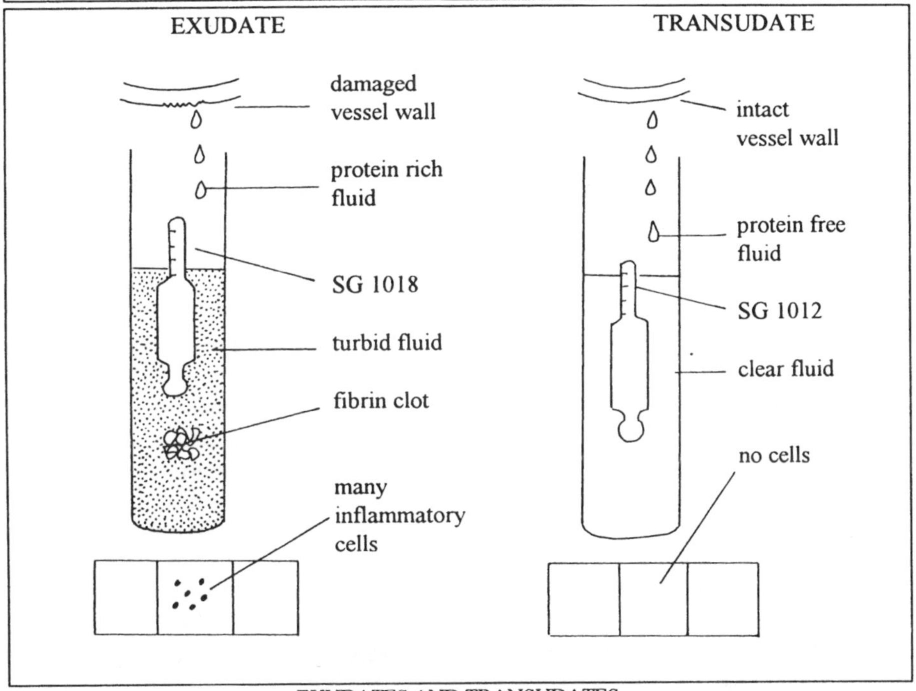

EXUDATES AND TRANSUDATES

Generalised oedema is somewhat more difficult to account for than local oedema in practice. The factors which may be involved include:

- **Increased hydrostatic pressure** (especially venous) of the blood; this may account in part for the generalise oedema of heart failure.
- **Decreased colloid osmotic pressure** of the blood due to a decrease in the circulating protein level. For example, protein, mainly albumin, may be lost via the urine in renal disease such as the nephrotic syndrome.

  In liver disease plasma protein may not be synthesised. In starvation lack of intake of proteins results in hypoproteinaemia. A serum albumin of 25g/l or less is generally associated with oedema of this type.

- **Changes in fluid balance** associated with an increase in the volume of extracellular fluid.

  Sodium retention automatically involves water retention and a combined sodium and water excess results. In both heart failure and renal disease there may be sodium retention.
- **Increased vascular permeability,** which may result from the action of bacterial toxins, chemicals, and most importantly anoxia, may play a part.

These factors may act together to cause generalised oedema.

Examples of generalised oedema include:

- **Cardiac oedema**: The oedema of congestive heart failure is a common form of generalised oedema.

  There are two factors at work. The obvious one is the congestion resulting from so-called backward failure of the heart. This causes increased hydrostatic pressure in the venous system and oedema results. There is also sodium retention and an increase in the volume of the extracellular fluid possibly due to reduced renal blood flow from forward failure. As a result there is a reduced glomerular filtration rate and hence increased sodium reabsorption by the tubules. There is also stimulation of the renin/angiotensin/aldosterone system which leads to further sodium (and of necessity, water) retention with an increase in plasma and interstitial fluid volume.
- **Renal oedema:** Acute glomerulonephritis is accompanied by oedema, particularly of the face and eyelids. Increased fluid retention, hypertension leading to heart failure and increased vascular permeability may be factors involved.

  The nephrotic syndrome is characterised by massive loss of protein from the kidney and hypoproteinaemia.

  In chronic nephritis oedema usually occurs as a result of hypertension and heart failure.
- **Famine oedema** is due to low protein levels. Vitamin B deficiency and heart failure may also play a part.
- **Oedema due to adrenal hormones**: In Cushing's syndrome or after administration of excess of cortisone there is sodium (and water) retention and there may also be hypertension resulting in some degree of heart failure.
- **Toxaemia of pregnancy:** There is hypertension, proteinuria and oedema.

	Increased venous pressure	Low plasma proteins	Sodium and water retention
Cardiac oedema	++	-	++++
Acute nephritic syndrome	+	+/-	++++
Nephrotic syndrome	-	++++	++++
Famine oedema	-	++	++
Adrenal steroids	+/-	-	+++
Toxaemia of pregnancy	+	++	++++

TABLE SHOWING FACTORS CONTRIBUTING TO GENERALISED OEDEMA

Oedema of special organs Severe oedema may have fatal consequences if it occurs in certain organs such as the brain or the lungs.

- **Oedema of the brain** is important because it may cause swelling of the brain which is contained within the non-yielding cranium leading to raised intracranial pressure and herniation of the brain through the dura or through the foramen magnum. The blood-brain barrier does not permit the free exchange of substances from the circulation into the brain substance in normal circumstances. Three types of oedema of the brain have been described:
  - Vasogenic oedema is the accumulation of fluid in the extracellular space following new vessel formation or damage to blood vessels which allow fluid to escape. This is the most common form and occurs in association with tumours, abscesses, infarcts, injury and in lead encephalopathy.
  - Interstitial oedema occurs in non-communicating hydrocephalus when there is obstruction to the circulation of cerebro-spinal fluid and when excess ventricular fluid permeates into the surrounding white matter.
  - Cytotoxic or intracellular oedema is not true oedema in terms of the definition. It is used to describe conditions in which the brain cells become waterlogged such as in cloudy swelling following ischaemia, or in water intoxication.
- **Oedema of the lungs** or pulmonary oedema may occur as a result of altered hydrostatic pressures, as may occur in left-sided heart failure, salt and water overload, with hypoproteinaemia or even as a result of blockage of draining lymphatics. Peripheral vasoconstriction is accompanied by a shift of blood to the lungs.

  Damage to the alveolar capillaries may occur in infections, and if diffuse may lead to the alveolar respiratory distress syndrome (ARDS), which is characterised by severe pulmonary oedema. This condition may follow on shock, oxygen toxicity, sepsis or aspiration of material into the lungs.

  Whatever the cause of pulmonary oedema, fluid first enters into the alveolar wall, and only when lymph drainage fails, into the alveolar space. Once the space is occupied the alveolar capillaries become hypoxic and capillary permeability increases allowing protein to escape into the fluid.

In this way pulmonary oedema may spread rapidly and contain an appreciable quantity of protein.

- **Hepatic oedema**. Oedema due to liver disease can occur when there is obstruction of the portal venous system, for example, in cirrhosis of the liver. This is usually manifested by ascites. The diseased liver may synthesise reduced amounts of albumin and plasma proteins, which may also contribute to the formation of oedema.

Examples of **localised oedema** include:

- **Acute inflammatory oedema** is a condition in which an exudate is formed at the site of inflammation due to an increase in vascular permeability. It is also present when there is increased hydrostatic pressure of capillaries (hyperaemia). There may also be increased osmotic pressure of the interstitial fluid and in increased fluidity of the ground substance.
- **Allergic or hypersensitivity oedema** occurs in hypersensitivity reactions (anaphylaxis) as a result of increased vascular permeability due to histamine release. Oedema may affect the eyelids, lips, tongue larynx and lungs, the latter two being especially dangerous.
- **Oedema of venous obstruction.** Increased hydrostatic pressure results in the formation of a transudate, for example, after prolonged standing, in patients with varicose veins or leg vein thrombosis. This may be generalised when venous obstruction is central, for example in congestive heart failure. As the venous pressure rises the circulatory flow is reduced and hypoxia may ensue resulting in an increased vascular permeability. There may also be associated overload of the lymphatics as a contributory factor.
- **Lymphatic oedema.** This is an oedema with a protein content intermediate between that of an exudate and a transudate (lymph contains a little protein normally). It is a firm brawny oedema which does not pit on pressure as does transudate. There may be some fibrosis in the oedematous tissue.

  Lymphatic obstruction can occur as a result of trauma, surgery (for example, following radical mastectomy with removal of lymph nodes in the axilla), following infection, tumour involvement or as a congenital malformation. Infection with the filarial worm *Wuchereria bancrofti* can cause lymph node obstruction, resulting in massive lymphatic oedema of the legs and genitalia known as elephantiasis.

CLINICAL FEATURES OF OEDEMA

Oedema has been defined as an excessive accumulation of fluid in the extravascular interstitial space. Considerable quantities of fluid must accumulate before the condition becomes clinically obvious, at least five litres or so. The development of generalised oedema can best be assessed by weighing the patient.

Classically in **cardiac failure** the distribution of the oedema is gravitational, that is oedema develops in the feet and ankles if the patient is up and about or in the sacral region of the back if bedfast. The oedema is due to transudate.

It is characterised clinically by pitting, that is when pressed firmly with a finger the oedematous subcutaneous tissue will be depressed and will take a minute or two to recover its normal shape.

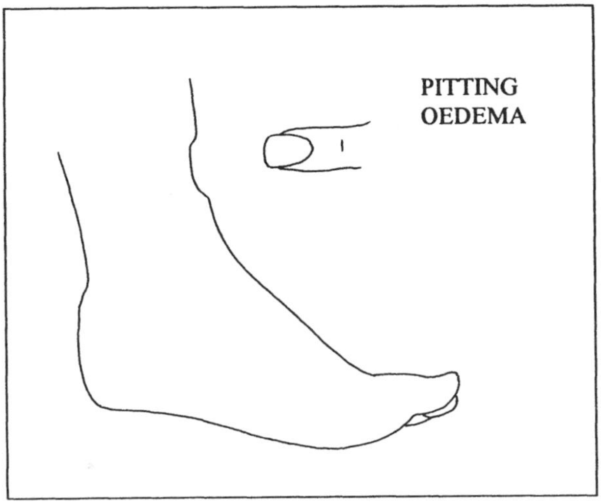

In **renal** oedema, the distribution may be somewhat different. There may be more fluid retained and this tends to be more generally distributed in the loose connective tissues of the body, often clinically obvious as a puffy face and eyelids.

In the more severe forms of generalised oedema fluid accumulates in the **serous cavities**, for example hydrothorax, ascites.

Oedema of the **brain** is characterised by a swollen brain with narrowed sulci and flattened gyri; the brain may feel soft and fluid will drip from the cut surface.

Oedema of the **lungs** makes the lungs firm and heavy. Frothy fluid will drip from the cut surface. This oedema fluid contains protein and a pink staining fluid may be seen within the alveolar spaces on microscopy. Oedema fluid in the lungs is prone to become infected and there may be an accompanying bronchopneumonia. Clinically the oedematous lungs sound as if air is bubbling through water when listened to through a stethoscope. These sounds are known as rales.

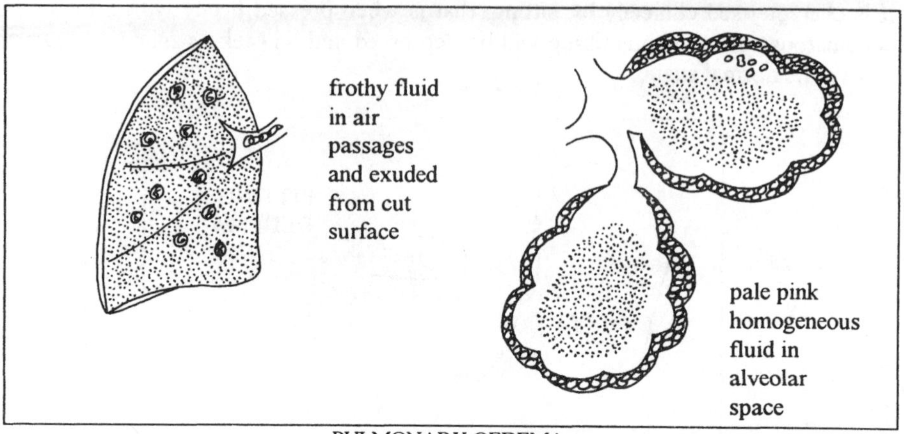

PULMONARY OEDEMA

Generally speaking, oedema under the microscope is unimpressive and can only be seen as a space separating collagen bundles and cells.

GENERAL PATHOLOGY - CHAPTER ELEVEN

HYPERAEMIA AND CONGESTION

In hyperaemia or congestion there is an increased content of blood in the vessels of a part, a tissue or an organ. This is closely related to the development of oedema.

Hyperaemia generally refers to a condition of increased blood flow to a part due to dilatation of arteries and arterioles. This may be referred to as **active hyperaemia**. There is a more rapid blood flow through the part, which becomes redder as a result. This may be brought about by the action of sympathetic nerves on vessels or by chemical substances which may be released in inflammation.

- Physiological hyperaemia may occur during strenuous exercise or when the body temperature is raised. Central nervous mediated vasodilation may occur in blushing.
- Pathologically systemic vasodilation occurs when there is an increase in pCO_2, in septicaemia or when there is increased cardiac output as may occur in anaemia or thyrotoxicosis. Local vasodilation is brought about by the inflammatory response, many of the features of which are due to locally increased blood flow.

Congestion usually refers to a condition where there is reduced venous drainage giving rise to what is called **passive venous congestion**. Blood flow is slowed.and the part becomes stuffed with slow moving blood, giving a reddish purple colour to the area. The part becomes more bluish purple as there is an increase of deoxygenated haemoglobin in the blood, a condition known as cyanosis.

- **Local passive venous congestion** may be produced by obstruction to venous return by thrombus or by pressure from without (for example, from tumours, from enlarged lymph nodes or in pregnancy). Back pressure as in varicose veins may also cause localised congestion.

  An important form of localised congestion is caused by cirrhosis of the liver. The intrahepatic portal venous system of the liver is deformed by fibrosis and so-called portal hypertension ensues, leading to passive venous congestion of, for example, the gut and spleen, and the development of a collateral venous return (with, for example, oesophogeal varices or haemorrhoids).
- **Generalised passive venous congestion** is found when there is heart failure. Backward heart failure is the term used to describe the build-up of blood in the venous systems when the failing heart cannot cope with the venous return.

There is passive venous congestion in the lungs when there is left ventricular heart failure, and in the systemic venous system when there is right ventricular heart failure. in many instances both left and right ventricular failure may be found together in generalised congestive heart failure.

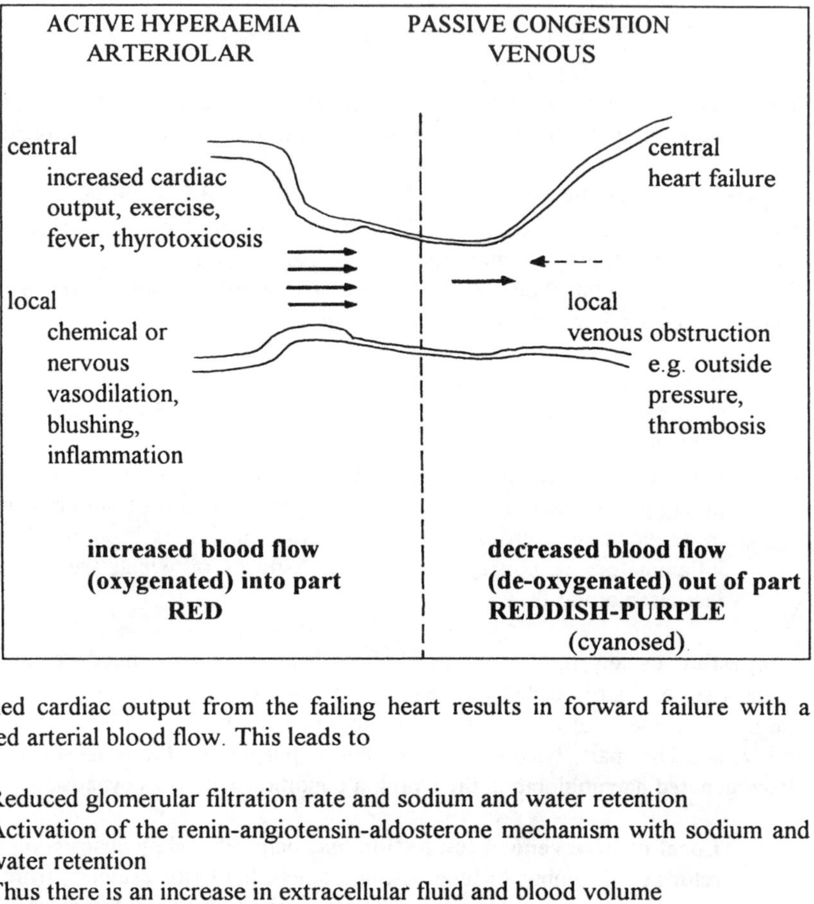

Diminished cardiac output from the failing heart results in forward failure with a diminished arterial blood flow. This leads to

- Reduced glomerular filtration rate and sodium and water retention
- Activation of the renin-angiotensin-aldosterone mechanism with sodium and water retention
- Thus there is an increase in extracellular fluid and blood volume
- Which may increase venous pressure and lead to congestive heart failure or backward heart failure.

THE PATHOLOGICAL EFFECTS OF CONGESTION

It must be remembered that venous congestion, if severe, may lead to the escape of fluid from the circulation to the tissues (oedema), or even escape of red cells, (haemorrhage). The reason for this is not only the increased pressure in relatively thin-walled post-capillary venules, but also because obstructive congestion, with slowing of the blood flow, leads to anoxia of the vessel wall which becomes more permeable. In the lungs particularly, congestion is likely to be accompanied by oedema and small haemorrhages because the lung capillaries and venules are particularly thin-walled, poorly supported and susceptible to hypoxia.

In many organs if congestion has been long standing, organs may show anoxic degeneration of parenchymal cells, small haemorrhages into the tissues and later, haemosiderin deposition and fibrosis. These changes are especially characteristic in the lungs, liver and spleen.

Chronic passive venous congestion of the lungs occurs when there has been left-sided heart failure. The most marked changes may be seen in cases of mitral valve stenosis.

Initially there is congestion in alveolar capillaries but these thin-walled vessels also allow fluid to escape. The fluid reaches the alveolar space and causes pulmonary oedema.

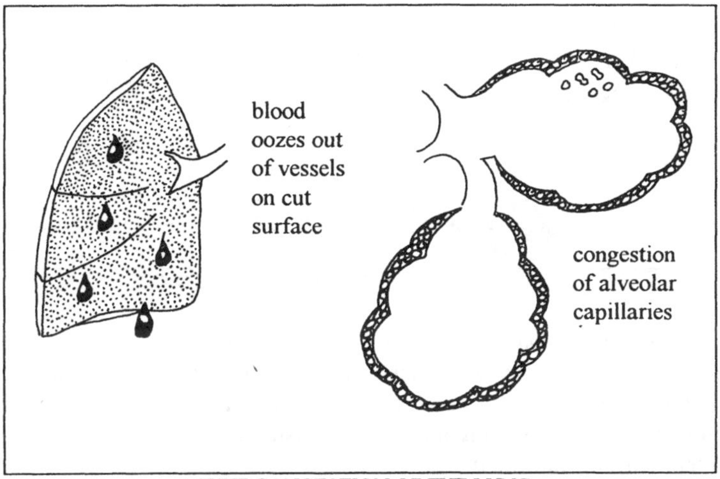

ACUTE CONGESTION OF THE LUNG

With fluid in the alveolar space oxygen from the inspired air can no longer reach the capillary, and the capillary wall suffers anoxic damage which permits the escape of small quantities of red cells. There is interstitial and intra-alveolar haemorrhage.

From the alveolar spaces red cells may be coughed up together with oedema fluid as a pink-tinged, frothy sputum which is characteristic of pulmonary congestion and oedema. Red cells may be broken down in the alveoli with the formation of haemosiderin, which is taken up by macrophages in the alveoli. These iron-laden macrophages are characteristic of more long standing or chronic congestion and are sometimes called 'heart failure cells'.

At this stage, grossly, the lungs may be heavy, dark red and bloody. In very severe long standing congestion the changes may progress further. Haemosiderin laden macrophages in the alveolar walls may accumulate with the release of haemosiderin, which can cause a fibrous tissue reaction. In addition, iron maybe laid down on the surface of elastic fibres, which then become thickened and bluish in colour. The altered elastic fibres may even excite a foreign body giant cell reaction. This process in areas of old haemorhage, where there is iron encrustation of elastic fibres and fibrosis, is known as **ferruginisation**.

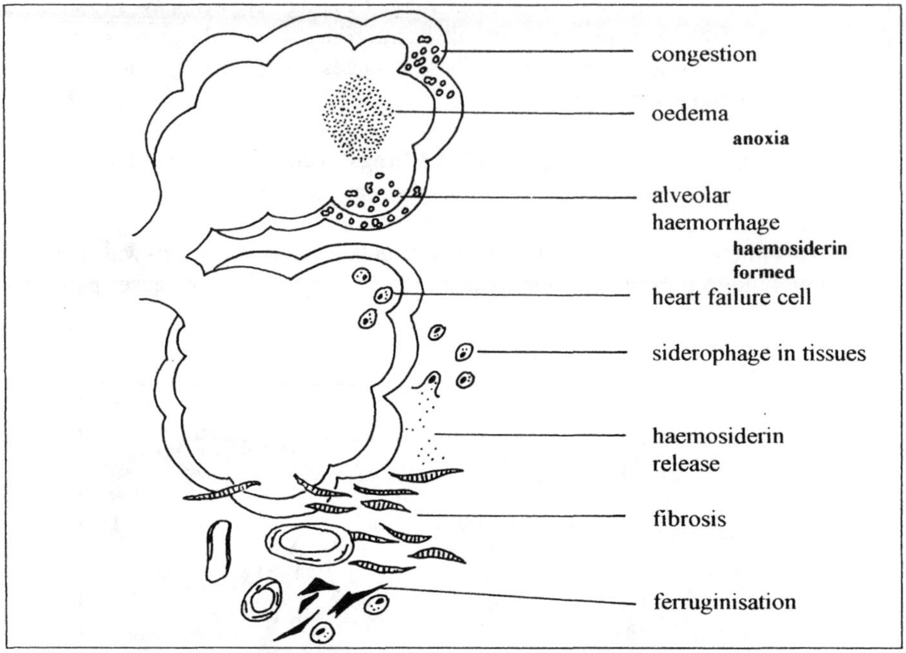

CHRONIC CONGESTION OF THE LUNG

Lungs in which there has been gross chronic congestion, marked haemosiderin deposition and fibrosis are particularly likely to be found in mitral stenosis. This combination of pathological changes in the lung is known as 'brown induration' as the haemosiderin gives the lung a rusty brownish colouration and the fibrosis makes the lung feel hard or indurated.

Chronic passive venous congestion of the liver is seen most frequently in cases of right-sided heart failure or more rarely in hepatic vein or inferior vena caval obstruction.

The liver is increased in size and bloody. On the cut surface the central veins appear red and are surrounded by yellowish brown uncongested liver, which may show fatty change. This appearance resembles the cut surface of a nutmeg, hence 'nutmeg liver' is the term used to describe it.

Microscopically blood accumulates in the central veins and the centrilobular vascular sinuses. The central liver cells may undergo atrophy and more peripheral ones show fatty change. In severe persistent congestion the red congested areas around central veins can extend and link up giving an appearance known as 'paradoxical lobulation', in which yellow brown areas of liver are surrounded by reddish areas. Where there has been necrosis of centrilobular liver cells there may be fibrous replacement, which can extend as bridging fibrosis between central veins, ultimately causing a diffuse fibrosis of the liver (which is sometimes referred to as 'cardiac cirrhosis').

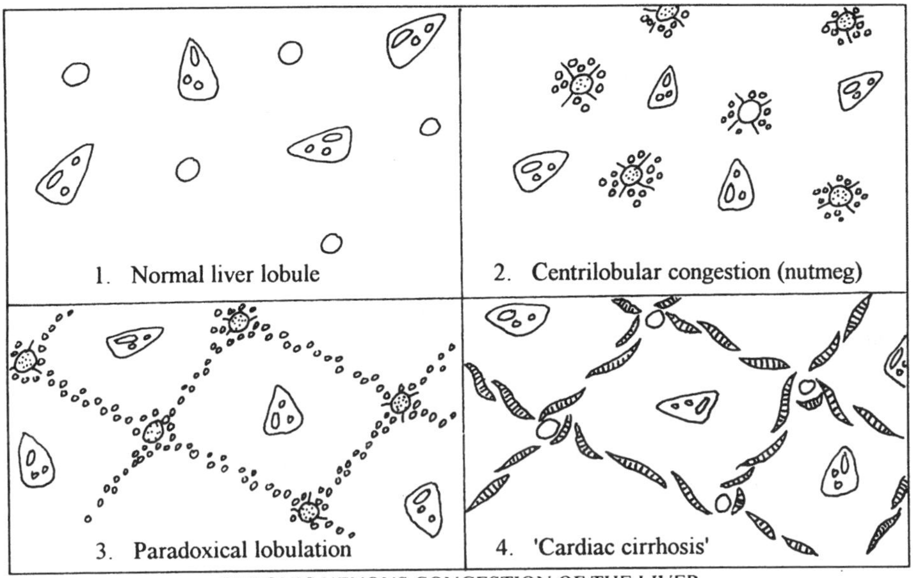

CHRONIC VENOUS CONGESTION OF THE LIVER

Chronic passive venous congestion of the spleen also occurs in right-sided backward heart failure. It may also be found in cirrhosis of the liver or where there is portal hypertension.

The spleen is enlarged and reddish purple in colour. It may be rounder and firmer than normal giving rise to the descriptive term 'cricket ball spleen'. The weight may be more than twice normal (more than 2x150g).

Microscopically there is sinusoidal congestion with thickening and fibrosis of the sinusoidal walls. In severe portal hypertension haemosiderin deposition and fibrosis may be particularly marked and ferruginised fibrous nodules (siderotic nodules, called Gamna-Gandy bodies in the spleen) may occur in the red pulp.

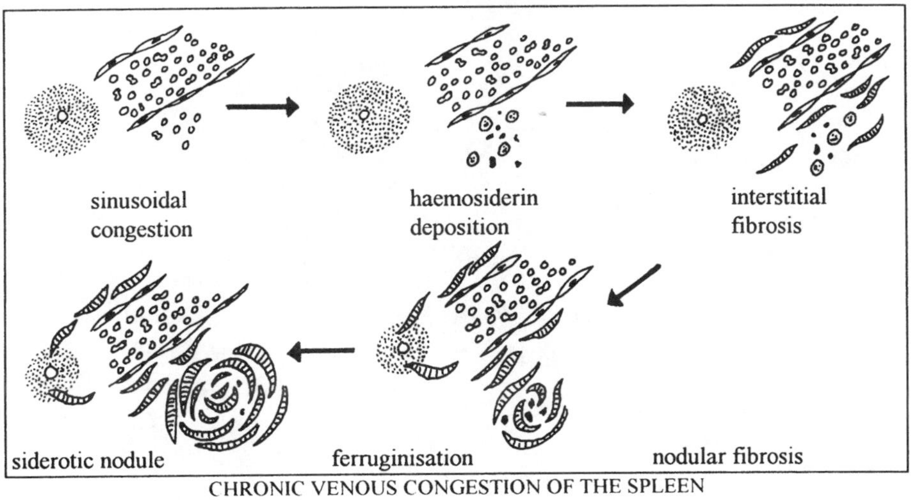

CHRONIC VENOUS CONGESTION OF THE SPLEEN

GENERAL PATHOLOGY - CHAPTER TWELVE

HAEMORRHAGE

Haemorrhage is the escape of blood from vessels within the circulatory system and implies that there has been damage to the integrity of the vessel in some way. Released blood may escape to the exterior, that is external haemorrhage as in bleeding from a wound or from the gut; or into the tissues where it may form a bruise or haematoma; or into a serous cavity, for example, the haemothorax, haemopericardium or haemoperitoneum.

Haemorrhages into the skin or mucous membranes may be classified according to size:

- **Petechiae** are pin point or very small haemorrhages
- **Purpura** describes areas of haemorrhage up to about one centimetre
- **Ecchymoses** are large and blotchy haemorrhages, greater than one centimetre across.

Chronic blood loss or repeated small losses, as may be found in bleeding from a peptic ulcer, or in heavy menstrual losses of blood, can usually be compensated for at first, but may eventually result in iron deficiency anaemia.

Acute haemorrhage, either arterial or venous, can lead to more serious immediate consequences.

The causes of haemorrhage are many and various but some of the more important causes include:

- **Trauma** for example motor vehicle accident, at surgical operation
- **Spontaneous** (in the apparent absence of trauma or in response to minimal trauma.
  - Damage to the vessel wall from without, for example by:
    Inflammatory lesions such as peptic ulcers or TB.
    Invasion by a malignant tumour such as cancer of the stomach
  - Disease of the vessel wall itself. Examples include:
    Congenital weakness, berry aneurysm, Marfan's disease
    Toxins from bacteria (meningococcal septicaemia)
    Anoxia (in small vessels)
    Primary inflammation, vasculitis
    Immune diseases
    Deficiency diseases such as Vitamin C deficiency or scurvy.
  - Problems within the lumen of the vessels:
    High blood pressure or hypertension increases the risk of haemorrhage, especially cerebral haemorrhage or bleeding from an aneurysm.

Disturbance of homeostatic mechanisms, such as decrease in platelets or clotting factors, or an increase in fibrinolysis or anticoagulants

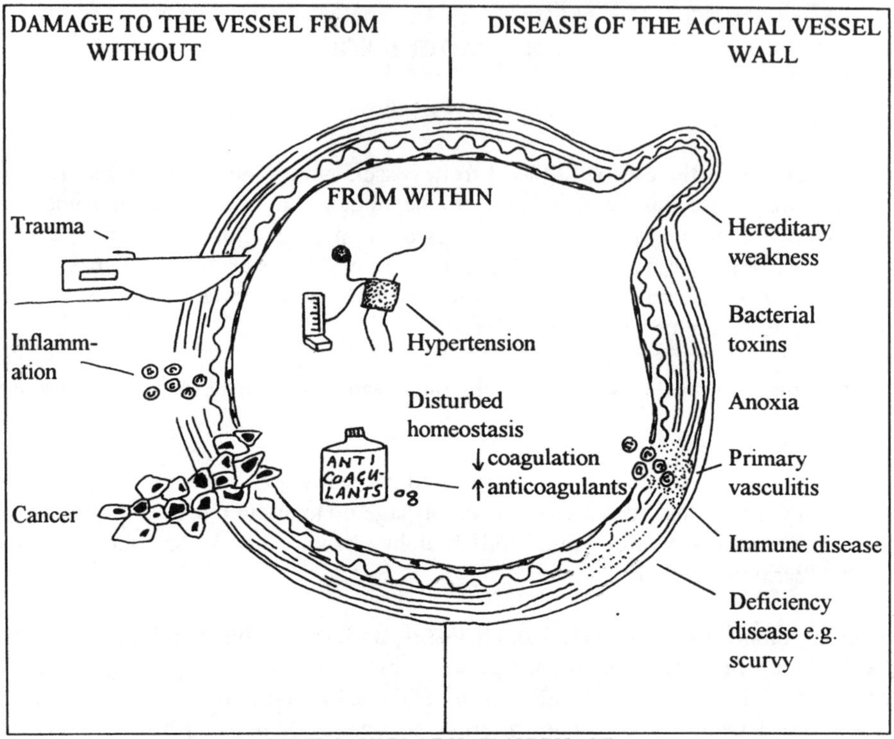

SOME CAUSES OF HAEMORRHAGE

The effects of haemorrhage depend on:

- The amount of blood lost
- The speed with which it is lost
- The site of the haemorrhage (more serious in the brain).

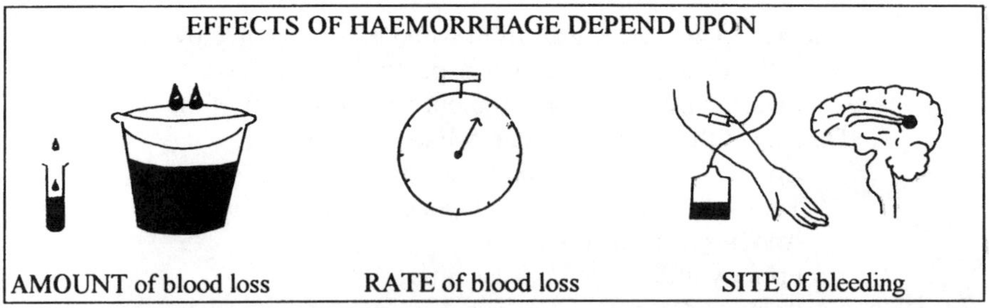

Generally a 20 per cent blood loss, up to about one litre, can be coped with by the body's compensatory mechanisms; a sudden loss of, say two litres, may cause death, but if the same haemorrhage takes place slowly over 24 hours or more, almost three litres or more than half the blood volume can be lost without necessarily fatal consequences.

The body's response to blood loss

- Initially there may be syncope (or fainting) with dilatation of the arteries to the muscles, a fall in blood pressure and a marked reduction in cerebral blood flow resulting in loss of consciousness. This may actually then help to improve the blood flow to the brain as the patient usually falls into a horizontal position and recovers consciousness.
- Then blood flows are altered so as to maintain the circulation to vital centres. As blood volume is lost the venous return to the heart is reduced and cardiac output is also reduced. Baroreceptors cause the vasomotor centre (VMC) in the brain to bring about vasoconstriction to maintain blood pressure. (Note that the coronary and cerebral vessels are not affected by this vasoconstriction). The heart rate is also increased.
- There are also hormonal mechanisms at work:

  - The adrenal medulla secretes adrenaline and noradrenaline
  - There is release of renin by the kidneys

  These mechanisms also cause vasoconstriction and the renin-angiotensin-aldosterone mechanism, together with diminished glomerular filtration rate, will bring about sodium, and hence water, retention with an increase in the cellular fluid to compensate for fluid volume loss. ADH secretion is also increased.

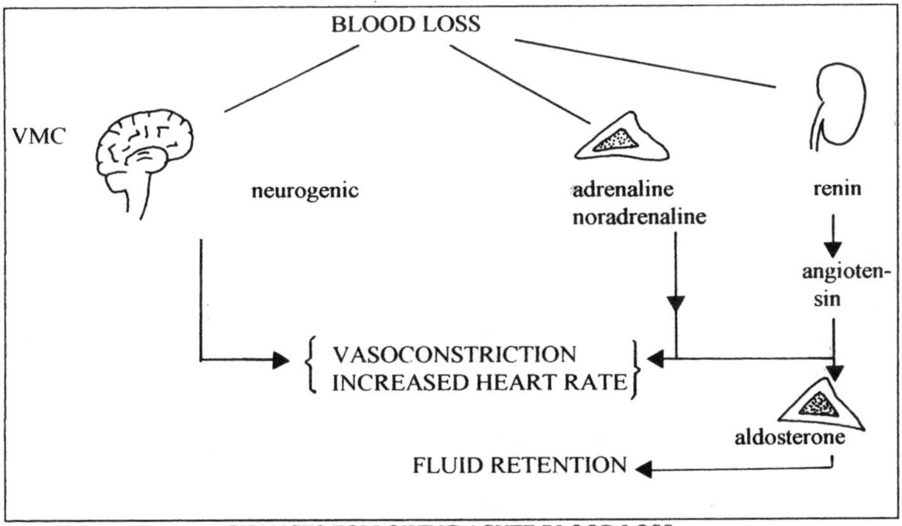

CHANGES FOLLOWING ACUTE BLOOD LOSS

- There is no change in the haemoglobin level, the red cell count or the concentration of blood constituents shortly after haemorrhage, but the platelet count may be increased and the clotting time decreased. The white cell count is generally increased due to a rise in the neutrophil count.

- Later changes include the restoration of blood volume by fluid being drawn into the intravascular compartment from the extracellular fluid. The body retains extra fluid as outlined above. Haemodilution takes two or three days. It is only after this period that haemoglobin levels will allow accurate assessment of the amount of blood loss.
- In the final stages of compensation, additional plasma proteins are manufactured by the liver over the course of the next few days and the red cells will eventually be replaced by the bone marrow although this may take up to six weeks. Treatment by blood transfusion will restore blood volume, protein and red cells almost instantly.

If the haemorrhage is too severe to be compensated for by these mechanisms a state of **haemorrhagic shock** ensues:

- Despite increased heart rate and selective peripheral vasoconstriction
- Blood pressure drops
- There is diminished blood flow to the tissues and ischaemic damage may result, particularly in susceptible tissue such as the brain and the kidneys.

Assessment of the patient's state may be difficult if the body manages to compensate for blood loss by maintaining the blood pressure. The amount of blood loss cannot be assessed by the estimation of haemoglobin or concentration of plasma proteins until haemodilution has taken place in a few days. Measurement of central venous pressure (vena caval or right atrial pressure) can be used as a guide to the need for fluid replacement. An indication of left atrial pressure may be gained from the wedged pulmonary arterial pressure. This may also be of use in calculating the requirement for fluid replacement to restore circulatory volume. Note that too much fluid can cause pulmonary oedema, which is definitely an unwanted complication.

GENERAL PATHOLOGY - CHAPTER THIRTEEN

SHOCK

The term **shock** can be regarded as a clinical term descriptive of a patient who is pale, has a cold clammy skin ('a cold sweat') with a rapid thready pulse, low blood pressure and rapid respiration. Cyanosis may be noted. The temperature may be subnormal. Little or no urine is passed. The patient is still and apathetic, and if the shocked condition is not relieved, may become confused, comatose and then die.

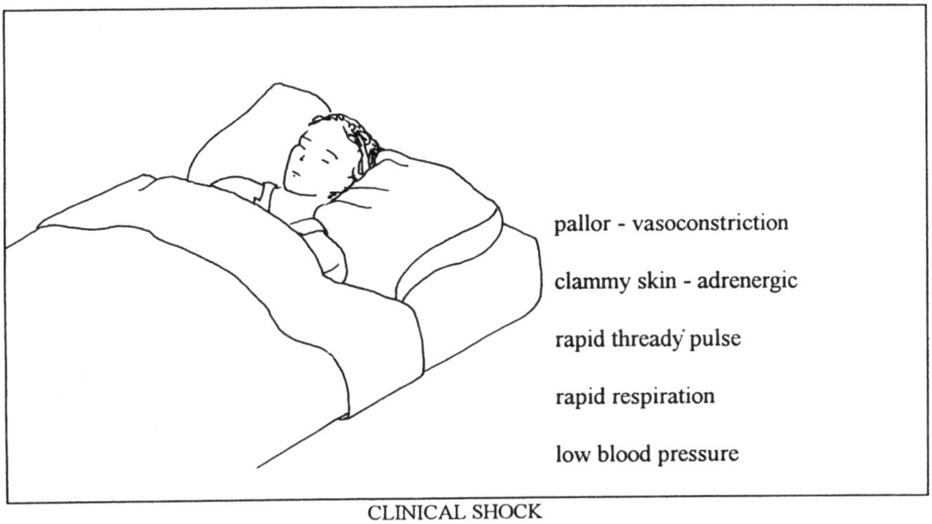

CLINICAL SHOCK

This pattern is well recognised, and few would argue with this description for a patient who had suffered severe blood loss followed by circulatory failure. Continued increase in pulse rate and drop in blood pressure are the telltale clinical signs, and are monitored at fifteen minute intervals or less in patients who are thought likely to be developing shock.

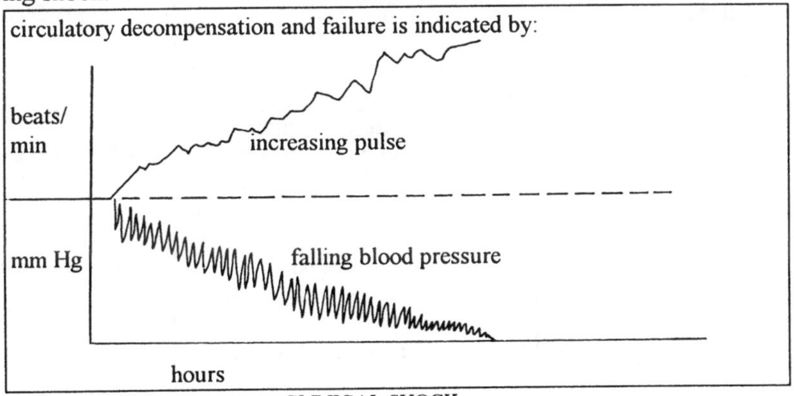

CLINICAL SHOCK

However, many different conditions can produce the clinical picture of shock, and not all of the symptoms may be present. In some cases of septic shock the patient may have a warm flushed appearance with a normal heart rate. Thus clinically although the term shock is useful and will doubtless be retained, it is necessary to qualify the diagnosis wherever possible by categorising the type of shock.

Almost all types of **shock can be defined as a state of circulatory failure resulting from some stress leading to inadequate cellular perfusion and an oxygen deficit in the tissues. It is this hypoperfusion of tissues which is the basic and important lesion.**

The arteries only contain about eight per cent of the total blood volume. The important metabolic changes take place at capillary level and the largest volume (60-70 per cent) of blood is contained in the venous system. It is observations of these vascular beds that are of most value.

The major problems are with blood volume, blood pressure and blood flow resulting in hypoperfusion. At cellular level there is widespread inadequate supply of oxygen to the cells, with anaerobic glycolysis and accumulation of lactic acid leading to metabolic lactic acidosis. Anoxia leads to cell damage and eventually to cell death.

TYPES OF SHOCK

Shock may be classified either on the basis of underlying clinical cause, or according to the presumed pathogenic mechanism:

Types of Shock

Clinical	Cause	Presumed Pathogenetic Mechanism
HAEMORRHAGIC	sudden loss 20%, slow loss 0-40% blood volume	LOSS OF BLOOD VOLUME (HYPOVOLAEMIC)
BURN	transudation and exudation of fluid	
TRAUMATIC] WOUND] SURGICAL]	haemorrhagic, neurogenic and possibly metabolic mechanisms	
LOSS OF FLUID & ELECTROLYTES	e.g. diabetic coma, sodium and water depletion	
CARDIOGENIC	M.I., myocarditis, rupture of papillary muscle, arrhythmias, cardiac tamponade etc.	PUMP FAILURE
SEPTIC	Gram negative endotoxic from bacterial lipoprotein. Gram positive enndotoxic *Cl. welchii*, α toxin of *Staph.* etc.	POOLING OF BLOOD
NEUROGENIC	Pain, anaesthesia, spinal cord injury	

Anaphylactic shock resulting from a hypersensitivity reaction, is a state resembling histamine poisoning. Asthma and oedema may be followed by circulatory failure due to pooling of blood and shock.

Other causes of shock which are difficult to classify exactly, include :

- Gastric or intestinal perforation
- Acute pancreatitis
- Incompatible blood transfusion

Note that these mechanisms need not necessarily act singly. A patient with traumatic shock has pain, that is, there are neurogenic factors, and may also develop infection. Bacterial infections may involve the action of toxins on the heart with peripheral pooling and pump failure.

- **Hypovolaemic Shock.** Haemorrhagic shock occurs when the compensatory mechanisms of the body are overwhelmed, thus loss of blood volume causes reduced cardiac output and lowered blood pressure in shock.

  The peripheral resistance begins to fall, perhaps due to anoxic damage to the vasomotor centre.

  Rapid treatment is essential. Severe blood loss with lowered blood pressure if rapidly restored by transfusion, is soon reversed. If the shock state is allowed to persist it may become irreversible.

  In traumatic shock there may be metabolic injury and pooling of blood mediated by metabolic and neurogenic mechanisms. There may also be plasma loss into the tissue.
- **Cardiogenic Shock** This is most commonly caused by extensive myocardial infarction, but other cardiac conditions, as listed above, or even massive pulmonary embolism may bring this about.

  The cardiac output and stroke volume are low. In cardiogenic shock there may also be some peripheral pooling of blood. Arteriolar vasoconstriction is followed by loss of tone with dilatation of small vessels due to metabolic acidosis caused by poor perfusion. Naturally this accentuates the shock state with inadequate venous return and loss of effective circulating volume.

  Cardiogenic shock is difficult to treat and carries a mortality of up to 80 per cent.
- **Septic Shock** is a physiological phenomenon in which blood is pooled in the peripheral circulation. It is commonly encountered in septicaemic states with Gram negative organisms and it is thought that endotoxins, lipopolysaccharides in the wall of the organisms, act as peripheral vasodilators.

  In this type of shock cardiac output is maintained at first and the skin may remain warm and even appear flushed; there is arteriolar dilatation followed by dilatation of the capillary bed and venous system resulting in peripheral pooling of blood. This will result in reduced venous return to the heart and hence reduced cardiac output.

The mechanisms by which changes are brought about are probably multiple. Damage to cell membranes of endothelial cells and white cells and platelets may activate the coagulation cascade sometimes with disseminated intravascular coagulation (DIC). Injured cells may release free radicals.

The complement system may also be activated and damaged platelets can release a wide variety of vasoactive substances, including histamine, serotonin, adrenaline, thromboxane, kallikreins.

Septic shock has a mortality rate of more than 50 per cent.

- **Neurogenic Shock** It is postulated that neurogenic mechanisms also can cause peripheral vasodilatation and shock, particularly in severe and painful injury.

MEDIATORS OF SHOCK

In shock the compensatory mechanisms listed as occurring following acute blood loss or haemorrhage are brought into play with ADH secretion and with sodium and water retention from reduced glomerular filtration rate and release of renin from the kidney. There is also (in most cases) vasoconstriction which may aggravate ischaemia if the blood pressure drops.

A whole host of chemical substances has been put forward as having a secondary role in the development and perpetuation of the shocked state. These include:

- Catecholamines
- Complement activation
- Interleukin-1
- Platelet activating factor
- Histamine
- Prostaglandins
- Interferon (alpha and gamma)
- Tumour necrosis factor-α (TNF-α)
- The kinin system
- Leukotrienes
- Endorphins
- Myocardial depressant factor

IRREVERSIBLE SHOCK

The condition of shock is self perpetuating and progressive if untreated. If treatment is delayed for too long, or is inadequate, it may be ineffectual. However, the term irreversible shock is only applied *post mortem*.

Treatment should be given energetically up to the last minute during life. In irreversible shock whatever the primary cause, cellular hypoxia will somehow bring about vasomotor paralysis and pooling of the blood. Obviously on the treatment of shock the first principle is **to treat the cause**.

This can be done effectively by the replacement of lost blood volume but is less easy in heart failure or in septic shock where even if organisms are killed by antibiotics their toxic products remain in the circulation.

Otherwise treatment is symptomatic - to restore blood volume, to maintain fluid and electrolyte balance, to promote blood flow to tissues, to correct any acidosis and to provide, for example, adequate oxygenation.

Note that:

- Skin vasoconstriction is a compensatory mechanism; the time honoured remedies of warming the patient or giving alcohol will tend to dilate skin vessels and worsen the patient's condition.
- With inadequate blood supply to the gut and tissues medicaments should be administered intravenously to be effective.

THE PATHOGENESIS OF SHOCK

The figure 'The Pathogenesis of Shock', on the last page of this chapter shows the action and interaction of the causative mechanism. It also illustrates how the effects on heart, brain, blood, small vessels and tissues tend to create a vicious cycle to perpetuate and worsen the condition of shock.

MORPHOLOGY AND PATHOLOGY OF SHOCK

The brain. Although most compensatory mechanisms tend to try and maintain circulation to the brain, if the circulation is slowed anoxia may result. Oxygen sensitive neurones in the cortex, especially the hippocampus, and cerebellum may die individually, or there may be small or large areas of infarction (softening).

Severe damage to the vital centres may cause death of the patient or irreparable brain damage which may persist even if the patient otherwise recovers.

The lungs are red and heavy due to congestion and oedema which occur after twenty four to forty eight hours. In severe cases the Adult Respiratory Distress Syndrome (ARDS) also known as 'diffuse alveolar damage' may ensue with marked pulmonary oedema, which is protein-rich, and the formation of hyaline membranes along the alveolar walls. Congestion and oedema interfere with gaseous exchange, hence there is an increased respiratory rate initially. There is also a predisposition to infection and bronchopneumonia may be a complication. Severe changes of this sort are found in shock following sepsis or trauma and the condition is sometimes referred to as 'shock lung' although there are many other causes such as diffuse infection (often viral), oxygen toxicity, inhalation of toxins and irritants (for example, nitrogen dioxide and smoke), narcotic overdose, hypersensitivity reactions, post cardiac surgery with extracorporeal pumps and aspiration pneumonitis.

The kidneys are particularly susceptible to ischaemic change in shock. The kidney suffers from acute tubular necrosis (ischaemic and toxic agents of many sorts can cause this). The ischaemic form of acute tubular necrosis may be known as 'shock kidney'.

In ischaemic acute tubular necrosis the kidneys show:

- Pale cortex, congested dark medulla and may be slightly enlarged.
- Focal renal ischaemic necrosis in the distal convoluted tubules and terminal parts of the proximal convoluted tubules.
- Rupture of the tubular basement membrane.
- Casts in tubules (hyaline, proteinaceous and pigment).
- Interstitial chronic inflammatory infiltration around necrotic areas. Early changes include interstitial oedema and leucocytes within dilated vasa recta.

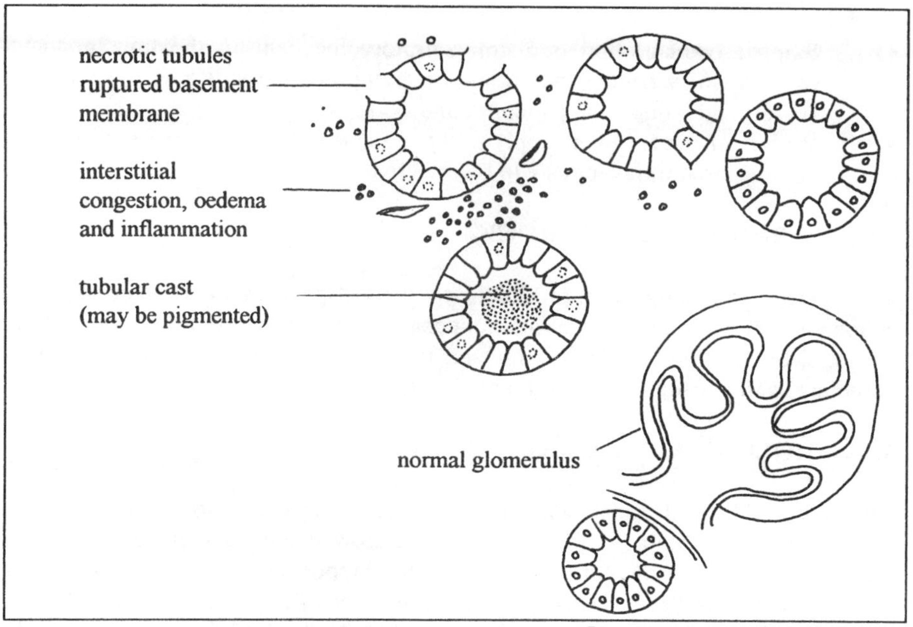

ACUTE TUBULAR NECROSIS - I

These changes occur within twenty four hours of the onset of shock and urine flow is reduced or may cease.

The glomeruli are usually unaffected but may contain fibrin thrombi as a manifestation of intravascular coagulation.

After seven days or so, if the patient survives, the oedema and inflammation subside and evidence of tubular regeneration with multinucleate tubular lining cells may be seen. At this stage there is heavy diuresis and profound electrolyte disturbances may occur (see diagram 'Acute Tubular Necrosis II').

Death from acute renal failure can follow recovery from the acute phase of shock.

The heart may show fresh subendocardial haemorrhages usually in the left ventricle in early fatal shock (twelve hours). Dark contraction bands may be noted within myocytes.

In later phases there may be microscopic fatty change after eighteen hours. Some cases show foci of necrosis.

The liver may also show fatty change. In severe shock there may be centrilobular necrosis.

The adrenals show lipid depletion as a stress response due to mobilisation of stored steroids after one to three days; there may also be small areas of focal necrosis.

Gastro-intestinal tract. Focal gastro-intestinal ulcration or diffuse haemorrhagic necrosis of the stomach or colon may occur in the gastro-intestinal tract, presumably as a result of ischaemia following on circulatory failure.

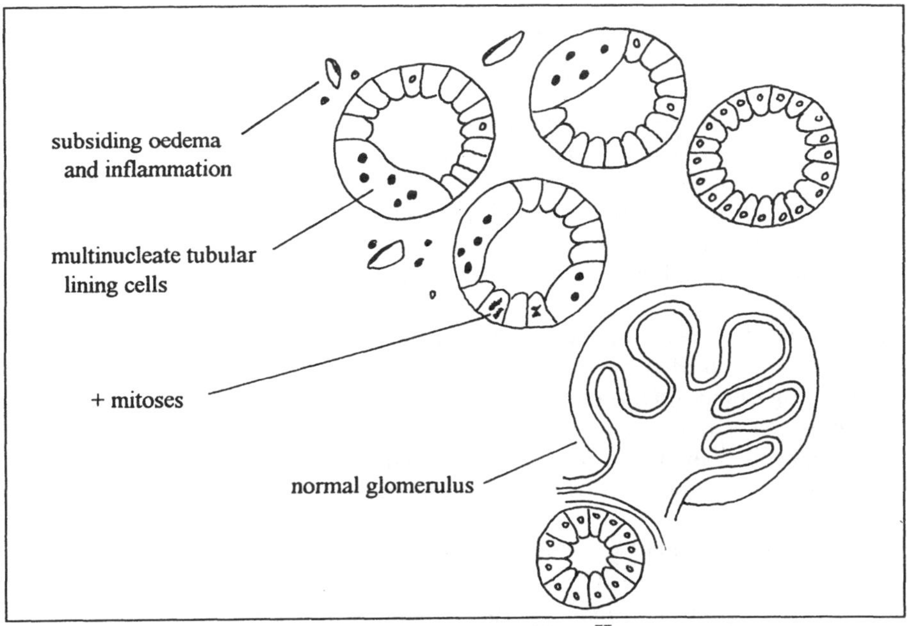

ACUTE TUBULAR NECROSIS - II

Blood: In areas of reduced flow and of pooling, aggregations of red cells may be seen in capillaries and vessels. This further reduces blood flow and may account for micro-infarcts in various organs. This is called sludging of the blood.

DISSEMINATED INTRAVASCULAR COAGULATION (DIC)

This may also take place. Activation of the coagulation complex leads to widespread microthrombus formation and hence to the depletion of platelets and fibrin (consumption coagulopathy). In addition, the fibrinolytic mechanism is brought into play. Fibrin degradation products may inhibit the coagulation mechanism. Thus the paradoxical situation where excessive coagulation is accompanied by haemorrhagic complications may develop.

Although DIC may be a feature of shock, DIC due to other causes may bring about shock.

Causes of DIC

- Release of tissue clotting factors into circulation as in severe trauma, fat embolism, surgery, obstetrical accidents (especially amniotic fluid embolism) and some malignancies.
- Endothelial damage as in heat stroke, vasculitis or shock.
- Infections such as meningococcaemia, any septicaemia and malaria.

The clinical and pathological effects may be due to microthrombi in many organs or only a few isolated areas. There may be:

- **Ischaemia** of organs affected, for example, brain, heart, lungs, kidneys and adrenals. Mostly ischaemia will not cause more than slight focal necrosis except in the adrenals where DIC may precede and contribute to massive haemorrhage (Waterhouse-Friderichsen syndrome).
- **Haemorrhage** may occur at any site, for example, into brain, skin and mucous membranes, serous membranes or lung.

Differential diagnosis must be made from a purely fibrinolytic state which may occur during and after surgery, in obstetrical conditions and in association with some malignancies. In this condition the fibrinogen/fibrin deficiency is pure and not accompanied by deficiencies of clotting factors or platelets.

Treatment is by transfusion of fresh blood to replace clotting factors and also with heparin to prevent further intravascular coagulation.

CAUSES OF DEATH IN SHOCK

In the early stages death can result from circulatory collapse, causing:

- Cerebral ischaemia
- Heart failure

At progressive stages, death may result from:

- Lactic acidosis and electrolyte disturbance
- Pulmonary oedema
- Acute tubular necrosis and renal failure

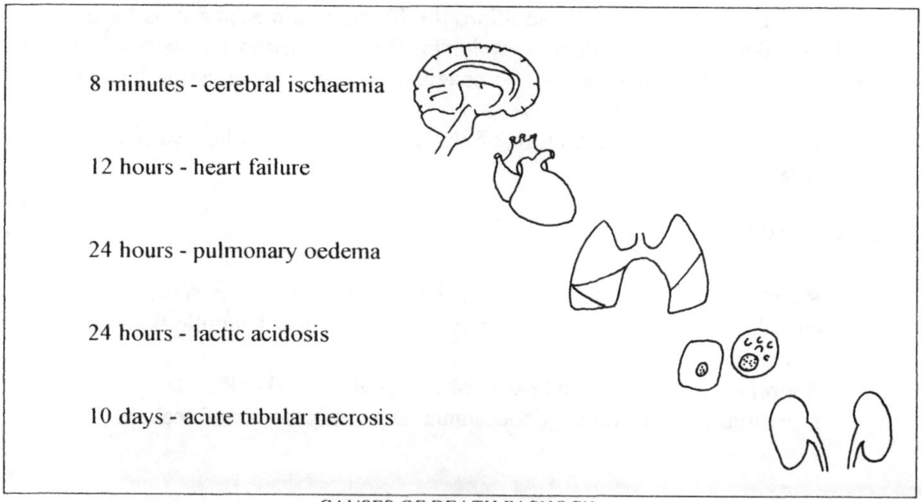

CAUSES OF DEATH IN SHOCK

THE PATHOGENESIS OF SHOCK

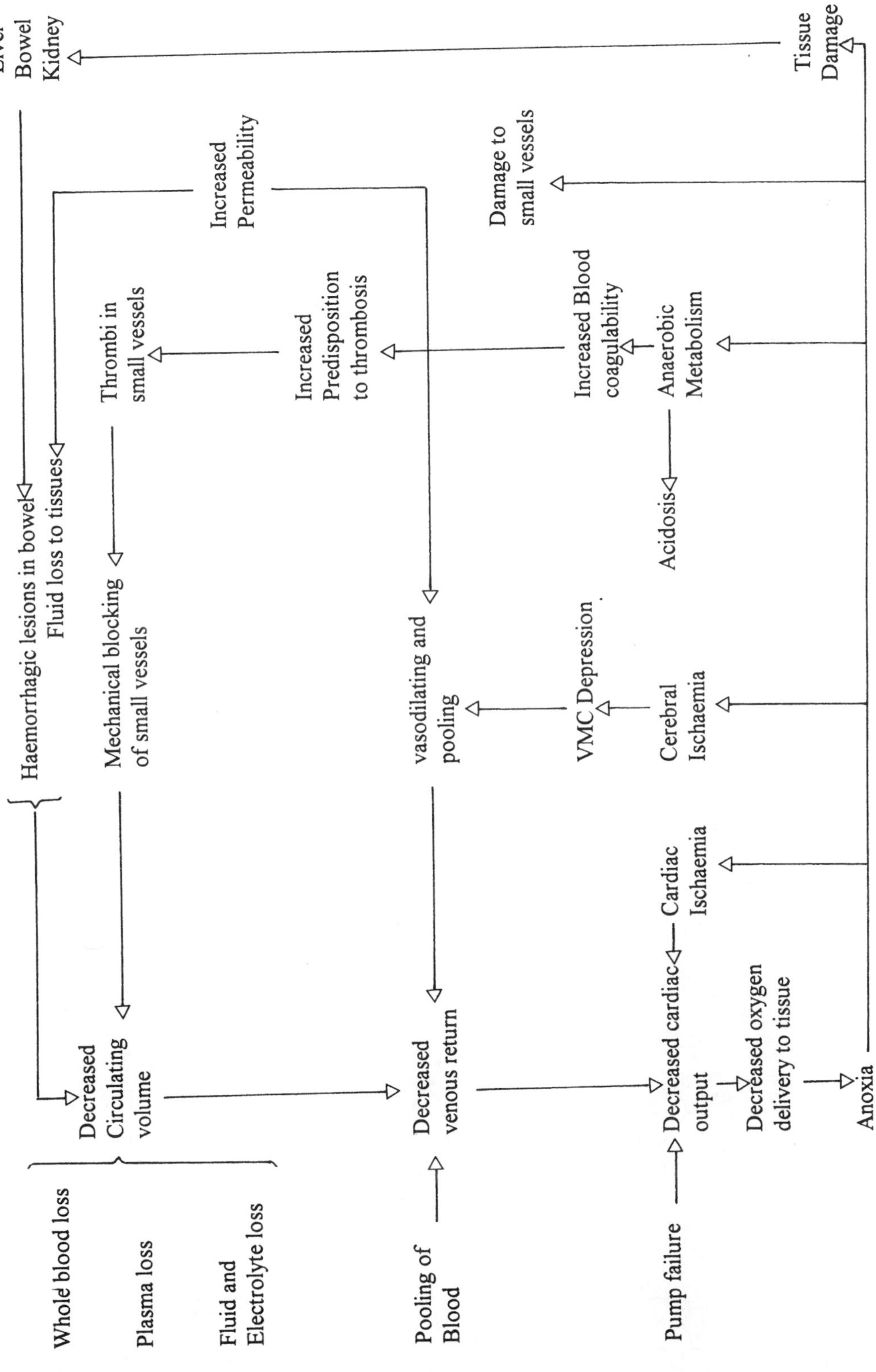

GENERAL PATHOLOGY - CHAPTER FOURTEEN

THROMBOSIS AND EMBOLISM

The blood remains fluid within the vessels due to a process known as **haemostasis**. This depends on the properties of:

- The vessel wall and its lining endothelium
- Platelets
- Coagulation factors and anticoagulants in the blood which may be activated by endothelial cell products and platelet secretions

In the case of damage to a vessel wall haemostatic mechanisms come into play to preserve the integrity of the vessel. Thus there is:

- **Vasoconstriction**.
- Formation of a **platelet plug** at the site of injury.
- Virtually simultaneous activation of the coagulation system to produce **fibrin**.
- Leading to **thrombus** formation which will fill the breach in the vessel wall.

A **thrombus** is a solid or semi-solid mass formed from the constituents of the blood within the vascular system during life.

This implies that it is formed within the flowing blood stream.

A thrombus is more than just a blood clot which can form outside vessels (as when blood is taken for pathological examination) or in the vessels after death. It differs because there is, in addition to blood coagulation, a contribution from the vascular endothelium and the platelets.

Some of the haemostatic mechanisms are shown in the following diagram:

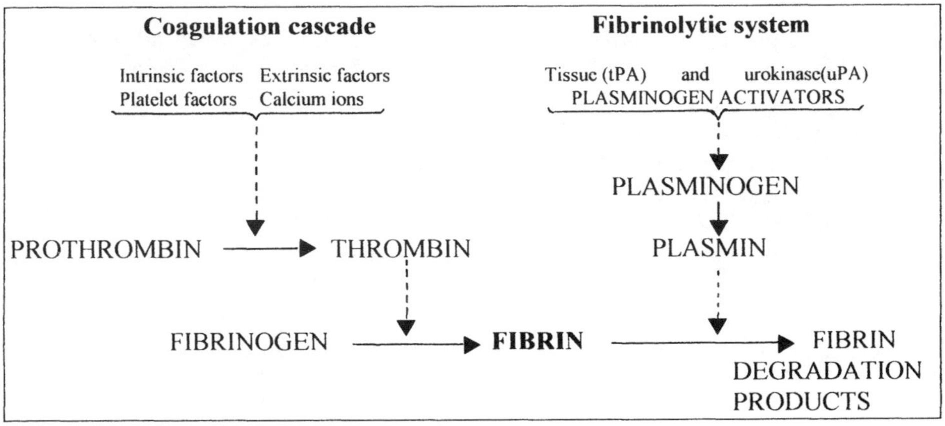

Thus there is one system promoting the coagulation of blood and one system devoted to clearing the product of coagulation, fibrin.

Another system which should be mentioned is the arachidonic acid cascade:

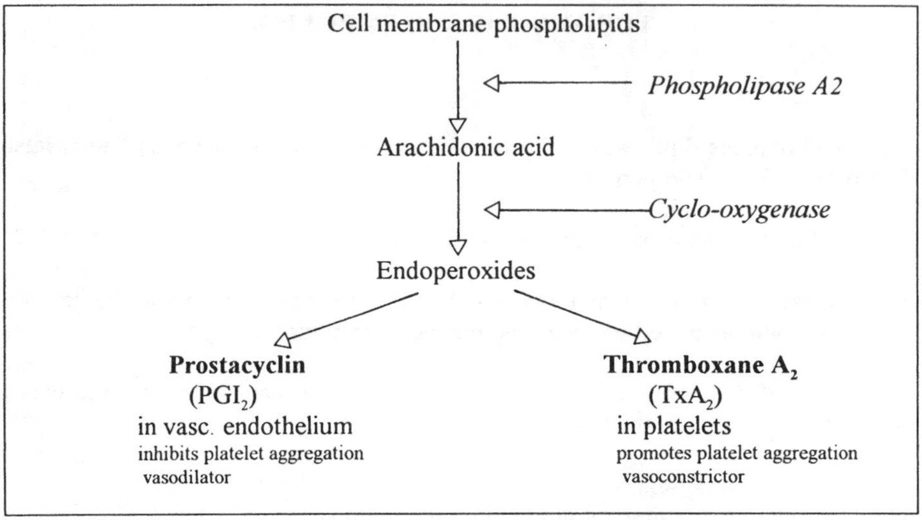

Here again in this system, which is important in haemostasis, there is a balance between factors promoting or inhibiting platelet aggregation or causing constriction or dilatation of vessels.

The kallikrein-kinin system is also activated at the same time as the coagulation and fibrinolytic systems.

Thus it would seem that within the blood circulatory system with regard to haemostasis that for every system or active substance there is an equal and opposite reactive system or substance, with vast numbers of active or potentially active factors occurring in the endothelial cells, the platelets and the plasma. Together these maintain the blood flow in normal circumstances.

THROMBUS FORMATION

- The process is started by the deposition of platelets which **adhere** to the usually damaged or breached endothelial surface and are activated
- To **secrete** (amongst other things) adenosine diphosphate (ADP) and to synthesise thromboxane A_2 (TxA_2)
- Which bring about platelet **aggregation.**
- At the same time the **coagulation process** is started (by platelet factor 3) and thrombin is generated.
- The aggregated platelets **fuse** and are bound down by fibrin which has been precipitated by coagulation.

There is now a firm platelet plug or platelet thrombus in position.

At this stage in many cases of minor injury to a vessel wall the process may stop. A small deposit of platelets and fibrin can rapidly be grown over by vascular endothelium and the vessel is then restored to normal with no further platelet deposition. The platelets are removed by phagocytosis and enzyme digestion and the fibrin is dissolved by fibrinolysin.

However, in cases of more severe damage to the vessel wall or in the presence of disturbances of blood flow, more platelets may aggregate, more fibrin may be precipitated and the platelet thrombus may continue to grow in size. This is now known as a **pale thrombus**. If the rate of fibrin formation is high then fibrin may enmesh red blood cells and further build up the surface of the thrombus. Another layer of platelets bound with fibrin may then form on the surface of this, followed by more fibrin and red cells to build up a laminated thrombus, demonstrating alternate pale and dark areas known as the lines of Zahn (which are often well seen in a thrombus in a cardiac or aortic aneurysm). If fibrin and red cells predominate then the appearances are those of a **red thrombus.** This is more often seen in veins where platelet and fibrin strands may be inconspicuous.

The whole process is under the influence of a number of control mechanisms. Robbins notes that sufficient thrombin is generated by the clotting of one millilitre of blood to coagulate all the fibrinogen in three litres of blood. But of course in practice this doesn't happen. Clotting factors may be washed away by the flowing blood and cleared in the liver or by phagocytosis. Clotting factors may become exhausted and used up. Anti-clotting factors such as antithrombin III (in plasma) and others activated by the endothelial cells may prevent further coagulation from taking place. Finally the fibrinolytic system via plasmin can digest and dissolve already formed fibrin.

FACTORS CONTRIBUTING TO THROMBUS FORMATION

There are three main factors which may predispose to thrombus formation; these are sometimes known as Virchow's triad. They are:

- **Damage to the endothelium of the vessel wall.** This is probably the most important factor; a breach in the endothelium may expose the collagen to which platelets adhere. Endothelial damage is particularly important in the causation of thrombi in the arterial system and heart, where there may be atheromatous ulceration of an artery or damage to the endocardium in myocardial infarction
- **Alterations in blood flow.** Turbulence in the arterial system and stasis in the venous system disrupt the laminar flow of blood and allow platelets to come into contact with the endothelium. Alterations in blood flow may prevent the clearing of clotting factors and restrict the access of inhibitors. Turbulence in arteries may also cause damage to the endothelium directly.

  Turbulence is most likely to occur in arteries where atheromatous plaques can obstruct the lumen.

  Stasis can occur in the leg veins (particularly in patients confined to bed), in the atria of the heart when flow is disturbed as in atrial fibrillation, or in arteries showing areas of dilatation (aneurysms).

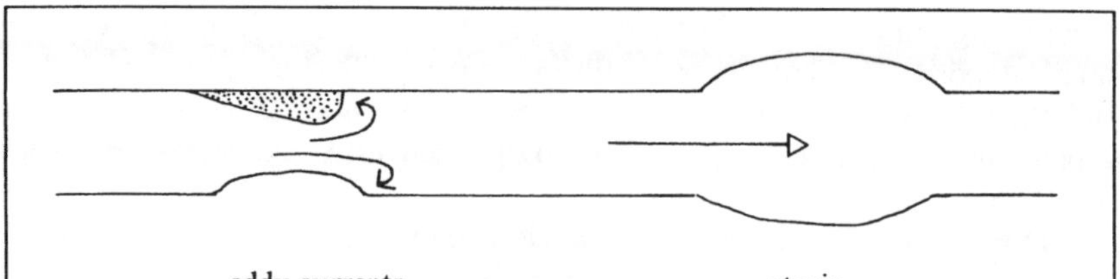

ALTERATIONS IN BLOOD FLOW

- **The composition of the blood.** There is an increased liability to thrombus formation after trauma, surgical operation, childbirth or in some cases of disseminated malignancy. There may be increased numbers of platelets and increased platelet stickiness as well as an increase in coagulation factors in these patients.

  There may also be inherited deficiencies of antithrombin substances (for example antithrombin III or protein C).

  There is increased risk of thrombus formation in older patients, cigarette smokers, obese patients and users of oestrogen-containing oral contraceptives, although it may be difficult to demonstrate any alteration in clotting factors.

SITES OF THROMBUS FORMATION

VENOUS THROMBOSIS

The venous system is a common site for thrombus formation, because it is a low pressure system and stasis is likely to occur, a condition known as **phlebothrombosis**. **Thrombophlebitis** is the name given to thrombosis occurring in an inflamed vein with damage to the wall.

Phlebothrombosis usually occurs in the deep calf veins of the legs, but may extend upwards to involve popliteal femoral and iliac veins. It is a common and important complication of surgery and is also common in patients who have had myocardial infarction. In these patients there is stasis and possibly also an increase in clotting factors. Stasis is particularly likely to occur:

- In heart failure with venous congestion
- In shock
- During bed rest when there is likely to be reduced venous return with muscle pumps not working and also with external pressure on these veins from the bed.

Nowadays it is the custom to keep patients in bed for as short a time as possible so as to reduce the likelihood of this complication. Some surgical patients may even be given anticoagulants to try and prevent phlebothrombosis

The importance of this sort of thrombus is that it may become detached from the veins of origin and transported through the venous circulation to the right side of the heart and thence to the lungs as a pulmonary embolism, which can be fatal.

An **embolus** is an abnormal mass of undissolved material which is transported from one part of the circulation to another. This process is called **embolism**.

To understand how this is brought about it is worth while considering how this sort of thrombus is formed:

- First of all a platelet thrombus is formed, and because of the stasis clotting factors can accumulate.
- Thus fibrin is laid down, firstly binding the platelets and then enmeshing red cells. A laminated thrombus is formed which causes a disturbance in flow as it grows, causing further thrombus formation until
- The thrombus occludes the lumen.
- Flow now stops, but loose red thrombus formation can be propagated on either side of the obstruction.
- Flow from tributaries can contribute further layered or laminated thrombus, until the lumen is again occluded.
- And so on until thrombus extends right up the vessel.

This thrombus is not well anchored to the wall of the vein, especially where the loose red thrombus has formed, and can easily break off and embolise to the lungs.

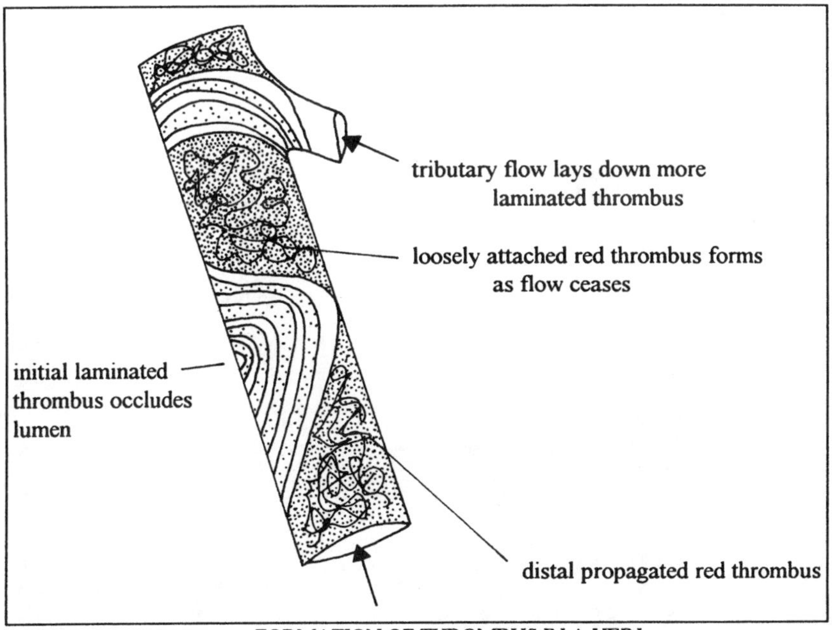

FORMATION OF THROMBUS IN A VEIN

Clinically the signs of deep vein thrombosis are often slight, but there may be pain when deep pressure is applied to the affected calf, pain on sudden forced dorsiflexion of the foot and some distal oedema.

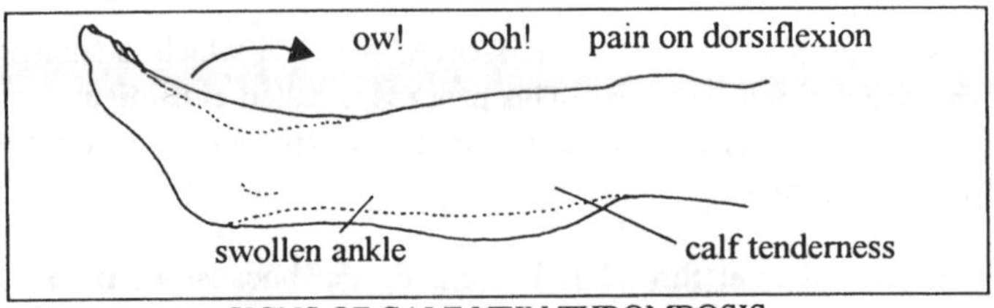

SIGNS OF CALF VEIN THROMBOSIS

Thrombophlebitis occurs when thrombus forms on the damaged wall of an inflamed vein. In this instance because of the endothelial damage, the thrombus is well anchored to the wall and as there is usually increased blood flow because of the inflammation, propagation is less likely. As a result, embolism from this type of thrombosis is uncommon.

Inflammation of a vein may be caused by:

- Damage to the vein wall for example by drip set needle, or by introduction of concentrated or damaging substances into vein.
- Severe local infection. In this instance the thrombus may be invaded by organisms which cause it to disintegrate and infected emboli may result, leading to pyaemia. For example acute appendicitis may result in the infection of the appendicular veins, which can lead to infected thromboemboli moving to the liver, causing what is called **portal pyaemia**.

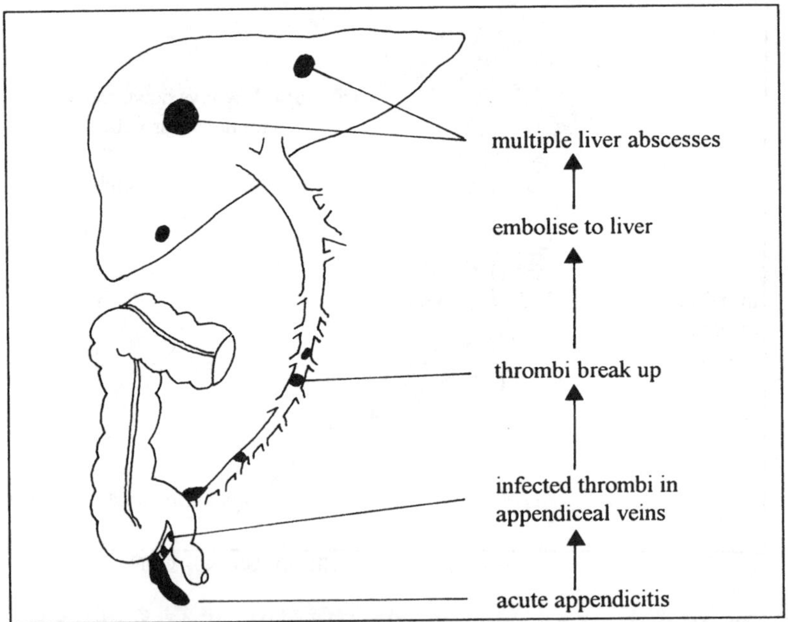

PORTAL PYAEMIA

Note: **Pyaemia**(literally pus in the blood) results from the presence of infected thrombi and emboli in the blood, and is accompanied by multiple abscess formation.

Bacteraemia (literally bacteria in the blood) is a transient phenomenon in which circulating organisms are usually soon killed off by the body's defence mechanisms.

Septicaemia is a severe illness with many organisms constantly being released into the blood stream or multiplying in the circulation.

Clinically thrombophlebitis is usually painful and obvious, especially if superficial.

Thrombophlebitis migrans is worth a mention. This is a condition in which there are recurrent thrombotic episodes in veins. Sometimes the cause is not known but it may be secondary to internal cancer of pancreas, stomach or lung.

ARTERIAL THROMBOSIS

This may occur where there is:

- Damage to the arterial wall as in atheroma.
- Where there are disturbances of flow, such as may be caused by:
  - Stasis in an area of local dilatation, that is, in an aneurysm
  - Pressure from without, for example, by a tumour
- When there is inflammation, either after trauma or due to arteritis as such, for example, in polyarteritis nodosa.

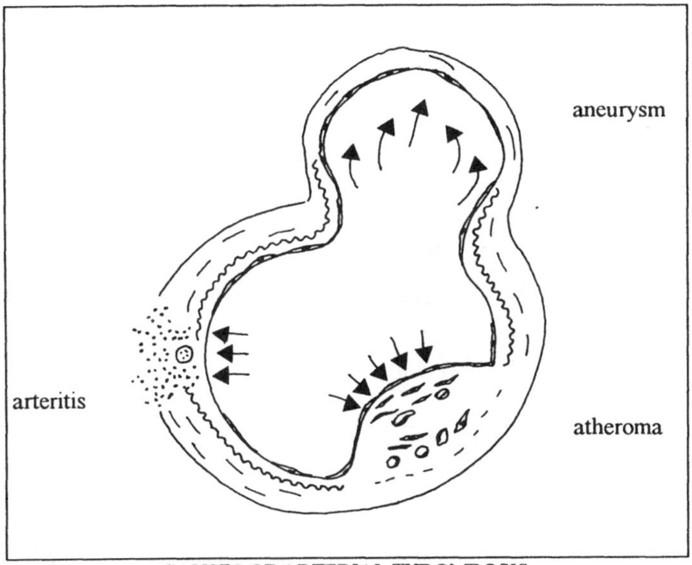

CAUSES OF ARTERIAL THROMBOSIS

CARDIAC THROMBOSIS

Thrombosis in the heart can occur in a variety of sites and for a variety of reasons:

- In the **atria**:
  - Stasis, which may be caused by obstruction (as in mitral stenosis), and atrial fibrillation may both contribute to thrombosis.
  - Atrial thrombi can become detached, leading to embolism.
  - Rarely a detached thrombus can remain in an atrium, growing in size until it can act as a ball valve thrombus and can bring about death by blocking the atrioventricular valve.

- In the **ventricles**: Thrombosis can occur on the damaged endocardium in myocarditis or following a myocardial infarct.
- On the **valves**: When inflamed in rheumatic fever or as infected thrombi in infective endocarditis.
- In the **coronary arteries**: Usually following atheroma with roughening of the wall and narrowing of the lumen.

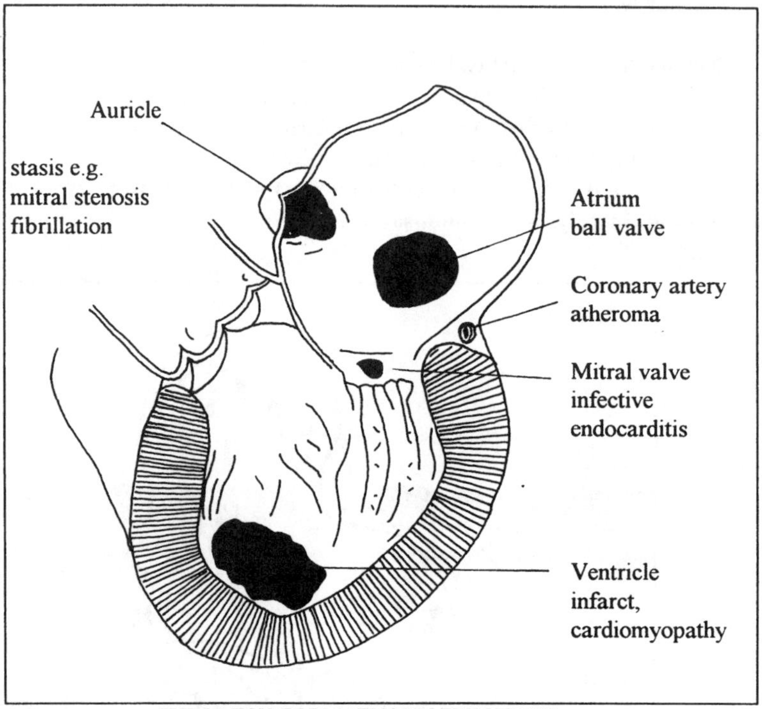

THROMBUS FORMATION IN THE HEART

THE FATE OF THROMBI

The development of ante mortem thrombi has been described. They are firm with a rough, dull, dryish surface. Distinctive ripple marks (lines of Zahn) are formed by the laying down of platelets.

Ante mortem thrombi take up the shape of the vessel in which they are formed and are adherent to the wall of the vessel (in differing degrees). Compare this again to the smooth post mortem clot already described. Most thrombi contain mixed pale and red areas. The emboli of venous thrombosis are usually bigger than arterial emboli or infected emboli.

A thrombus once formed does not remain as such:

- Some thrombi can be dissolved by fibrinolysin.
- The thrombus may retract, opening or widening a channel which is then lined by endothelium and flow is re-established.

- The thrombus may undergo organisation, that is ingrowth of fibrous tissue and new capillaries from the wall of the surrounding vessel. If the thrombus has totally obstructed the vessel in which it is situated, the capillaries tend to join up with each other and dilate and communicating channels can be formed which could allow flow through the previously obstructed vessel. This is a form of recanalisation.
- Sometimes recanalisation fails to take place and the thrombus is replaced by fibrous scar tissue, which may contract and remain as a permanent obstruction to flow. Occasionally these fibrous scars calcify to form **phleboliths** (literally vein stones). This seems most likely to occur in pelvic veins.
- Some thrombi may become detached to form emboli.
- Infected thrombi are likely to break up and form multiple emboli. Infected lesions or abscesses may occur at the site of impaction. Mostly these are of venous origin but infected thromboemboli may arise following pulmonary infection or from thrombotic vegetations on the heart valves in infective endocarditis.

Disseminated intravascular coagulation has already been discussed as a feature of shock. In this condition many microthrombi are formed within vessels. If the patient survives, these will generally be lysed or dissolved and disappear.

EMBOLI

An **embolus** is an abnormal mass of undissolved material which is transported from one part of the circulation to another. This process is called **embolisation.**

It invariably ends up by encountering vessels which are too small for it to pass through. This process of meeting small vessels and causing a blockage is known as **impaction**.

A number of undissolved materials can act as emboli including:

- **Thrombi:** 95 per cent or more of all emboli are thrombi (thromboemboli) which become loosened form their site of formation and are carried through the circulation until they impact, which may be anywhere in the circulatory system, depending on the original site of formation.
- **Other substances** such as:
  - Atheromatous material
  - Gas (air or nitrogen)
  - Fat
  - Bone marrow
  - Tumour fragments
  - Bacteria, parasites etc.
  - Foreign bodies
  - Amniotic fluid

PULMONARY EMBOLISM

At post mortem thrombi may be found in the pulmonary arteries and it should always be assumed that these are embolic in origin, as primary thrombus formation in a pulmonary artery is most uncommon.

Pulmonary emboli are common and may be found in up to ten per cent of patients dying of acute medical and surgical conditions who come to autopsy.

It is said that in hospital patients only one third of fatal pulmonary emboli are diagnosed during life. The clinical diagnosis may be difficult to make and is often incorrect.

Most pulmonary emboli (95 per cent) arise in the deep leg veins, mostly from thrombus which has propagated into the popliteal, femoral and iliac veins. Loose thrombus detaches and passes through the right side of the heart to the lungs. Very rarely it may pass through a septal defect to the left side of the heart to give rise to embolism in the systemic arterial system; this is known as paradoxical embolism.

The thrombus may detach in a large mass or in pieces (often the latter occurs, as multiple pulmonary emboli are common).

A very large embolus can block the main pulmonary artery at its bifurcation. This may be known as a 'saddle embolus'.

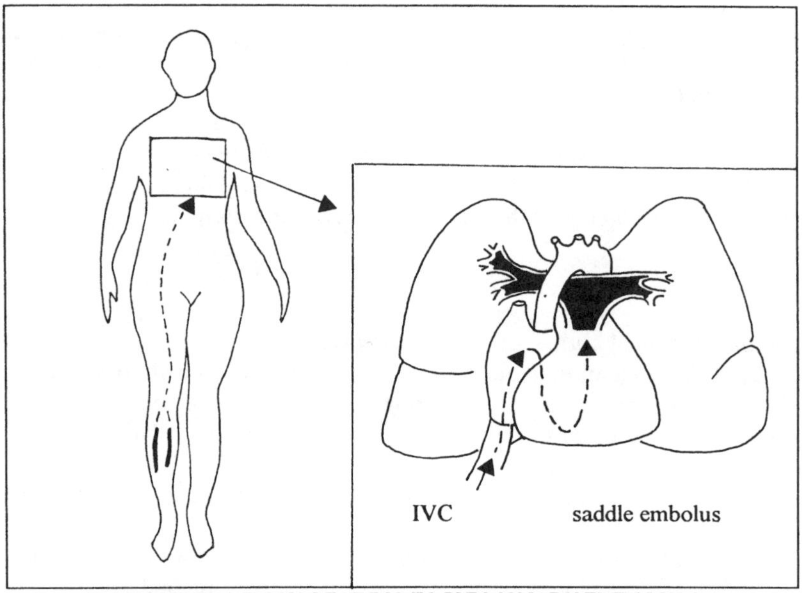

LEG VEIN THROMBOSIS LEADING TO PULMONARY EMBOLISM

Somewhat smaller emboli can block a major pulmonary artery. Blockage of the pulmonary circulation entirely will lead to right heart failure with cessation of cardiac output. Blockage of a major artery may also lead to right heart failure, partly by obstruction to the circulation, but also because of reflex vasoconstriction and possibly also constriction of both vessels and bronchi due to release of serotonin and thromboxane A_2 from platelets. The end result is sudden death, or acute right heart failure and the development of cardiogenic shock.

Pulmonary embolism can cause *immediate* sudden death. Classically the patient is said to sit up in bed, call urgently for a bedpan and then drop back dead. If death is not immediate the symptoms may mimic a myocardial infarct (heart attack) with chest pain, difficulty in breathing and the development of heart failure and shock.

However this sudden death is not the usual effect of pulmonary embolism and is likely in only a small percentage of cases. Most often pulmonary emboli are absolutely silent, because they are small.

There are no symptoms and the thrombus is rapidly dissolved by fibrinolysis after impaction. Because the lung has a double blood circulation via the bronchial arteries, tissue death due to anoxia is not likely if there is no other circulatory disorder such as heart failure. Multiple small silent emboli may present later with pulmonary hypertension.

If emboli do not dissolve completely they may either undergo fibrosis and be incorporated into the wall of the pulmonary artery (possibly contributing to the later development of pulmonary hypertension) or they may partly resolve, leaving only a fine fibrous web to be seen in the lumen of the affected artery.

In patients who have an impaired pulmonary circulation, for example in pulmonary congestion from heart failure then smaller emboli can cause **infarction** with chest pain, difficulty in breathing and sometimes slight **haemoptysis** (coughing up of blood).

ARTERIAL SYSTEMIC EMBOLISM

Most arterial emboli arise from the heart, mostly from the wall of the left ventricle following myocardial infarction. They can also arise from the atria in cases of mitral stenosis following rheumatic fever, or from infected thrombi on valves in infective endocarditis. In the arteries thrombi may form on atheromatous plaques or in aneurysms.

Emboli impact in smaller arteries and are likely to give rise to serious ischaemic effects and infarction, which will be described in detail later. Brain and gut are sites at which emboli can cause serious, potentially fatal infarcts. Limbs, kidneys and the spleen are other common sites for emboli of this sort to lodge

Atheromatous emboli are a form of systemic arterial embolus. They consist of lipid debris which may arise from an ulcerated atheromatous plaque. They can be recognised histologically by the presence of cholesterol crystals in the lumen of small arteries, a fairly common finding at autopsy in the kidneys of elderly patients.

GASEOUS EMBOLI:

Air embolism: Air can be introduced into a vein during child birth or abortion, when air may enter venous sinuses in the uterus. Injury to the lung or chest wall may allow a large bolus of air to enter a vein during inspiration, or a similar event can occur during cardiothoracic surgery. Small volumes of air are usually absorbed without noticeable effects but if there is a considerable volume of air in the circulation, over 200ml, the air will make the blood froth in the right ventricle, an no blood will be delivered to the lungs. Frothy blood in the heart is reputed to make a churning noise which is very characteristic.

Decompression sickness: A sudden reduction in pressure, as can occur in divers ascending too rapidly to the surface, can result in bubbles of gas (usually nitrogen) coming out of solution in the tissues. Muscle cramps known as the 'bends' and acute respiratory distress known as the 'chokes' are presenting features, but there may also be serious cerebral damage resulting in coma.

In chronic decompression sickness gaseous emboli seem to persist in bone causing foci of ischaemic necrosis.

FAT EMBOLI

These can occur following fractures or severe trauma. In a small proportion of cases symptoms occur known as 'fat embolism syndrome', with symptoms of respiratory distress, mental confusion, and petechial haemorrhages in skin and conjunctiva. It is thought that fat globules in the circulation become impacted in the lungs, but some may squeeze through to enter the systemic arterial circulation to impact in, for example, the brain and skin. It is not just the presence of fatty droplets in the circulation which causes the disease, as in most cases few ill effects can be seen. Additional features that have been postulated include toxic injury to capillaries (for example from fatty acids) and activation of the coagulation system to produce disseminated intrvascular coagulation.

The symptoms occur one to two days after injury. At autopsy, slices of lung tissue under saline may release fat globules on light pressure. In addition to skin petechiae small haemorrhages may be seen in the white matter of the brain. The lungs may show well marked oedema with features of the adult respiratory distress syndrome.

TUMOUR EMBOLI

Most malignant tumours invade local blood vessels and isolated cells frequently enter the circulation. Most are destroyed but a very small proportion may impact and produce secondary deposits. Some tumours actually grow into larger blood vessels, (for example renal cell carcinoma) and large tumour emboli may be produced.

AMNIOTIC FLUID EMBOLISM

This is fortunately a rare condition, occurring in about one in 50,000 deliveries. It carries a high maternal mortality of about 80 per cent. Amniotic fluid containing squamous cells from the skin of the foetus, and occasionally bile from meconium, enters the maternal circulation following rupture of the membranes, probably via a torn uterine vein. Clinically there is severe respiratory distress, shock and coma. This is followed by pulmonary oedema and bleeding due to disseminated intravascular coagulation. It is thought that material contained within the amniotic fluid causes marked pulmonary vasoconstriction and impairs cardiac function; there may be some obstruction of pulmonary vessels by solids within the amniotic fluid. Contained thromboplastin-like substances are thought to give rise to disseminated intravascular coagulation.

FOREIGN BODIES

Foreign bodies such as pieces of polythene derived from intravenous catheters, may act as emboli. **Parasites**, such as Schistosomes and their ova may also be carried by the blood stream.

TIMING OF EMBOLI

Type of embolus	Time after incident
Gas, amniotic fluid	immediate
Fat	1-2 days
Pulmonary thromboembolism	7-14 days

GENERAL PATHOLOGY - CHAPTER FIFTEEN

ISCHAEMIA AND INFARCTION

To commence this chapter let us review some definitions:

Ischaemia is a condition of inadequate blood supply to an area of the body. Ischaemia may be complete or partial. Complete ischaemia is likely to result in an **infarct** which is a localised area of ischaemic necrosis.

ISCHAEMIA

The most important consequence of ischaemia to a part is **hypoxia** or **anoxia**, that is too little or no oxygen. As has already been discussed, hypoxia or anoxia are very important causes of cell injury and cell death.

Hypoxia may be of a number of different types:

- **Ischaemic hypoxia** results from an inadequate blood supply.
- **Hypoxic hypoxia** when the blood is not properly oxygenated due, for example, to respiratory failure
- **Anaemic hypoxia** occurs when there is too little haemoglobin available in the blood to carry sufficient oxygen to the tissues
- **Toxic or histotoxic hypoxia** is brought about by cell injury or poisoning (for example cyanide poisoning) which prevent uptake of oxygen by the cells.

Most of the following discussion will be devoted to ischaemic hypoxia and the results thereof, but it is worth while to remember that ischaemic hypoxia may be aggravated in the presence of inadequate oxygenation of the blood or anaemia.

In addition to hypoxia there are other consequences of ischaemia, such as a reduction in the supply of nutrients to ischaemic tissues and the accumulation of waste products or metabolites in ischaemic areas.

Ischaemia may be **generalised**, affecting the whole body (for example in a case of cardiac arrest when the circulation suddenly ceases), or **localised,** affecting only a part of the body. It may be **complete** if the blood supply is totally cut off, or **partial** if the blood supply is reduced.

The effects of ischaemia may be modified according to the nature of the tissues affected. Certain cells are particularly sensitive to anoxia, and others are less so, for example

Cell type	Period of anoxia leading to cell death
Brain cell (neurone)	4-8 minutes
Heart muscle, liver and kidney epithelial cells	1/2-2 hours
Connective tissue cells, and skin	many hours

CAUSES OF ISCHAEMIA

- In most instances, ischaemia results from the failure of the arterial supply. It is usually inadequacy of arterial blood flow which leads to clinically important anoxia.

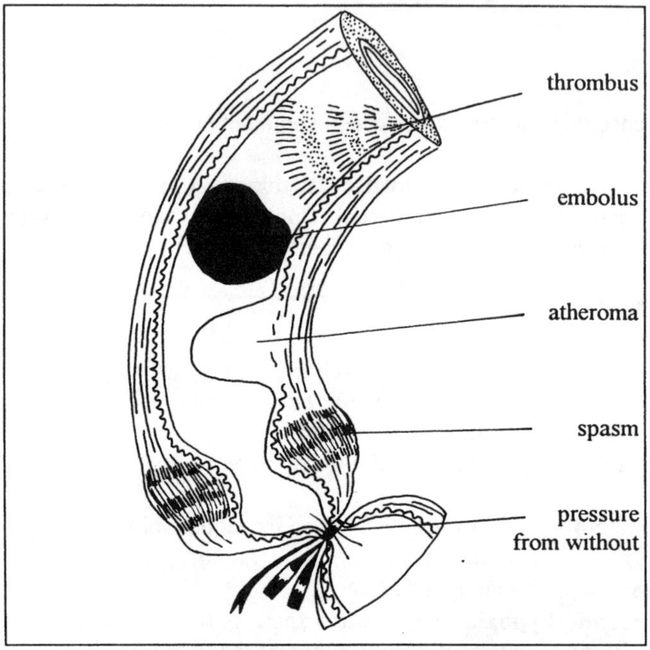

SOME COMMON CAUSES OF ARTERIAL OBSTRUCTION

Arterial obstruction, if complete, is likely to cause ischaemic necrosis, that is infarction. The common causes are obstruction due to thrombosis or embolism. However degenerative arterial disease such as atheroma can cause partial ischaemia. This is characterised by ischaemic fibrosis, with patchy death of specialised cells and proliferation of more hypoxia resistant connective tissue, resulting in some loss of function.

- Venous obstruction can also cause a stoppage of blood flow, which will then prevent the arterial flow from entering the part. This may best be seen when there is complete cut-off of venous return by twisting of the blood vessels supplying a part. For example, there may be torsion (twisting) of the testis which will cause occlusion of the veins of the spermatic co.d resulting in testicular infarction. Similar occurrences may take place in the ovary and in the gut.

  In these circumstances the low pressure venous system is obstructed before the arterial system, and it is venous obstruction which brings about infarction. Veins are more likely to be obstructed by external pressure, but obstruction of a vein by thrombus can also cause infarction, for example, mesenteric venous thrombosis.

- Small vessel obstruction may cause micro-infarction, as has been noted in fat embolism and in decompression sickness. Vasculitis affecting small vessels and external pressure (causing vascular obstruction leading to bedsores) are other examples.

It can be noted at this stage that ischaemia in muscle may give rise to pain before there is cell death of any degree. Probably *angina pectoris* is the best example of this where chest pain occurs on exertion when degenerate coronary arteries cannot increase the blood flow sufficiently to meet the added demand by the heart muscle.

FACTORS DETERMINING THE EXTENT AND SEVERITY OF ISCHAEMIA include:

- **The blood supply to the part**: Some organs, such as the liver and the lungs, have a double blood supply and hence arterial obstruction must be severe before they are rendered ischaemic. In other parts there is a well developed collateral circulation, for example, in the brain the circle of Willis will maintain the arterial supply even if there is complete occlusion of one of the carotid or vertebral arteries. In the lower limbs there is a good collateral circulation with duplicated arterial supply and in the gut there are plentiful arterial anastomoses. However some arteries are true end arteries with no anastomotic arteries, for example, the central retinal artery and others such as the coronary arteries, the cerebral vessels (distal to the circle of Willis) and the arteries to the kidneys and the spleen are virtual end arteries with little collateral supply . Blockage or obstruction to an artery which has a naturally poor collateral circulation will result in severe ischaemic effects.
- **Rate of onset of ischaemia:** If the onset is sudden, for example due to embolism, there is no time for a collateral circulation to develop. If the onset is slow, such as when there is a gradual build up of atheroma in a coronary artery, there may be time for collateral vessels to become opened up.
- **The degree of vascular obstruction:** This may be partial or complete. Atheroma in a coronary artery, for example, may lead to partial obstruction, while a thrombus in the same artery may cause complete obstruction and usually infarction.
- **The nature of the tissue:** Some tissues are more susceptible to hypoxia and anoxia than others, as was illustrated in the table earlier in this chapter.
- **Other contributory factors**: Anaemia or respiratory insufficiency may lead to anaemic or hypoxic hypoxia. In the presence of these complicating factors infarction can occur with a lesser degree of ischaemia. Thus a patient with a just adequate coronary supply may develop infarction because of a sudden drop in blood pressure or the onset of anaemia for some reason.

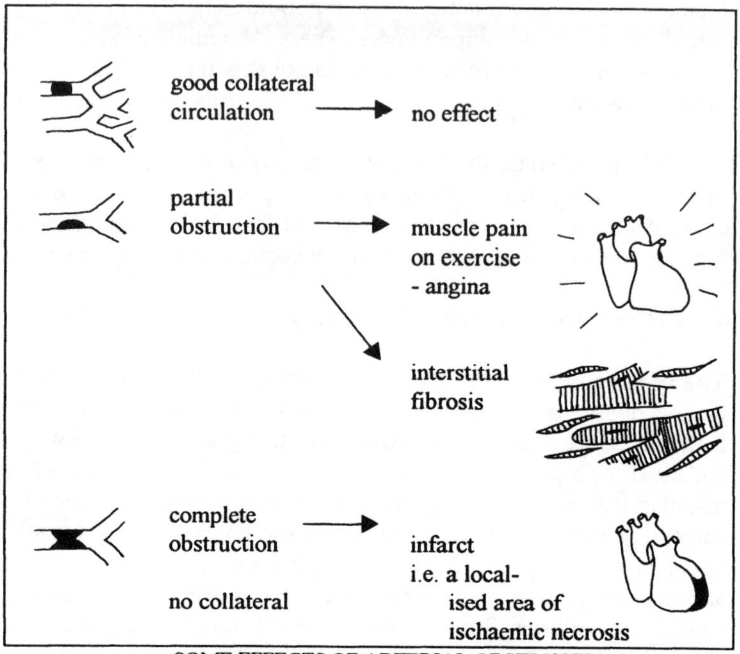

SOME EFFECTS OF ARTERIAL OBSTRUCTION

INFARCTION

An infarct develops when there is sudden, complete arterial (or less often venous) obstruction in an area where the collateral circulation is poor. The process of infarct formation is known as **infarction**.

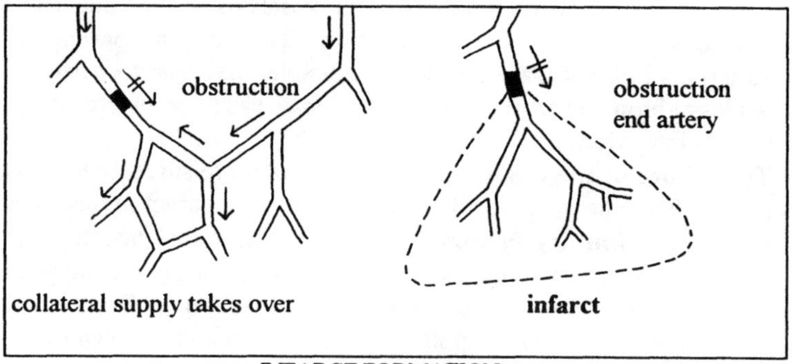

INFARCT FORMATION

THE DEVELOPMENT OF AN INFARCT:

- Initially there is death of cells due to the cutting off of the blood supply. There is, seemingly paradoxically, an accumulation of blood in the area from anastomotic vessels and from venous reflux. Red cells may escape into the tissue from anoxic capillaries.

Venous infarcts are usually intensely haemorrhagic and more blood will accumulate in loose tissues such as lung and intestine, than in solid organs such as kidney or heart. The blood in an infarct is, of course, deoxygenated.

- The dead tissue undergoes necrosis, that is recognisable structural changes of cell death. It can then truly be called an infarct. This takes from twelve to twenty four hours. Usually infarcts show typical coagulative necrosis in which the necrotic cells swell. In solid organs blood will be squeezed out of the area and the infarct will become pale. After a few days nuclei disappear but the tissue may retain a 'ghost' architecture for many weeks. (Note that in infarcts of the brain there is colliquative necrosis.)

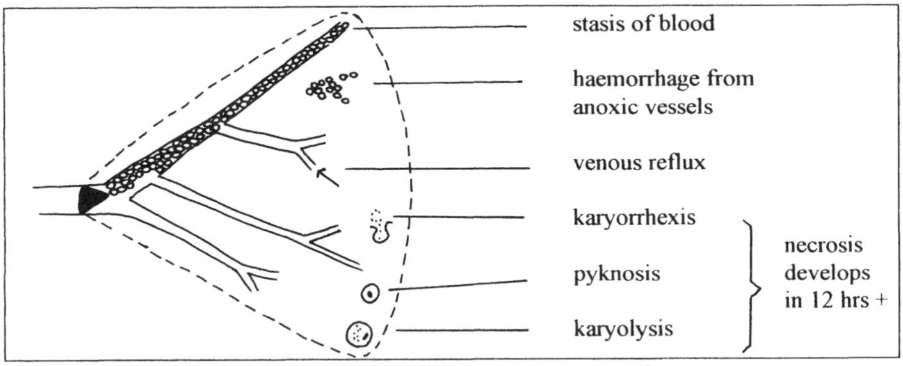

THE PROCESS OF INFARCTION 1 AND 2

- There is surrounding inflammatory change. Polymorphonuclear leucocytes gather first followed by mononuclear phagocytes. These macrophages phagocytose and remove any breakdown products together with bilirubin and haemosiderin from haemolysed red cells. These changes can best be seen one week or so after the ischaemic episode, when a pale yellowish area of infarction may be seen to be surrounded by a red inflammatory zone.

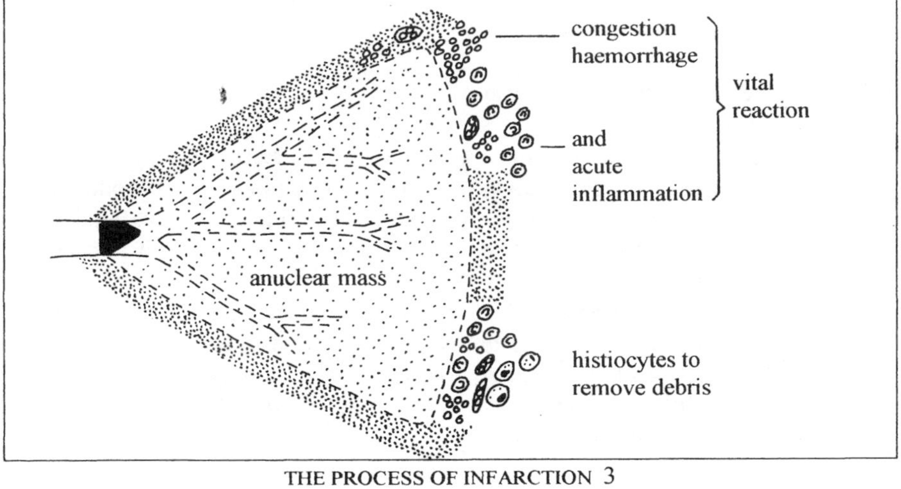

THE PROCESS OF INFARCTION 3

- Slowly reparative fibrosis takes place with the ingrowth of capillaries and fibroblasts from the edges of the lesion. Eventually the dead tissue is replaced and a fibrous scar is left. Fibrosis may be complete in six weeks in the heart or may take months to be completed in, for example, the kidney. The scar, once formed, may then undergo hyaline change and dystrophic calcification.

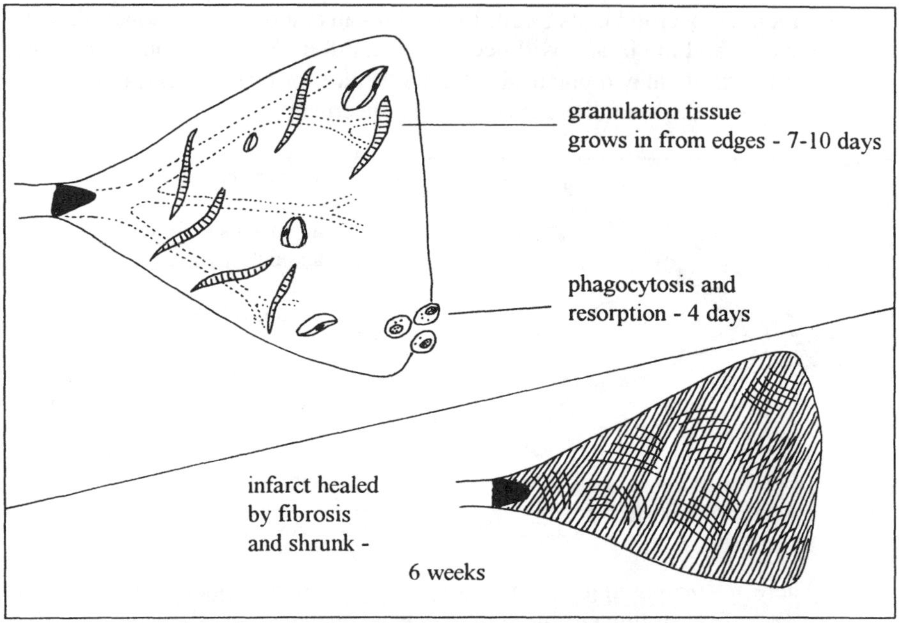

THE PROCESS OF INFARCTION 4

That summarises the events which take place in most infarcts. As has been illustrated infarcts are usually wedge shaped with the apex at the site of arterial obstruction and the base on the surface of the organ. The surface of the organ may be inflamed and covered with a fibrinous exudate.

CLASSIFICATION OF INFARCTS:

Infarcts can be classified in various ways:

- **On the basis of colour**: Although most start out by being **red** (haemorrhagic), in solid organs the blood is soon squeezed out and the infarct becomes **white** or **pale** (anaemic). The infarct remains red when it occurs in loose tissues (such as lung or gut) and also when it is caused by venous occlusion.
- **On the basis of infection**: The area of necrosis may be infected if there are already organisms present, for example infarcts of the gut soon become gangrenous. Infection may also follow on infarction caused by an infected thrombus. Infected infarcts are referred to as **septic**, while non-infected infarcts are **bland**.

THE BODY'S REACTION TO INFARCTION:

The body as a whole reacts to infarction as it does to any large area of necrotic tissue, with fever, raised erythrocyte sedimentation rate (ESR) and raised white blood cell count (WBC). Enzymes may be released from the dead and dying cells into the circulation, for example, aspartate transaminase (AST), lactic dehydrogenase (LDH) and creatine kinase (CK) from infarcts of the heart. This can be useful diagnostically.

INFARCTS OF PARTICULAR ORGANS:

Myocardial infarction is of particular importance as it is a common cause of death in developed countries, accounting for some half a million deaths per annum in the United States alone. Atheromatous narrowing of the coronary arteries with superadded thrombus formation is the most frequent cause. The most common artery to be affected is the anterior descending branch of the left coronary artery. Virtually all infarcts are in the left ventricle and inter-ventricular septum (even when there is obstruction of the right coronary artery). Thrombi may not always be found at autopsy and sometimes infarcts may be attributed to relative ischaemia due to an episode of hypotension or anoxic anoxia. Coronary artery spasm may also be implicated.

Myocardial infarcts may present in two ways:

- As a localised infarct in the wall of the left ventricle. This is a pale infarct which follows the pattern of coagulative necrosis. If the infarct involves the full thickness of the heart wall there may be an overlying pericarditis and also inflammation of the endocardium (sometimes with thrombus formation). If the patient survives, this will heal with a fibrous scar in about six weeks.

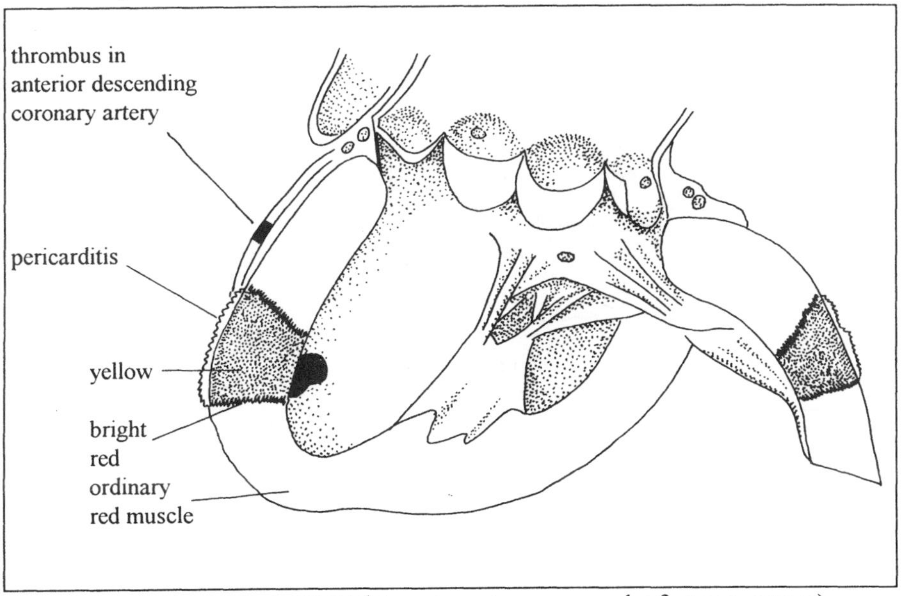

MYOCARDIAL INFARCTION (at post mortem one week after occurrence)

- A circumferential subendocardial infarct may occur involving only the inner third of the left ventricular wall. There is generally advanced atheromatous narrowing of all coronary arteries but thrombus formation may not be found, suggesting that this type of infarct is caused by an additional strain on a heart with a severely compromised blood supply, such as an episode of lowering of blood pressure.

Infarction of the brain is usually due to thrombus developing in an atheromatous cerebral artery or due to embolism (most often following thrombus formation in the heart). Brain infarcts may be pale or haemorrhagic. Haemorrhagic infarcts may occur when a large thromboembolus blocks an artery, giving rise to an area of necrosis. It then breaks up and moves on, allowing blood to flow into the primary area of necrosis.

Ischaemic brain undergoes colliquative (rather than coagulative) necrosis. Grossly, the brain tissue first shows softening, becoming so soft that it may be washed away by a stream of running water. The softened, liquefying tissue is removed by the action of macrophages over a period of weeks, eventually leaving a small cystic space. This may show some yellow staining of the wall due to haemosiderin. The cyst is surrounded by fibrillary material produced by the supporting cells of the brain. This is called gliosis of the brain (a repair process more or less equivalent to fibrosis elsewhere in the body).

Brain infarcts most commonly involve the area supplied by the middle cerebral artery, often including the internal capsule, and hemiplegia may be the clinical result. Brain infarcts are a common cause (along with cerebral haemorrhage) of the clinical condition known as 'stroke'.

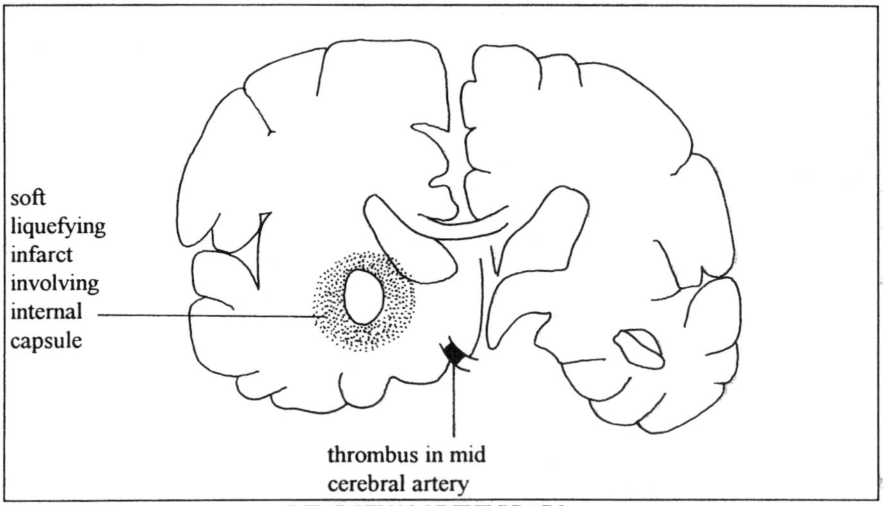

INFARCTION OF THE BRAIN

Pulmonary infarction: Pulmonary embolism is a common post mortem finding, but pulmonary embolism in healthy people does not cause infarction because of the double blood supply to the lungs.

There needs to be some degree of circulatory defect or heart failure before pulmonary embolism can cause infarction, for example, due to mitral stenosis when there is congestion of the lung and a reduced cardiac output and hence reduced bronchial artery supply,

Infarcts, when they occur, are wedge-shaped, firm and deep red in colour at the periphery of the lung. There is inflammation of the overlying pleura (pleurisy) and sometimes even pleural effusion. Histologically pleural infarction can be distinguished from pulmonary haemorrhage by the ischaemic necrosis of lung substance which has taken place.

Because of the relatively good blood supply, healing by fibrosis is rapid and a healed infarct may eventually only be represented by a small insignificant depressed scar.

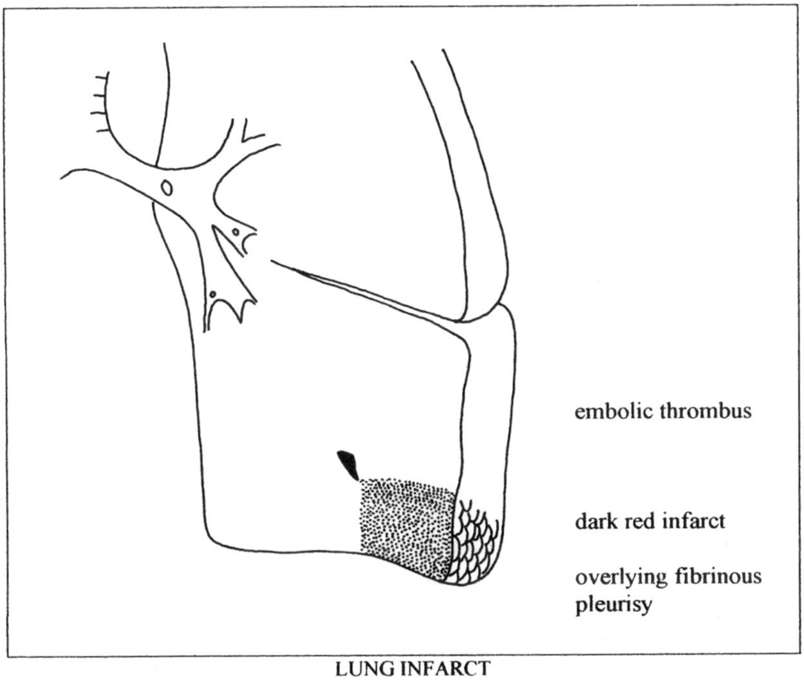

LUNG INFARCT

Infarct of the kidney: Local ischaemia to a part of the kidney occurs when there is occlusion of one of the branches of the renal artery; these branches can be considered as end arteries. This is usually the result of embolism.

The infarct is a typical wedge shaped infarct, pale after a few days, and shows the changes of coagulative necrosis. The 'ghost' structure of kidney preserved in an infarcted area can persist for months. The capsule of the kidney is not usually involved, as capsular vessels can maintain flow to the extreme periphery of the kidney. Eventually healing by fibrosis will take place.

It is worth noting that relative ischaemia or generalised ischaemia due to shock will cause acute tubular necrosis and not infarction, although in very severe shock bilateral cortical necrosis can take place.

Intestinal infarcts: may be due to arterial or venous obstruction. In both instances the loose tissue of the gut appears red. Twisting of the vessels to the gut in **volvulus** or compression of the vessels in a strangulated hernia are some causes. The affected ischaemic gut dies and becomes necrotic, and if not treated (by surgical removal) gangrene will follow when putrefactive organisms from the faeces invade the necrotic wall.

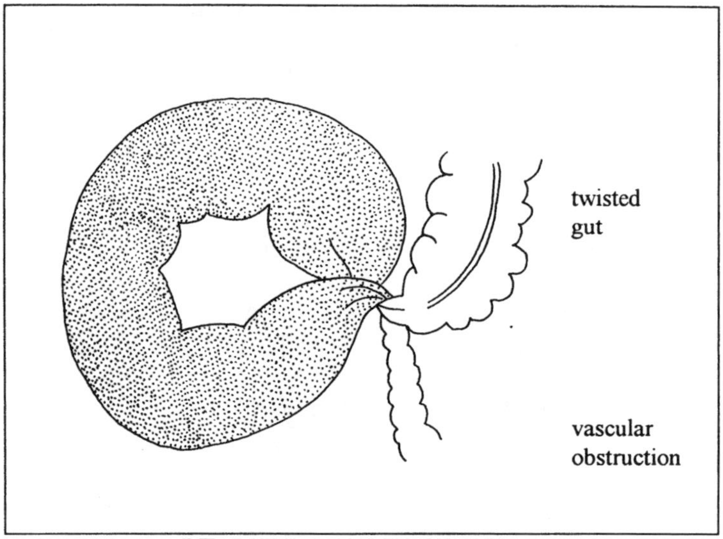

INFARCTED GANGRENOUS GUT

Splenic infarcts: These are usually post embolic in the normal spleen. Enlarged spleens are more susceptible to infarction perhaps because the blood supply is outgrown and there may be more local stasis.

The infarcts appear as pale areas of coagulative necrosis after a few days. They are wedge shaped. The capsule is involved with a fibrinous peritonitis overlying the spleen and causing pain which is a presenting symptom of this condition.

These examples of infarction have been described in some detail so as to allow visualisation of the application of general principles in practice. These important conditions will be dealt with again in lectures on systematic pathology.

For the present note once again that myocardial infarcts ('heart attacks') and brain infarcts ('strokes' or cerebro-vascular accidents) are amongst the most important causes of death in the developed world, together accounting for about one third of all deaths in the United States.

GENERAL PATHOLOGY - CHAPTER SIXTEEN

ACUTE INFLAMMATION

Most pathology books have a description of the changes found in acute inflammation somewhere near the beginning, because acute inflammation is part of the response of living tissue to injury. It is in fact the initial response, and may be followed by other changes such as regeneration of destroyed cells and repair by proliferation of fibrous tissue.

The changes following injury at cellular level have already been described, as have the changes of necrosis. Now it is proposed to examine the changes that occur at tissue level in vascularised tissue. In fact the definition is given as

Inflammation is the reaction of vascularised living tissue to local injury (Robbins)

It is a process designed to neutralise the injurious agent in some way (for example, by diluting it, by sealing it off or by actually destroying it).

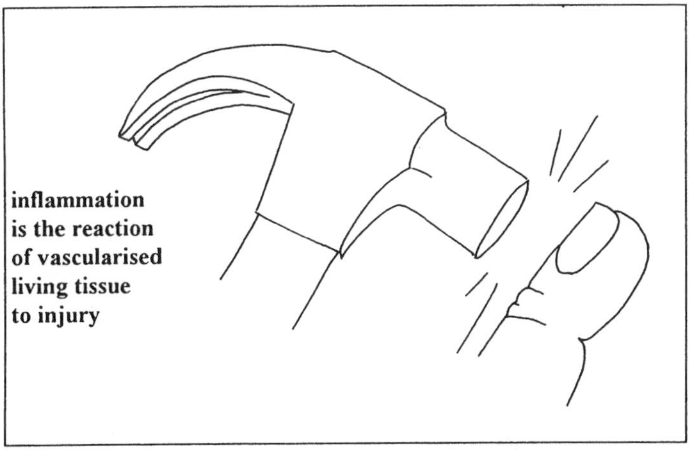

It must be remembered that the inflammatory reaction is a response on the part of living tissue and cannot occur in dead tissue. Thus, in an injury severe enough to cause complete destruction and death of an area, the changes of inflammation will not be seen - only those of necrosis. In these circumstances inflammation will be present at the *edge* of the destroyed or necrotic area. This is the **vital reaction** to necrosis.

THE CAUSES OF INFLAMMATION

are the same as those for any damage to cells:

- **Mechanical injury** for example by cutting or crushing
- **Chemical injury** for example. by acids, alkalis, bile within tissues
- **Physical injury** for example by heat, light electricity, X-irradiation
- **Injury due to living organisms** for example by bacteria, viruses, fungi and parasites
- **Reaction to necrosis** for example physiologically to menstrual endometrium, around infarcts or in the centre of tumours
- **Immunologic reaction** for example when there is an antigen/antibody reaction in the body such as in autoimmune disease or in hypersensitivity such as contact dermatitis.

In practice infective causes are the most important of those listed.

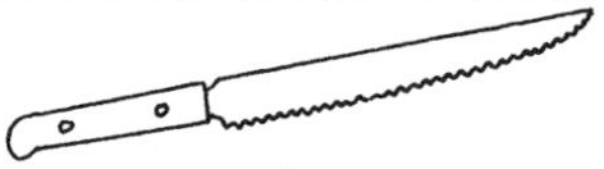	MECHANICAL e.g. cutting crushing
	CHEMICAL e.g. acids, alkalis . bile etc.
	PHYSICAL e.g. sun, heat electricity
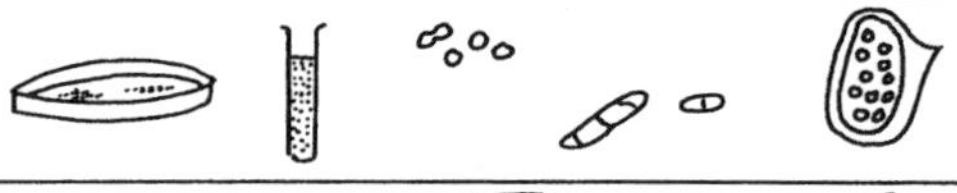	INFECTIVE e.g. bacteria, viruses, fungi, parasites
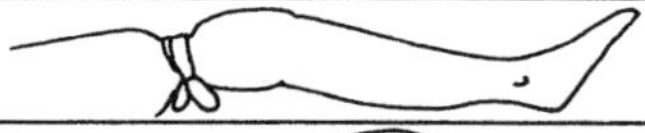	NECROSIS e.g.around infarcts
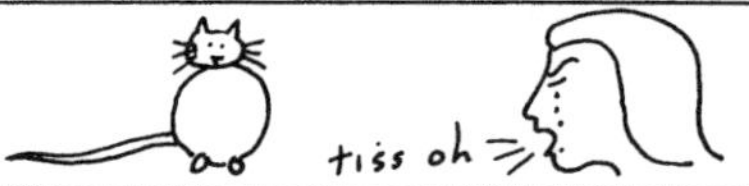	IMMUNE OR ALLERGIC e.g. rhinitis, dermatitis

CAUSES OF INFLAMMATION

As can be seen, this list includes most of the headings given earlier on as the causes of disease in general. Thus the causes of disease are similar when considered at cellular level (as in cell injury), at tissue level (as in inflammation) and when looking at the individual as a whole.

The features of acute inflammation following injury are relatively constant and a general description of the sequence of events can be applied to most parts of the body.

THE CLINICAL SIGNS OF INFLAMMATION

Certain so-called cardinal signs of inflammation have been known since described by Celsus in the first century of this era.

- **Redness** *(Rubor)*
- **Heat** *(Calor)*
- **Pain** *(Dolor)*
- **Swelling** *(Tumor)*, and a fifth added by Virchow last century
- **Loss of function**

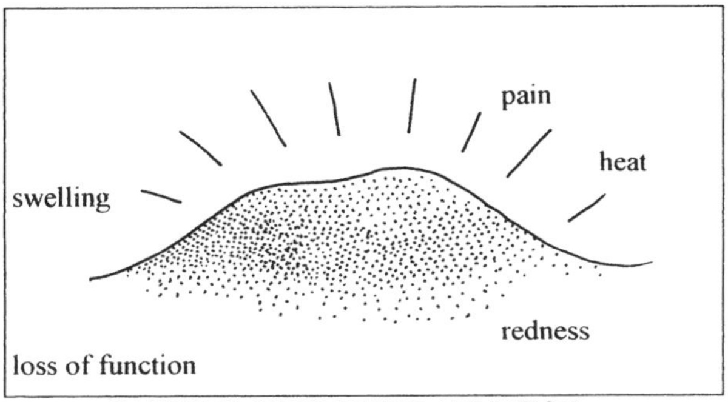

CARDINAL SIGNS OF INFLAMMATION

This collection of symptoms and signs is not a disease in itself, but a response of the body to many different types of injury.

MICROSCOPIC FEATURES OF INFLAMMATION

The microscopical features were originally described by Cohnheim in the 1880s. It is thought that these changes are brought about initially by the release of vasoactive substances.

In order the changes are:

- Arteriolar dilatation (sometimes preceded by transient vaso-constriction).
- Increased blood flow through arterioles, capillaries and vessels.
- Increased permeability of microvessels allows for the
- Exudation of fluid through permeable vessels.

} Triple response of Lewis

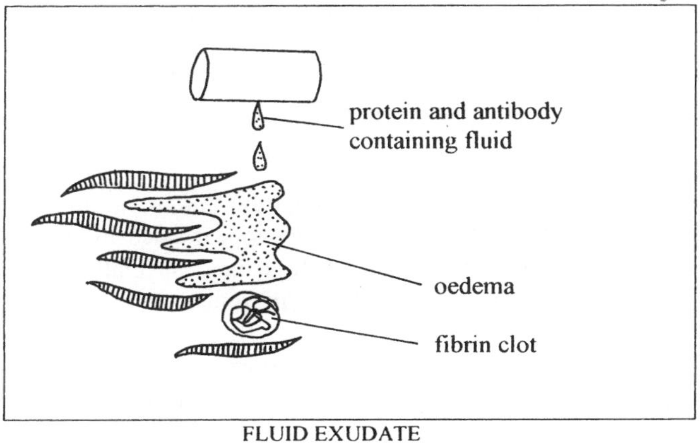

FLUID EXUDATE

- Concentration or packing of red cells in vessels due to fluid loss. This results in increased viscosity of the blood and
- Slowing or complete cessation of blood flow.
- Peripheral margination or pavementing of white cells which stick to the endothelium in micro vessels.
- Emigration of white cells by **active** passage through vessel wall into the inflammatory focus; first polymorphonuclear leucocytes, then monocytes together with the **passive** escape of red blood cells forced out with them by a process known as diapedesis.

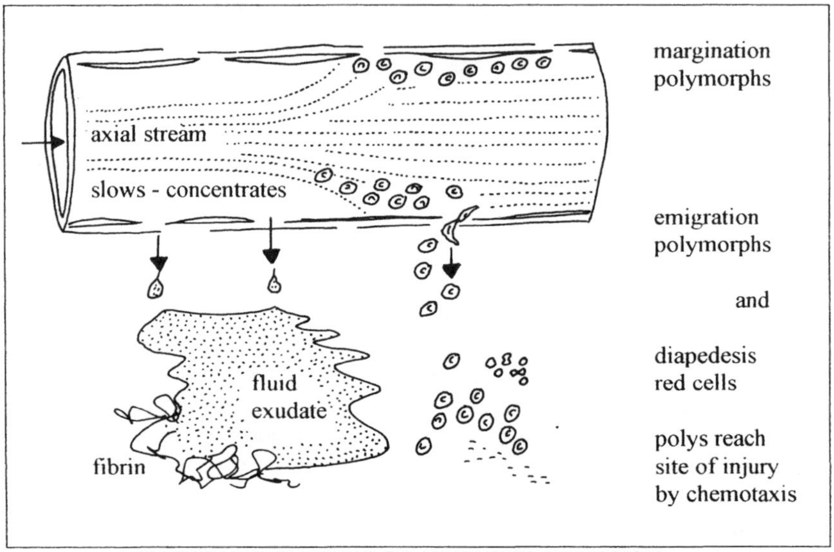

MICROSCOPIC FEATURES OF ACUTE INFLAMMATION

These changes can be summed up as:

- **Vascular**
- **Exudative** - fluid or cellular exudate

The changes can be seen down the microscope in a section from a tissue or organ which is the site of acute inflammation as:

- Vascular: Active hyperaemia.
- Exudative: Oedema, space occupied by fluid separating fibres and cells.
- Fibrin precipitation in the exudate may appear as eosinophilic strands
- There is an accumulation of white cells, especially polymorphonuclear leucocytes in and on vessel walls and in tissues.

It is the polymorphonuclear leucocytes (polymorphs or PMNs) which are the most characteristic histological finding in acute inflammation. These cells may show phagocytosis.

This description has summarised the initial changes that may occur in acute inflammation. Thereafter the inflammation may progress or resolve (get better) depending on the nature of the injury and the host resistance. The whole process is dynamic and variable. The timing of the development of changes can also vary, depending on the intensity of the injurious stimulus.

Some of these events can now be considered in more detail

Vascular permeability: Experimentally three types of vascular response are described.

- **The immediate-transient** response which usually comes on immediately after injury, lasts for fifteen to thirty minutes and is brought about by histamine and other chemical mediators. Leakage occurs from small venules between gaps in endothelial cells.
- **The immediate-sustained** response follows more severe injury to all small vessels (that is arterioles, capillaries and venules) that are damaged directly, for example, by a severe burn. There is necrosis of endothelial cells and leakage may last for several days until damaged vessels are thrombosed or repaired.
- **Delayed-prolonged** leakage may take some hours to develop and lasts for hours or days. It affects capillaries and venules and follows on injuries such as moderate heat, sunburn or X-ray. Other causes include bacterial toxins and delayed hypersensitivity reactions.

The histamine response or triple response evoked by mild injury was described by Lewis in 1924 and consists of:

- A red line, due to localised capillary dilatation.
- A red flare, which is a blotchy flush due to more diffuse arteriolar dilatation.
- A weal, which is a raised area along the red line due to oedema.

Exudate formation depends upon the raised intracapillary hydrostatic pressure in the first instance, although changes in permeability soon take place:

	Hydrostatic pressure	
	Arterial end	Venous end
Normal	32mm Hg	12mm Hg
Inflammation	60mm Hg	30mm Hg

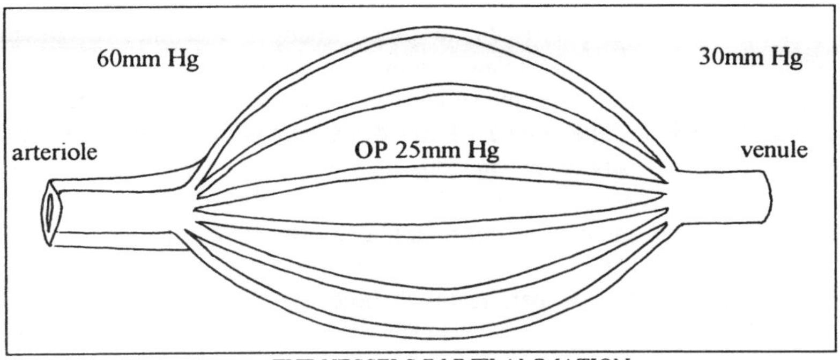

THE VESSELS IN INFLAMMATION

In addition to the increased hydrostatic pressure, fluid may leave more permeable vessels either by a process of pinocytosis across the endothelial cell, or, as seems more certain in small venules, through gaps between the intercellular junctions of endothelial cells which are opened up by the chemical mediators of inflammation.

The fluid part of the exudate is virtually equivalent to plasma in composition. It will act by:

- Diluting toxins
- Transporting antibodies and drugs to the site of inflammation
- Precipitating fibrin to wall off the area and to aid phagocytosis

The lymphatics dilate and drain oedema fluid and waste products from the inflamed area. An immune response may be evoked by the transport of macrophages containing antigenic material to draining lymph nodes. If lymphatics are involved in the inflammation there may be lymphangitis seen as red streaks radiating from the primary area along the course of the lymphatics. Draining lymph nodes may be enlarged and tender.

The clinical cardinal signs of inflammation can be related to the vascular changes:

- Redness and heat are due to the increased blood supply with vascular dilatation.
- Swelling is due to the oedema caused by the exudation of fluid from vessels into the tissues
- Pain may be due to oedema causing increased tissue tension and pressure on nerve endings. Release of chemical mediators of inflammation (such as bradykinin) may also cause pain.
- Loss of function may occur following the swelling and the pain or following inflammatory destruction of tissue.

The cellular part of the exudate

As the white cells fall out of the axial stream when the circulation slows **margination** or **pavementing** of leucocytes takes place.

These cells then **adhere** to the vascular endothelium. Adhesion takes place by virtue of cell adhesion molecules (CAMs) on leucocyte or endothelial cell surfaces which are activated by the chemical mediators of inflammation.

Leucocytes then **emigrate** by active extension of pseudopods through gaps in the endothelial cell junctions. They may then move along the vessel wall between the endothelial cell and the basement membrane until they manage to penetrate the basement membrane and enter the tissues. Movement of pseudopods is carried out by means of actin filaments associated with the contractile protein myosin within the cell. Emigration may start within minutes of the onset of injury and can be well established in a few hours. The cells involved in the first six to twenty four hours are the neutrophil polymorphonuclear leucocytes (polymorphs or PMNs), but after twenty four hours they are replaced by monocytes. A few red cells may passively follow the white cells through the vessels wall in the process of **diapedesis**.

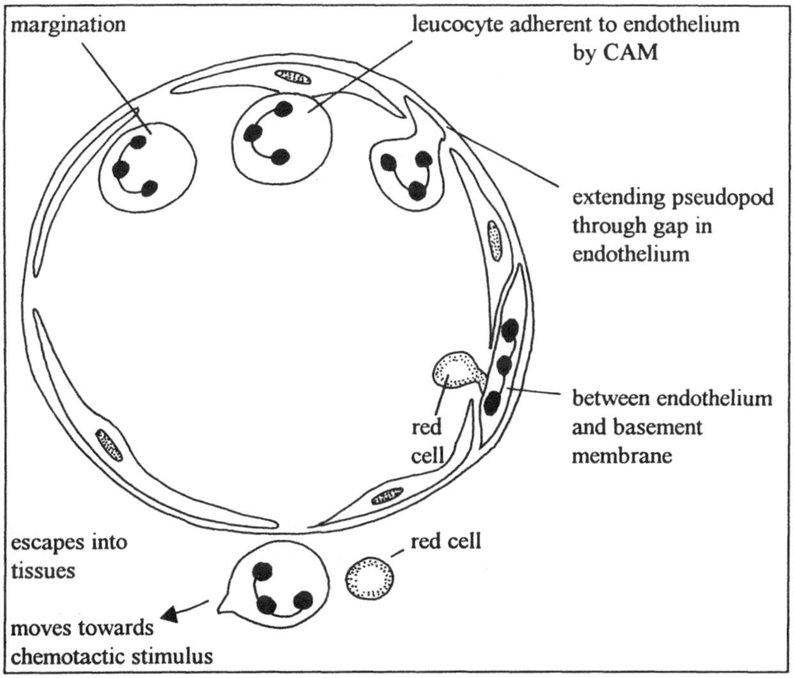

MARGINATION, ADHESION AND MIGRATION OF WHITE CELL

White cells find their way to the site of injury by a process of **chemotaxis**, which is a positive movement by white cells towards an attracting chemical stimulus, involving movement along a chemical gradient.

Chemotactic agents for neutrophils include:

- Bacterial products
- Components of complement system, particularly C5a
- Components of the lipoxygenase pathway of arachidonic acid metabolism especially leukotriene B_4 (LTB_4).

Chemotaxis takes place through specific receptors for chemotactic agents on the leucocyte cell surface. By a series of complex chemical reactions on the cell membrane, calcium ions are admitted to the cell to stimulate the cell movement mechanisms and also to activate the arachidonic acid cascade.

Once in the tissues leucocytes proceed to **phagocytosis** to try and deal with noxious particulate material such as bacteria. **Degranulation** of leucocytes with the release of enzymes also takes place.

Phagocytosis consists of the stages of **recognising** the material to be phagocytosed, **ingestion** of the material into the cell cytoplasm and denaturation including **killing** and/or **digesting** the material.

Opsonins are substances in the serum, including immunoglobulin (IgG) and a fraction of complement (C3b) which can coat the matter to be phagocytosed and bind on to the surface of white cells allowing the subsequent stages of phagocytosis to take place.

Cell cytoplasm then extends around the material to be ingested until it is surrounded, and a piece of cell membrane breaks off to leave an ingested membrane bound particle within the cell cytoplasm - a **phagosome**. The surrounding membrane rapidly fuses with that of a lysosome to form a **phagolysosome**. Controlling chemical mechanisms for this process resemble those involved in chemotaxis.In addition to opsonins, phagocytosis may also be assisted by specific antibodies. Precipitated fibrin may form a framework trapping the material to be phagocytosed.

The final stages of the killing of bacteria and digestion of phagocytosed material depend on intracellular metabolic events. There is a sudden increase in respiratory activity with the formation of free radicals of oxygen (superoxide and hydrogen peroxide), which may act through the Hydrogen Peroxide-Myeloperoxidase-Halide system, which makes HOCl· (hypochlorous acid), a powerful antibacterial agent. The free radicals may also act independently. There are also a number of non-oxygen dependent mechanisms whereby substances within leucocyte granules can kill and digest organisms.

Not all organisms may be disposed of in this fashion. Some, such as the tubercle bacillus, may be resistant to these intracellular processes and survive, perhaps to be transported throughout the body via the lymphatics and to be released when the cell containing them dies. In other circumstances, very virulent organisms may resist phagocytosis.

Neutrophils contain two sorts of lysosomal granules:

Smaller specific granules	Larger azurophil granules
• Lactoferrin	• Myeloperoxidase
• Lysozyme	• Lysozyme
• Alkaline phosphatase	• Cationic proteins
• Collagenase	• Acid hydrolases
• Leucocyte adhesion molecules	• Neutral proteases (elastase)

The contents of these lysosomal granules are released not only into phagolysosomes, but also into the surrounding tissues, together with oxygen derived free radicals and products of the arachidonic acid cascade. These then act to potentiate the inflammatory reaction.

Lysosomal enzymes may be released by a process of regurgitation if the cell membrane is not completely closed off during phagolysosome formation, by exocytosis or after cell death.

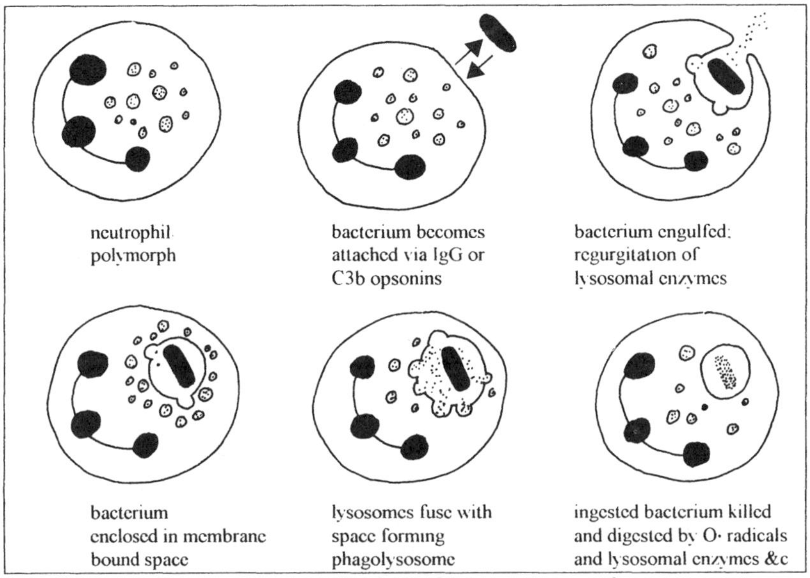

PHAGOCYTOSIS

CHEMICAL MEDIATORS OF INFLAMMATION

There are vast numbers of substances to which a role in the causation of inflammation can be attributed, and some of these have been mentioned in the previous discussion. Once the inflammatory process has started it may be self-perpetuating inasmuch as some of the mechanisms involved (for example lysosomal granule contents) may themselves act as mediators of inflammation. However, just as in so many of the microsystems of the body, for example, those concerned with haemostasis, there are very elaborate systems of control, with agonists and antagonists active at every stage in the process.

The major groups of chemical mediators include:

- **Vasoactive amines,** such as histamine and 5-hydroxytryptamine, may be released from mast cells in the tissues and basophils in the blood and platelets following certain stimuli, such as trauma, IgE hypersensitivity, release of fragments of complement (C3a and C5a) and other substances from the blood cells. It has been noted that histamine is important in the early stages of inflammation.
- **The complement system** is a complex cascade system in the blood which is important in the immune and inflammatory reactions. The system is activated by antigen-antibody complexes (the classic pathway) or by a variety of substances such as bacterial toxins and aggregated globulins (the alternate pathway).

Fractions of complement increase the vascular permeability (C3a and C5a), exert chemotactic influence (C5a), act as opsonins (C3b), and lyse the membranes of cells and organisms.

- **The kinin system, the coagulation system and the fibrinolytic system**. These are related systems which may be activated together by contact activation factor (Factor XII or Hageman factor).

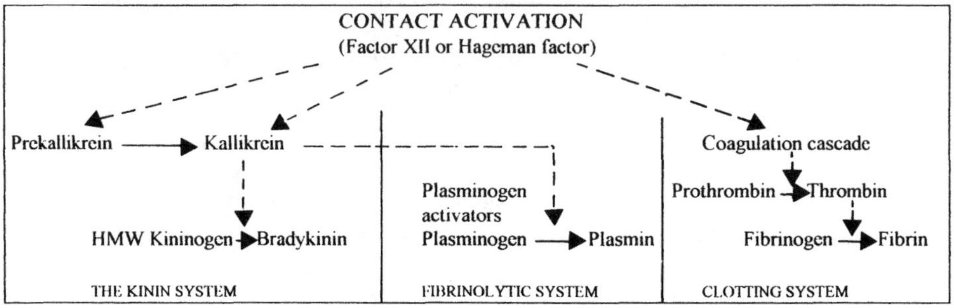

Bradykinin increases vascular permeability and is a vasodilator and may cause pain.

Plasmin activates C3 to C3a which increases vascular permeability and also splits fibrin into degradation products that may increase vascular permeability. It lyses any fibrin formed.

In the conversion of fibrinogen to fibrin substances are formed which increase vascular permeability and have a chemotactic effect

- **The Arachidonic Acid Metabolites**

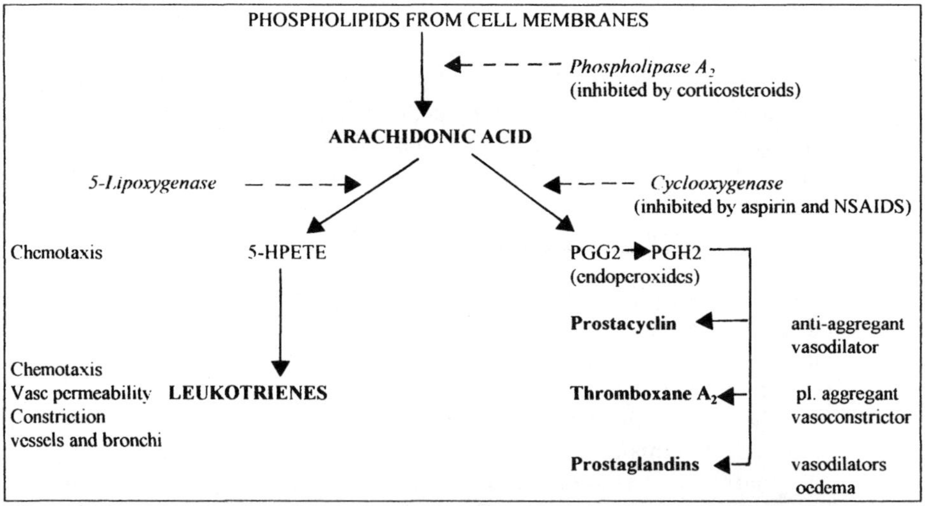

Platelet-activating factor may be derived from endothelial cells or mast cells. It brings about the release reaction from platelets, whereby the contents of their granules are released. These contain vasoactive amines (in the dense bodies) and lysosomal enzymes and cationic proteins (in the alpha granules) that can increase vascular permeability. Activation of platelet factor 3 aids in the coagulation process.

Platelet activation factor in low concentrations will bring about vasodilatation and increased permeability. It will also increase leucocyte adhesion and is a chemotactic agent.

- **Cytokines,** interleukin-1 (IL-1) and tumour necrosis factor (TNF), are produced by macrophages and many other cells. They have local effects on endothelial cells, increasing adhesion of leucocytes, stimulating release of prostacyclin and platelet activating factor and promoting coagulation. They stimulate fibroblast proliferation and collagen production. They are also responsible for some of the system effects of inflammation, such as fever and increased polymorph count.
- **Lysosomal contents** and **free radicals** mostly derived from oxygen, have already been discussed when describing the inflammatory response.

Thus there are a multitude of factors at work in bringing about the changes of inflammation. They are not necessarily sequential as has been described. The process may be at different stages in the same area, and there are many modifying factors such as the type of noxious stimulus, the anatomical site involved and the general status of the person affected, including their immune status.

The majority of cells found in the inflammatory exudate occur in the peripheral blood. They include:

POLYMORPHONUCLEAR LEUCOCYTES (PMNs or polymorphs), consisting of neutrophils, eosinophils and basophils.

- **Neutrophils** are so called because they contain faintly staining granules or lysosomes of granular appearance in the cytoplasm. These are of two types - the smaller specific granules and the larger azurophil granules, containing a variety of enzymes that have been listed previously.

  They are 10-12μm in diameter with a distinctive lobed nucleus which stains deeply; the number of lobes increases with the age of the cell. In the circulating blood neutrophils number 2,500-7,500 per c mm and constitute 40-75 per cent of total white blood cells. They have a short life of only a few days; they are actively motile and intensely phagocytic and are the first cells to arrive when there is acute inflammation.

  The cytoplasm contains glycogen to provide energy for the activities of the cell. Released lysosomal enzymes are important in the formation of **pus**, which is a semi-liquid substance composed of living and dead polymorphs, digested dead tissue cells and fluid exudate. Living and dead micro organisms are generally present in pus, but it may rarely be sterile as in chemically induced inflammation.

Neutrophil polymorphs are by far the most numerous and the most important of the cells participating in the initial stages of the acute inflammatory reaction.

- **Eosinophils** are multilobed (often bilobed) less actively phagocytic cells with prominent red-staining granules (with eosin). There are 40-440 eosinophils per c mm of blood (one to six per cent of total white cells). Their functions are not well understood but they are thought to ingest immune complexes and produce substances that inhibit other mediators of inflammation.

  In inflammation due to parasites and in allergic conditions eosinophils accumulate and they are also attracted by degranulating mast cells (which secrete eosinophilic chemotactic factor). They may destroy antibody coated larval parasites, particularly schistosomes by virtue of toxic substances contained within their granules; they may also cause lysis of epithelial cells, all by virtue of a major basic protein (MBP) contained within their granules.

  Eosinophils may modulate the inflammatory reaction and are often seen at a later stage of acute inflammation when healing and repair are commencing.

- **Basophils** in the blood are similar to **mast cells** in the tissues. **Mast cells** are characterised by granules which contain histamine and 5-hydroxytryptamine, heparin, eosinophil-chemotactic factor of anaphylaxis and platelet-activating factor. These granules may be released in response to stimuli such as IgE hypersensitivity reactions, stimulation by activated complement factors, injury, and cytokines. Their activation may also lead to the operation of the arachidonic acid metabolic pathways.

  In the circulating blood basophils number 0-100 per c mm and form less than two per cent of circulating white cells.

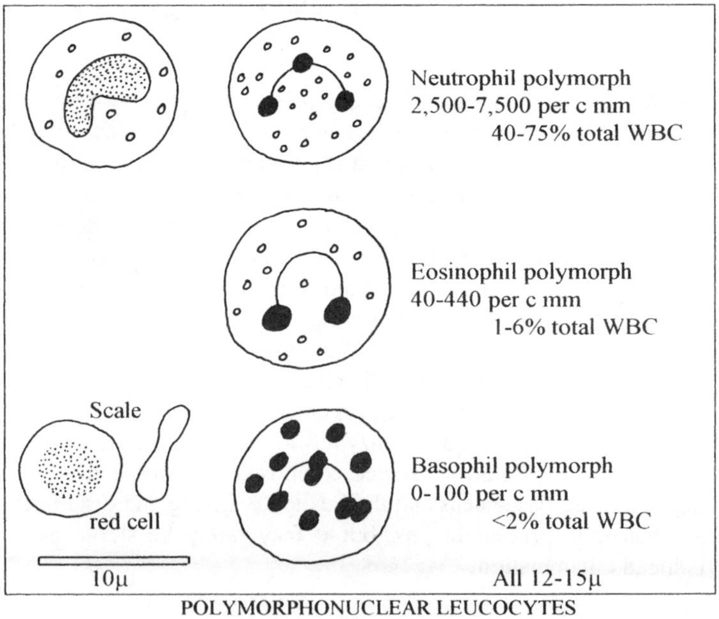

POLYMORPHONUCLEAR LEUCOCYTES

MONONUCLEAR PHAGOCYTES

These are the scavenger cells of the body, responsible for most of the phagocytic activities which clear up foreign matter from the blood and from the tissues and the debris of inflammation. They are prominent in the inflammatory reaction after the first twelve hours or so. They also have an immune function as presenters of insoluble antigen to T cells for development of cell-mediated immunity; they produce cytokines such as interleukin-1 (IL-1) and tumour necrosis factor (TNF) which are mediators of inflammation as well as T cell activators; they have a role in the delayed hypersensitivity reaction.

Mononuclear phagocytes are called **monocytes** in the blood and **macrophages** elsewhere. They are derived from precursor cells in the bone marrow which give rise to blood monocytes (200-8,00 per c mm or from two to ten per cent of white cells in the blood). Macrophages are also present in tissues:

- In the sinuses or sinusoids of lymph nodes, spleen, liver and bone marrow
- In connective tissues (histiocytes); in bone as osteoclasts.
- In lung alveoli; lining serous cavities; in the nervous system (microglia).

They are capable of phagocytosis and pinocytosis. They carry cell surface receptors for opsonins (the Fc fragment of IgG and C3b) to aid phagocytosis. They contain a large variety of lysosomal enzymes. They can become 'activated' with increased levels of enzymes and enhanced capacity to kill bacteria or tumour cells; activation may be brought about by lymphokines from sensitised T lymphocytes (γ interferon) or by bacterial endotoxins or other causes of inflammation. Activated macrophages produce a great number of biologically active agents including:

- Plasminogen activator
- Coagulation factors
- Complement
- Arachidonic acid metabolites which act as mediators of the inflammatory reaction
- Cytokines
- Chemotactic factors for other white cells
- Free radicals of oxygen
- Collagenase and elastase which break down connective tissue
- Growth promoting factors

Most of these are themselves mediators of inflammation, so that the presence of activated macrophages can perpetuate inflammation. However there are also substances produced by cells in inflammation which will switch the macrophages off or de-activate them, for example, transforming growth factor β.

Macrophages are less active than polymorphs, but are none the less mobile and able to reach all parts of the body. They become prominent later in the course of inflammation and being long lived may persist for weeks or months at the site of inflammation. Macrophages can fuse together to form multinucleate giant cells in certain conditions for example, TB, fat necrosis and foreign body reactions.

LYMPHOCYTES

These are a small darkly staining cells 8-10μ in diameter with little cytoplasm. They recognise foreign antigenic material in the body. The **T lymphocytes** may be cytotoxic or exercise a regulatory function, often by the secretion of lymphokines, which amongst other things will activate macrophages and cause monocyte chemotaxis. **B lymphocytes** synthesise uniquely specific immunoglobulins after antigenic stimulation.

Lymphocytes number 1,500-3,500 per c mm in the peripheral blood and form 20-45 per cent of all white blood cells. They are mostly T cells with about 10-20 per cent B cells in the blood.

They occur most often later in inflammation after there has been time for an immune response to develop.

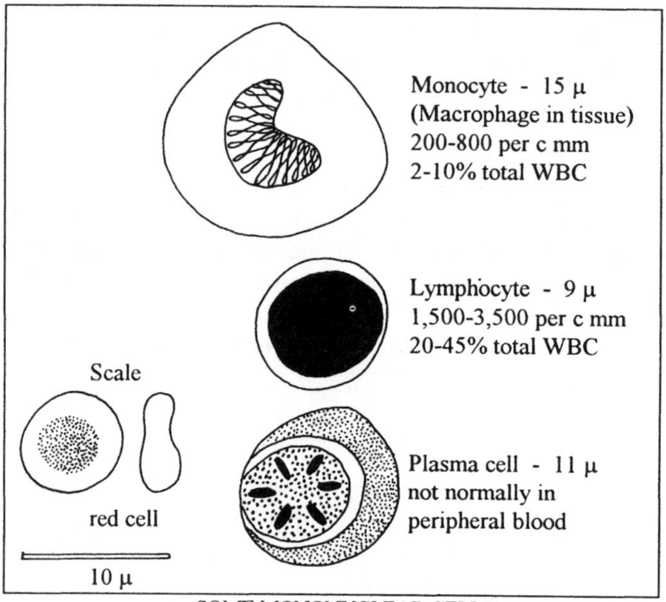

SOME MONONUCLEAR CELLS

PLASMA CELLS

Plasma cells are normally not found in the peripheral blood but only in tissues, especially lymphoid tissue and the spleen. They are somewhat larger than lymphocytes measuring 10-12μ in diameter. They have a distinctive appearance. The nucleus is dark and shows chromatin clumping, often around the periphery in a clock face pattern. The nucleus is eccentric, that is situated at one edge of the round cell. It is separated from the dark reddish purple staining cytoplasm by a pale peri-nuclear halo. These cells may be derived from transformed B lymphocytes and synthesise antibody globulins. They occur, with lymphocytes, late on the scene in inflammatory areas and are more often evidence of chronic rather than acute inflammation.

GENERAL OR SYSTEMIC REACTION TO ACUTE INFLAMMATION

There are a number of systemic effects which affect the body as a whole:

- **Fever:** A raised body temperature with shivering as the temperature rises and sweating as it remits may occur in association with inflammation, particularly if the inflammation is due to an infective cause. his may be due to the effect of bacterial endotoxins or may be brought about by interleukin-1 (IL-1), a cytokine produced by macrophages and other cells, which acts on the thermosensory centre. IL-1 acts via the formation of a prostaglandin; aspirin, which inhibits the action of cyclooxygenase in arachidonic acid metabolism, reduces body temperature by blocking prostaglandin formation. Patients with fever also feel unwell, with general symptoms such as headache, aches and pains in the muscles and joints, lethargy, weakness, loss of appetite and difficulty in concentration.
- **Leucocytosis**: There is usually an increase in the number of neutrophils circulating in the peripheral blood in inflammatory conditions. This is due to the effect of cytokines (IL-1 and TNF) and also to the release of colony stimulating factors (CSFs) from macrophages and T cells. In certain infections, mostly viral, there may be a lymphocytosis and parasitic infestations lead to an eosinophilia.
- **Changes in protein synthesis** are brought about by IL-1 and TNFα and involve the synthesis of proteins by the liver, including C-reactive protein, serum amyloid A complement and coagulation proteins including fibrinogen. An increase in fibrinogen is responsible for a rise in the erythrocyte sedimentation rate (ESR), in which red cells in anticoagulated, citrated blood settle or sediment more quickly in most inflammatory condition.

In prolonged inflammation there may also be **weight loss**, also thought to be due to the action of cytokines; **amyloidosis** is another possible late complication, but these last two are more associated with chronic inflammation than with acute inflammation.

GENERAL PATHOLOGY - CHAPTER SEVENTEEN

THE CLASSIFICATION OF INFLAMMATION

There are a number of factors which may modify the inflammatory response by tissues of the body. These include:

- The nature of the injury
- The site of the injury
- The general health status of the person affected

INJURY factors	HOST factors
Strength, Dose, Duration, Invasiveness or penetration, Site } of causative agent Bacterial virulence Bacterial toxin production	General state of health State of immunity White cell count Local factors such as blood supply and site of inflammation

There are endless possibilities depending upon the balance of these factors. They range from an overwhelming rapidly fatal infection resulting from infection with an organism of relatively low virulence, as may occur in a patient with a deficient immune system, to subclinical or inapparent infection caused by a highly virulent organism in a patient with a high level of immunity and good general health.

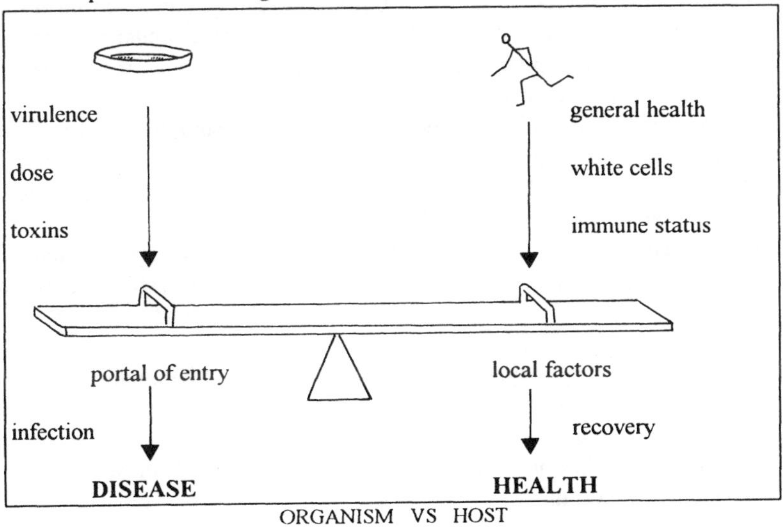

ORGANISM VS HOST

The different types of response are best considered by **classifying** inflammation.

Classification can be based upon the following three points:

- The nature of the inflammatory exudate.
- The duration of the inflammation.
- The site of the inflammation.

CLASSIFICATION ACCORDING TO THE NATURE OF THE INFLAMMATORY EXUDATE.

The fluid exudate is most characteristic of acute inflammation but may persist into the chronic stages. Exudates can be classified according to the type of cells present. This classification is most useful when considering inflammation of serous cavities such as pleura and pericardium. Inflammation can also be classified as to the form the exudate takes, for example, an abscess.

Inflammation with:

- **Serous exudation** is characterised by much fluid exudate either into the tissues or into a body space. This is an early response to mild or moderate injury. Examples include skin blisters which form after burns, and pleural effusion in tuberculosis
- **Fibrinous exudation** in which the inflammatory exudate contains considerable amounts of fibrinogen which precipitates as fibrin. This tends to be a feature of inflammation of the body cavities such as pleura, pericardium or peritoneum.

  It may be combined with a large outpouring of fluid, that is sero-fibrinous inflammation or exudation. An example may be seen in rheumatic fever involving the pericardium, when the pericardial sac may contain much fibrin deposited between the two layers. When these surfaces are separated or pulled apart they may show a rough sticky appearance - the so-called 'bread and butter' appearance of fibrinous inflammation of a surface, because it resembles the surface of a slice of buttered bread (when it has been dropped on the floor butter side down!). In uraemic patients a similar pericarditis may occur and in lobar pneumonia of pneumococcal origin a fibrinous exudate forms in the alveolar spaces.

  Microscopically, fibrin appears as strands of eosinophilic material.

  Clinically fibrinous inflammation of serous surfaces in the pericardium or pleura may be noted because of the sound or 'rub' made by roughened surfaces moving against each other.
- **Haemorrhagic inflammation** occurs when there is severe vascular damage as part of the inflammatory process and there is considerable loss of red cells into the exudate. Examples include influenzal pneumonia, meningococcal septicaemia and anthrax
- **Catarrhal inflammation** occurs when there is inflammation involving a mucus producing epithelium which may be stimulated to produce a profuse secretion of mucus, for example, in inflammation of the nasopharynx, bronchi, gut. Chronic bronchitis is actually defined in terms of mucus production as sputum.

However the best known example is the common cold in which inflammation of the nasal mucosa is accompanied by the secretion of mucus and inflammatory exudate which runs from the nose.

- **Pseudomembranous* inflammation** is said to occur when there is inflammation of a mucosal surface, for example, throat, respiratory passages or gut, in which a false membrane consisting of necrotic epithelium, fibrin and inflammatory cells is formed. [* Note: a true membrane is said to consist of **living** cells]

  A well known example is diphtheria, where a pseudomembrane can form in the throat or larynx. This membrane may cause respiratory obstruction and necessitate immediate tracheotomy to save the patient's life.

  Pseudomembranous colitis is most often due to the toxin of *Clostridium difficile*, when this organism is able to proliferate in the gut, for example, after broad spectrum antibiotic therapy. There is severe diarrhoea and a grayish yellow membrane forms on the surface of the bowel.

- **Suppurative inflammation** is a condition in which inflammation is accompanied by the production of pus.

Pus is a semi-liquid substance composed of living and dead polymorphs, digested dead tissue cells, and fluid exudate and which may contain living or dead micro-organisms.

Pus producing inflammation is known as **suppurative** and the process of pus production may be known as **suppuration**. The adjective derived from pus is **purulent**.

Suppurative inflammation may be localised or diffuse.

A localised collection of pus is known as an **abscess**.

An abscess forms when the destructive inflammatory process is walled off at first by fibrin and inflammatory cells, and later by newly formed connective tissue. If pus production continues after this walling off it can be extremely painful. Whilst the pus remains *in situ,* healing cannot easily take place. The pus must be removed and this often happens naturally because with increasing tension the pus will tend to track between tissue planes to a surface.

A boil is a common form of superficial abscess (formed in an infected hair follicle) and those who have experienced this form of infection will know something of the behaviour of an abscess. Surgeons tend to anticipate the bursting or tracking of an abscess by incising it and draining the pus. This tends to minimise tissue damage and relieve the pain, as well as reducing the chances of spread of infection due to the tracking or 'pointing' of the abscess.

Abscesses which burst into the body spaces, for example, from the brain into the meninges, from the lung into the pleural cavity (giving rise to empyema) or from the appendix into the peritoneum, are particularly dangerous.

Occasionally the abscess may not drain and healing can still take place. The pus thickens as fluid is absorbed. Eventually it becomes solid and dystrophic calcification can take place.

Diffuse suppurative inflammation of subcutaneous tissues is called cellulitis. This sort of spreading inflammation is caused by organisms which produce spreading factors. Examples of this are Streptococci which can give rise to a deep subcutaneous spreading inflammation called erysipelas and gas gangrene which is a condition where inflammation spreads through tissue and up and down muscle sheaths.

- **Gangrenous inflammation** has already been described together with necrosis, and occurs when there is superadded infection with putrefactive organisms in an area of necrosis. Typically the exudate is foul smelling and there is also gas production from bacterial fermentation of carbohydrates.

Note that inflammation which involves an epithelial surface may cause an **ulcer** which may be defined as a localised full thickness defect of epithelium covering a surface.

CLASSIFICATION ACCORDING TO THE DURATION OF THE INFLAMMATION.

This forms one distinction between acute and chronic inflammation:

Acute inflammation is the term used to denote the vascular and exudative changes that have been described in some detail. But this is far from the end of the story. In acute inflammation, polymorphs are the predominant cells, but with the passage of time and neutralisation of the injurious agent, macrophages accumulate and the demolition phase commences. In this phase the macrophages remove the dead cells and their breakdown products. The macrophages themselves then either die or are drained through the lymphatics. Then follows healing and repair.

However this course of events can only take place when the original cause of the inflammation has been removed. If the noxious agent persist then there may be simultaneous acute inflammation and attempts at demolition and healing. It is at this stage that the inflammation becomes chronic.

Chronic inflammation can be defined as 'a prolonged process in which destruction and inflammation are proceeding at the same time as attempts at healing' (Walter and Israel).

In terms of time, acute inflammation can be assumed to be entering a chronic phase after it has persisted for a week or two. Acute inflammation may go on to become chronic for a number of reasons:

- Persistence of the injurious agent
- The presence of foreign bodies
- Delayed healing for some reason
- The development of hypersensitivity
- Repeated bouts of acute inflammation as may occur in the kidneys, especially when there is a predisposing cause such as stasis.
- Lowered host resistance

In some instances there is no obvious acute inflammatory phase and a low grade chronic inflammation may appear be present from the outset, for example, in tuberculosis.

In chronic inflammation although some polymorphs may be present, macrophages, lymphocytes and plasma cells are the main types of cells present. In addition, fibroblasts and new capillaries are visible evidence of attempts at healing. The fluid part of the exudate is generally less conspicuous.

CLASSIFICATION OF INFLAMMATION ACCORDING TO SITE

Inflammation may be classified according to the tissue involved, such as appendicitis, tonsillitis and osteomyelitis, generally by adding the suffix '-itis' to the anatomical name for the part affected. Thus inflammation can be classified under a series of headings according to duration, nature of the exudate and the site

Duration	**Nature of exudate**	**Anatomical site**
Acute	fibrinous	pericarditis
Acute	haemorrhagic	pancreatitis
Chronic	suppurative	osteomyelitis
Chronic	ulcerative	colitis

THE OUTCOME OF ACUTE INFLAMMATION

The following diagram illustrates some of the possible outcomes of acute inflammation. Perhaps it is worth stating at this stage that acute inflammation is not always advantageous. For the most part it is beneficial in destroying or neutralising noxious agents that may have come into contact with the tissues and clearing up the debris. However there are systemic effects which are unpleasant, such as fever. There are also conditions which are caused by an aberration in the immune mechanism, such as hypersensitivity reactions and certain forms of glomerulonephritis and connective tissue disease (for example, rheumatoid arthritis), in which unwanted and unnecessary inflammatory reactions are the basis of the disease.

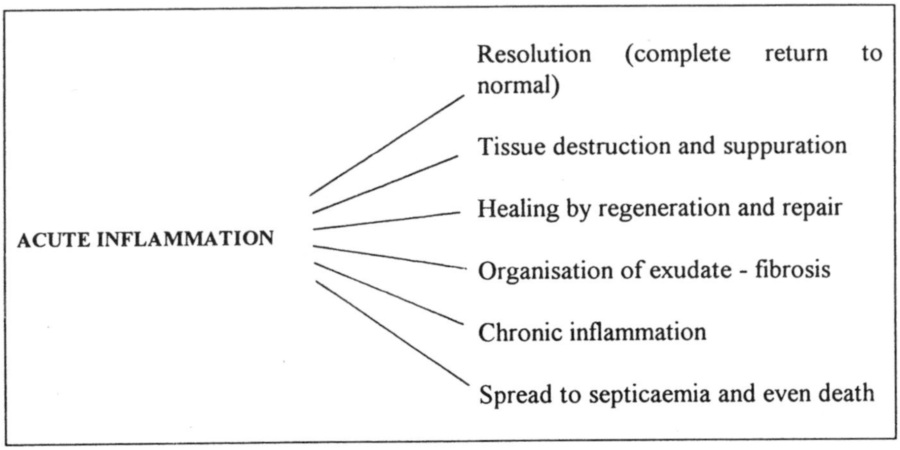

GENERAL PATHOLOGY - CHAPTER EIGHTEEN

CHRONIC INFLAMMATION

As has been noted, chronic inflammation may succeed acute inflammation if the noxious stimulus persists, if foreign material is present, if there is delayed healing for any reason, if hypersensitivity develops, or if there is lowered host resistance. It may also follow on repeated bouts of acute inflammation. In some instances inflammation may have the features of chronic inflammation, with attempts at healing and repair and with a mononuclear cell infiltrate, apparently from a very early stage. Tuberculosis was cited as an example of this.

To go into the microscopical changes in somewhat more detail:

HISTOLOGICAL FEATURES OF CHRONIC INFLAMMATION

These fall into three major groups:

- **Exudation:** This is the persistence of acute inflammatory changes with polymorphs, fibrin, fluid exudate and pus formation, due to the persistence of the injurious agent. There is likely to be continued tissue destruction.
- **Macrophage accumulation:** As part of the demolition phase macrophages may be present in very large numbers. They can continue to reach the area of inflammation from the blood. In addition, in some cases, there may be local proliferation of macrophages at the site of inflammation, although this is not common. As macrophages are long lived cells, they may persist at the site of inflammation in considerable numbers for an appreciable period of time. In some circumstances, for example, foreign bodies and tuberculous infection, macrophages may fuse to form multinucleate giant cells.

  The presence of macrophages may contribute to the continuation of the inflammatory process as these cells produce many of the mediators of inflammation.
- **Repair**. Reparative granulation tissue can be formed. This consists of :

  - Newly formed capillaries and budding endothelial cells
  - Proliferating fibroblasts and collagen
  - Lymphocytes, plasma cells and other inflammatory cells

  Granulation tissue is so called because, when seen at the bottom of a healing wound, it has a red granular appearance.

  Granulation tissue may mature to form proper fibrous tissue which will form a scar. At this stage neighbouring vessels may show fibrous thickening of the wall, with narrowing of the lumen that is known as endarteritis obliterans.

Problems may arise due to contraction of the fibrous scar tissue, which can lead to serious complications such as stenosis or narrowing of scarred heart valves, or obstruction of the pylorus of the stomach following on healing and scarring of a chronic gastric ulcer.

The lymphocyte and plasma cell accumulation is due to an immune reaction of either antibody-mediated or cell-mediated type. Lymphocytes can be stimulated by contact with antigen and can then produce lymphokines which will attract and activate macrophages, which in turn produce substances that modify the behaviour of both T and B cells.

The exudative, proliferative and reparative phases of chronic inflammation may all be present at once; usually the exudative area is central at the site of continued tissue destruction. However this is not always the case; sometimes there is central necrosis, as in the central caseous necrosis seen in tuberculous infection. Generally fibrosis will be most marked at the periphery of the lesion.

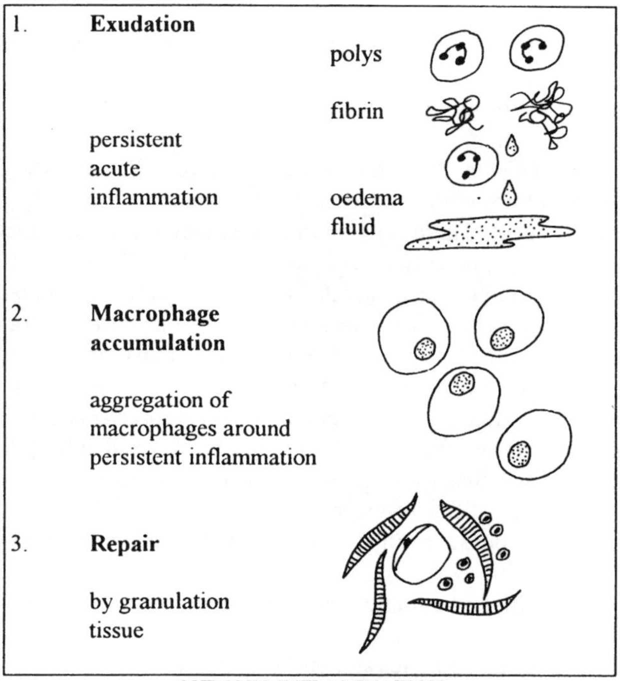

CHRONIC INFLAMMATION

CLASSIFICATION OF CHRONIC INFLAMMATION

Chronic inflammation may be classified as either 'non-specific' or 'granulomatous'.

NON-SPECIFIC CHRONIC INFLAMMATION

This may occur in a variety of circumstances due to organisms of various sorts or in response to foreign material. Many conditions due to this sort of inflammation will be described in systematic pathology.
Examples include chronic osteomyelitis and chronic bronchitis:

Chronic osteomyelitis: This usually occurs in the long bones of children or young adults. Chronic osteomyelitis follows acute inflammation in the bone. The increased tissue tension due to the exudate is exerted against unyielding bone, and there is compression of blood vessels leading to ischaemia and necrosis of the actual bone. Pus is formed, which tracks under the periosteum, further interfering with blood supply. This pus may eventually be discharged from the surface penetrating muscles and skin with the formation of **sinuses**.

The dead bone acts as a foreign body in which organisms can continue to grow. At the same time it prevents adequate drainage of pus and also delays healing. In chronic osteomyelitis the dead bone is known as the **sequestrum**.

Pieces of sequestrum are broken down by osteoclasts and may be discharged through the sinuses. Usually, however, the body cannot manage to remove all the dead bone in this way, a process which would be necessary for complete healing.

So attempts at healing continue and there is new regenerated bone formed by the periosteum which encases the sequestrum of dead bone. This new bone is known as the **involucrum** and this presents a further obstacle to proper drainage.

The end result is chronic supppurative inflammation of bone or chronic osteomyelitis. This can cause death due to persistence of infection. Amyloidosis may also occur.

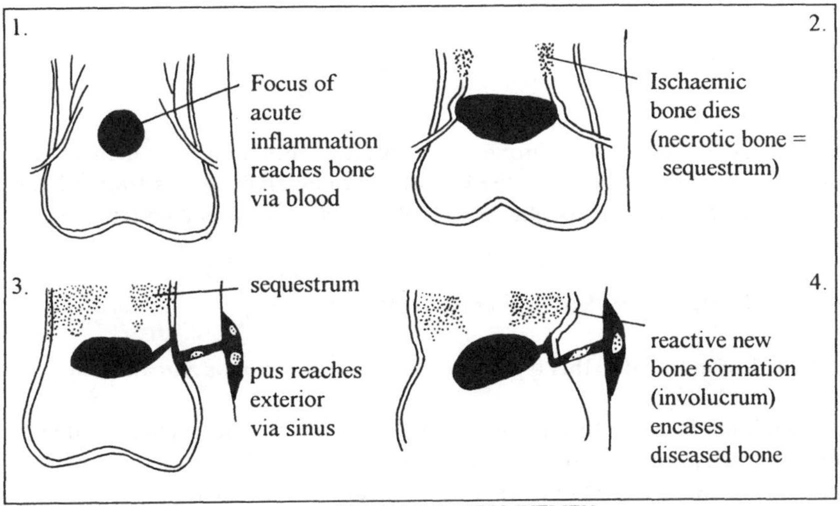

CHRONIC OSTEOMYELITIS

Fortunately with antibiotics in the early stages and with surgical treatment to bring about drainage, this disease is no longer the danger that it used to be. None the less it does still occur and late cases may only be cured by amputation of the affected limb.

Chronic bronchitis may be defined clinically as a disease in which there is a cough, productive of sputum, every day. These symptoms must be present for at least three months in at least two consecutive years.

The major cause is cigarette smoking which irritates the air passages and predisposes to infection.

The major histological feature is an increase in the number and the size of mucus secreting glands in the bronchial wall. These glands may come to occupy more than half the distance between the epithelium and the cartilage in the bronchial wall (whereas they normally occupy less than one third of the thickness of the wall). There is also an increase in the number of goblet cells in the epithelium. Hyperaemia and oedema of the submucosa with lymphocytic infiltration may be present. Later there may be scarring of the bronchial wall.

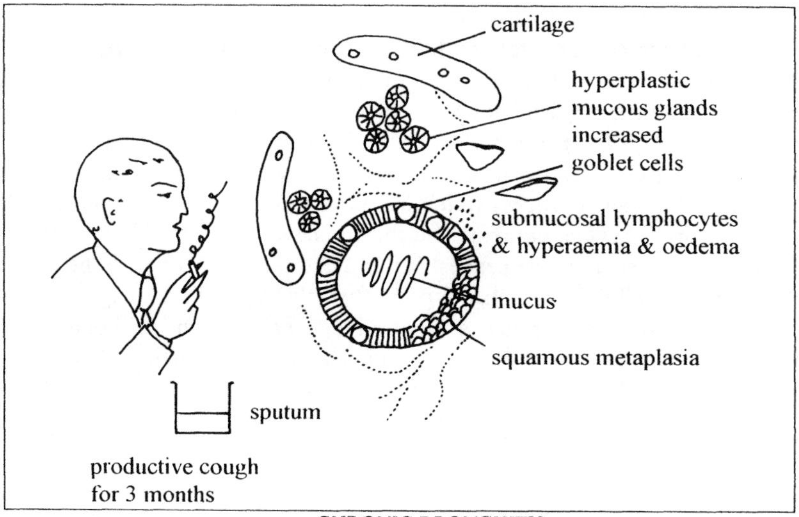

CHRONIC BRONCHITIS

Other types of common chronic non-specific inflammation include chronic pyelonephritis (inflammation of the kidney), chronic cholecystitis (inflammation of the gall bladder, chronic peptic ulcers of stomach or duodenum and ulcerative colitis.

GRANULOMATOUS CHRONIC INFLAMMATION.

The basic lesion found in this type of chronic inflammation is a **granuloma.**

A granuloma is a localised collection of modified macrophages, often called epithelioid cells, which usually contains multinucleate giant cells. [*Note: Do not confuse granuloma formation or granulomatous inflammation with granulation tissue, which is healing connective tissue.]

Many small separate or larger fused granulomas may be found at the site of granulomatous inflammation.

Granulomas may be surrounded by lymphocytes. Giant cells are formed from the fusion of macrophages or epithelioid cells in the presence of poorly soluble, indigestible material.

The modified macrophages have a large amount of pale pink cytoplasm and resemble epithelial cells, hence the name epithelioid (which means epithelial-like) cells. These cells contain, for example, many organelles, endoplasmic reticulum and Golgi apparatus, but contain few phagosomes and do not appear to be well equipped for phagocytosis. It is thought that they develop:

- In response to material which is difficult to break down and digest.
- In association with a T cell-mediated delayed hypersensitivity type of immune response. It would appear that lymphokines such as γ-interferon may stimulate the transformation of monocytes or macrophages to epithelioid cells and giant cells.

A description of a tubercle, which is a granuloma, will give an idea of the appearances found in granulomatous inflammation.

In the tubercle there is often a central area of caseous necrosis - and this is *almost* diagnostic of a tuberculous granuloma. Around this are masses of epithelioid cells, among which may be found numbers of multinucleate giant cells. In tuberculosis giant cells characteristically have the nuclei arranged around the periphery of the cell in a horse shoe shape - the so-called Langhans type of giant cell. There may be an outer rim of lymphocytes, and external to these there may be some fibrous tissue proliferation.

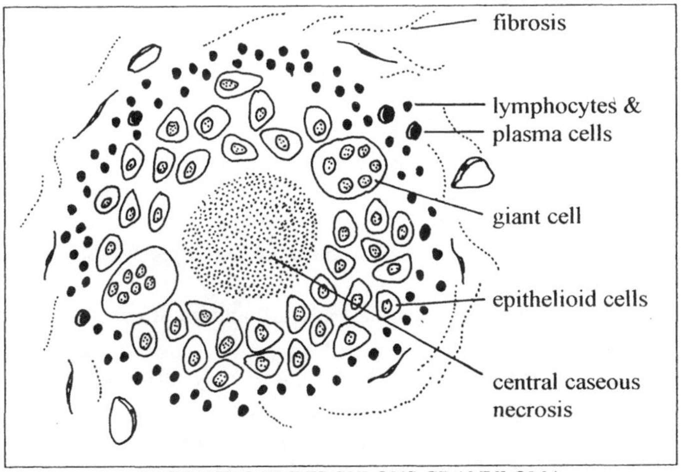

CASEATING TUBERCULOUS GRANULOMA

The caseating granuloma of tuberculosis may be known as a tubercle. Initially small distinct granulomas may fuse together, together with fusion of the central caseous necrosis. Granulomas without caseation may occur in many other conditions, but the specific appearance of the inflammatory infiltrate may aid in diagnosis.

Some causes of granulomatous inflammation:

- **Infection by micro organisms:**

  - Bacteria:

    Tuberculosis - caseating granulomas
    Tuberculoid leprosy - no central necrosis but involvement of nerves
    Yersinia enterocolitica - mesenteric lymphadenitis
    Cat scratch disease
    Lymphogranuloma venereum

  (Note: in these last three conditions there may be a central polymorph-containing abscess surrounded by epithelioid cells).

  - Spirochaetes. In tertiary syphilis there is granulomatous inflammation around gummatous necrosis. Plasma cells are prominent in the infiltrate.
  - Fungi, for example histoplasmosis and cryptococcosis
  - Parasites, particularly around schistosome ova.

- **Exogenous foreign material** in the tissues may be inorganic or organic:

INORGANIC	ORGANIC
Silica and talc Beryllium and zirconium	Starch Catgut sutures

- **Endogenous insoluble material** in the tissues includes:

  - Keratin from ruptured epidermoid cysts
  - Elastic lamina of arteries in giant cell arteritis
  - Faecal material in fistulas
  - Sodium biurate in gout
  - Rarely in association with tumours for example Hodgkin's disease and seminoma
  - Fat in fat necrosis - foamy macrophages and giant cells

- **Cell-mediated hypersensitivity** reactions may be related to granuloma formation by the release of lymphokines (for example, γ-interferon). This is probably of importance in tuberculosis and leprosy, but may also play a part in the development of granulomas in the lung due to inhaled antigens, for example, farmer's lung and bird fancier's lung. The rheumatoid nodule is a palisaded granuloma occurring in rheumatoid arthritis which is an inflammatory disease of immune aetiology.
- **Unknown causes,** for example, sarcoidosis and Crohn's disease (regional ileitis).

Recognition of inflammation as granulomatous limits the possibilities and may help in coming to a more exact diagnosis.

TYPES OF GIANT CELLS

Fused macrophages may form giant cells in a number of different ways as indicated in the diagram:

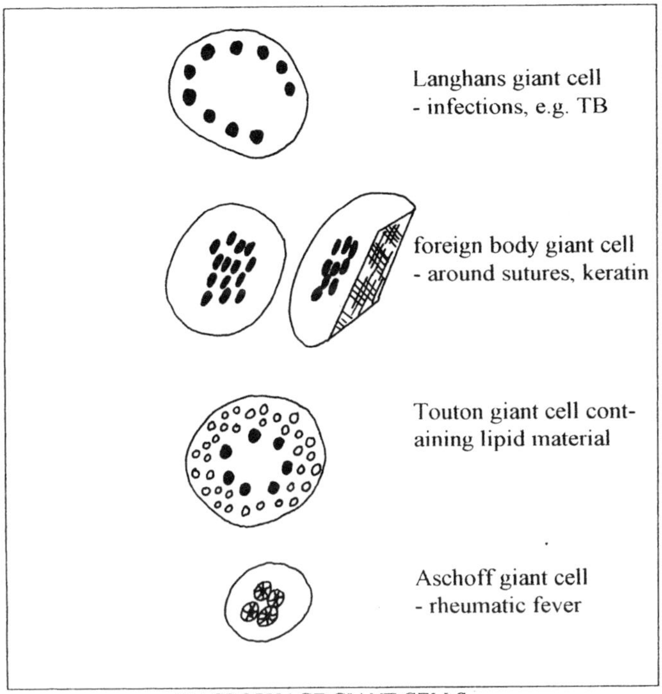

MACROPHAGE GIANT CELLS

THE EFFECTS OF CHRONIC INFLAMMATION

There may be systemic effects from chronic inflammation similar to those in acute inflammation, and due to the same causes. The fever tends to manifest itself by night sweats rather than by shivering attacks and chills. Leucocytosis is more likely to involve mononuclear cells than polymorphs. There will be increased protein synthesis by the liver, and increased C reactive protein and fibrinogen with raised ESR. Raised γ-globulin levels may be present also as part of an immune response with increased antibody production.

Loss of appetite and weight loss are common features of chronic illness; the loss of weight may be due to increased catabolism resulting from the actions of IL-1 and TNFα.

Non specific symptoms such as tiredness, headache and a general feeling of being unwell are difficult to explain. Although they may be attributed to 'toxaemia', no specific toxins have been identified. Anaemia may develop.

Finally amyloidosis is likely to occur in cases of long standing chronic inflammation.

GENERAL PATHOLOGY - CHAPTER NINETEEN

HEALING AND REPAIR

Healing is the process by which the body replaces destroyed or dead tissue by living tissue. This may take place in a number of ways:

- The tissue may be completely restored to normal, both structurally and functionally; this is known as **resolution**.
- There may be proliferation of parenchymal cells with some loss of normal architecture, known as **regeneration**.
- Lost tissue may be replaced with granulation tissue and fibrous repair will take place; this is known as **organisation**.

For resolution to take place the underlying framework of the tissue must have been preserved. Regeneration and organisation may occur together, although if the underlying framework has been destroyed, normal architecture may not be restored.

Different cells of the body have different capacities for regeneration. There are:

- **Labile cells** which are constantly multiplying during life to replace cells that are being destroyed. In this group are all the cells of surface epithelia that cover the body or line hollow organs. The cells of the bone marrow are continually making new blood cells and lymphoid tissue also multiplies throughout life.
- **Stable cells** have the capacity to regenerate if the need should arise. Most glandular epithelial cells and connective tissues fall into this category. When there is cellular destruction it is essential that some normal cells are left for regeneration to take place. As will be seen when healing of the skin is described. If there has been total destruction of hair follicles or sweat glands, then these cannot grow back again. Fibrous tissue has great regenerative capacity and is tough and resistant, being able to grow in relatively poor conditions where there may be ischaemia and anoxia and for this reason fibrous repair may predominate in some circumstances.
- **Permanent cells** have little or no powers of regeneration. These include nerve cells and skeletal and cardiac muscle. Although nerve fibres (axons) can regenerate if severed, if cell bodies are destroyed those cells are lost permanently and there will be replacement by fibrosis (or, in the central nervous system, gliosis).

  Muscle cells have extremely limited powers of regeneration, although this may be of little consequence if there is only minor damage as surviving muscle cells may undergo hypertrophy to maintain function; muscle mass will also be preserved. However, once there has been loss of cardiac muscle (as in myocardial infarct), it will be replaced by a fibrous scar.

Some examples of healing in terms of resolution, regeneration and organisation into a fibrous scar can be considered:

The lung can be infected with pneumococci giving rise to lobar pneumonia. In this condition there is consolidation or solidification of the whole lobe of a lung, with outpouring of fluid and cellular exudate into the alveolar spaces. However the alveolar walls, the framework of the lung, are usually not destroyed. Digestion of the exudate by lysosomal enzymes may be followed by resorption or coughing up of the debris, with the restoration of the lung to normal. This is a case of healing by resolution.

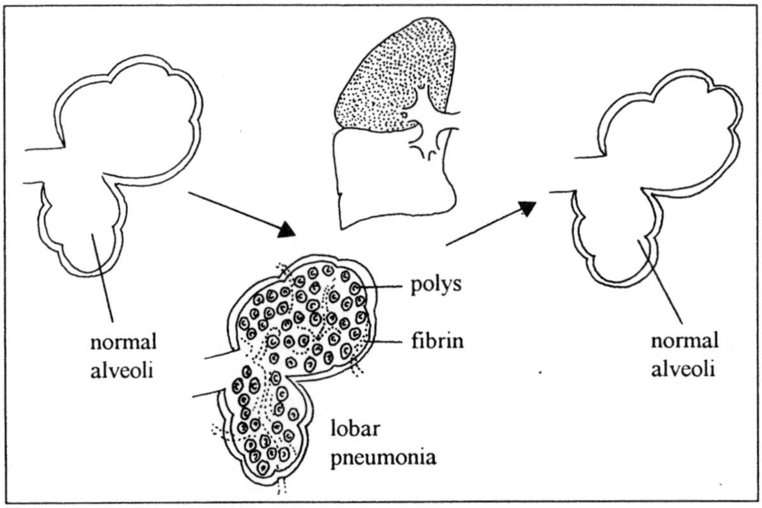

HEALING AND REPAIR - RESOLUTION

A more virulent organism or a different type of pneumonia such as bronchopneumonia will damage and cause necrosis of alveolar walls resulting in organisation with fibrous scarring.

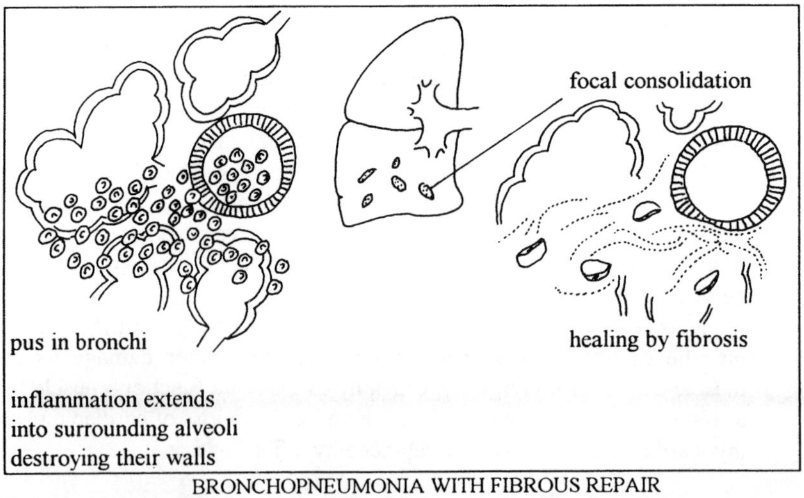

BRONCHOPNEUMONIA WITH FIBROUS REPAIR

The liver is an organ in which the parenchymal cells have great powers of regeneration. Generally speaking, if the underlying framework survives, healing will be by resolution, with complete restoration to normal both structurally and functionally. This can occur when necrosis is focal, (for example, in infective hepatitis) or when it is zonal, that is, restricted to one part of the liver lobule, as in certain types of poisoning or in severe centrilobular congestion.

Massive necrosis of the liver in very severe infective hepatitis or in poisoning (if not immediately fatal) can result in destruction of large areas of liver, including the framework. Regeneration occurs in an irregular nodular fashion, but there is also much fibrous scarring. If these changes are diffuse the end result may be a coarsely scarred nodular liver showing post necrotic macronodular cirrhosis. Liver function will be impaired thereafter.

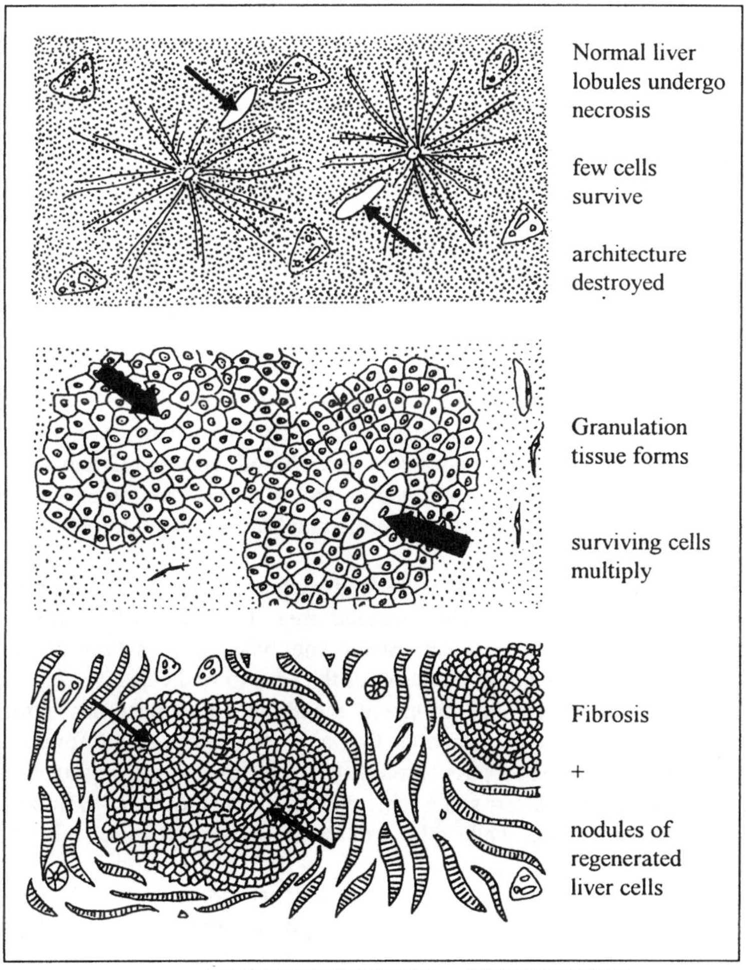

HEALING AND REPAIR - REGENERATION

Note that micronodular portal cirrhosis can be found in the liver in alcoholics, where smaller areas of necrosis diffused widely throughout the liver result in the formation of multiple small regenerative nodules.

Heart muscle cannot regenerate. After necrosis, such as may occur in myocardial infarction, the dead muscle tissue is replaced by granulation tissue, collagen is laid down and a fibrous scar results. Thus there is healing by organisation or fibrous repair.

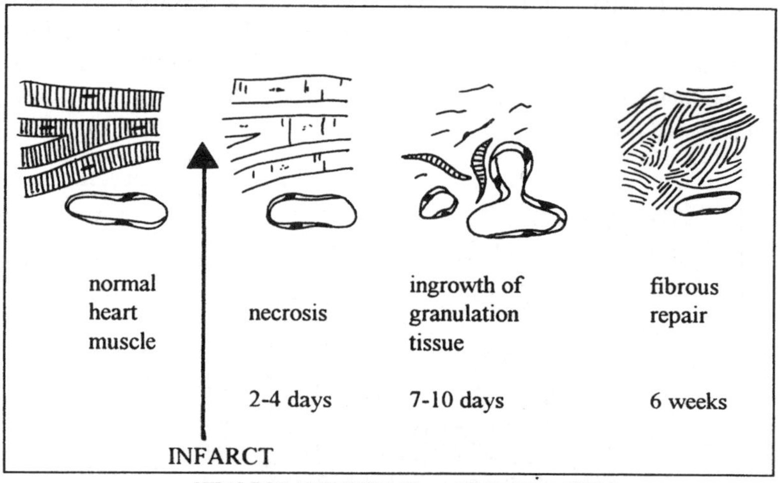

HEALING AND REPAIR - ORGANISATION

THE MECHANISM OF HEALING

Granulation tissue is the red granular tissue that is formed when a wound heals. It is composed of proliferating fibroblasts and new blood vessels. Granulation tissue formation commences quite early on in the inflammatory process, where its presence indicates the commencement of chronic inflammation, at the time of macrophage accumulation.

An important part of the process is the ingrowth of new small blood vessels from the normal tissue at the edges of the damaged area. These bud off existing vessels, which develop a defect in the basement membrane, which allows endothelial cells to bud off, multiply and move in the direction of the area to be healed. A solid capillary bud soon develops a lumen and basement membrane, and becomes a functional capillary. Function may not be perfect at first and often these newly formed vessels are leaky, allowing proteins to pass through the wall; as a result granulation tissue may be oedematous. Thereafter, some remain as capillaries, others develop muscular coats and become arterioles while others dilate to become venules. Fibroblasts, at first plump and active, grow in from the edges too. These produce proteoglycan ground substance initially, and then lay down collagen fibres. Some fibroblasts may form contractile proteins within their cytoplasm and are then known as myofibroblasts.

At this stage the granulation tissue is very red due to its vascularity. The rich blood supply also means that it is quite resistant to infection.

Lymphatics may form in the same way as blood vessels, and nerve fibres will also grow into the new tissue, often alongside arterioles. Macrophages persist, continuing to remove debris from the site, and other inflammatory cells may also be seen.

As collagen is laid down and fibrosis proceeds, many of the new vessels will thrombose and degenerate, and the cells will die and be resorbed Others will have joined up with the circulation, anastomosing with each other and with existing vessels.

At this stage there is a pale and relatively avascular scar. The collagen fibres tend to become oriented in a direction to resist stress. Contraction of the scar will follow. This may be an undesirable complication of scarring in some instances. Burn scars in particular seem to undergo marked contraction, with resultant deformity which may be unsightly and restrict movement .

Granulation tissue may be found wherever there is healing, in inflammation, in healing wounds, in thrombi and infarcts and in the stroma of neoplasms.

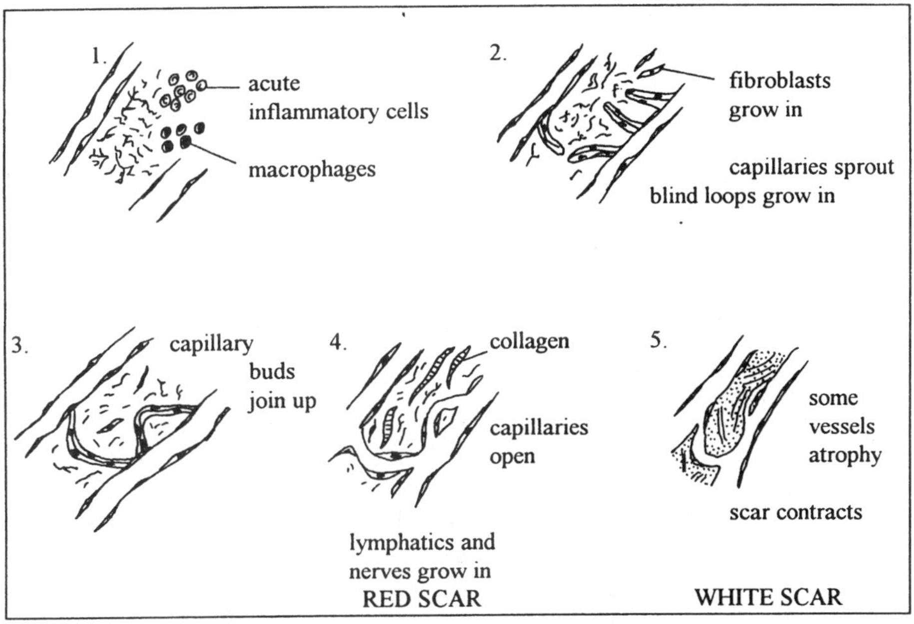

HEALING AND REPAIR - ORGANISATION

EXAMPLES OF WOUND HEALING

The healing of the two types of skin wound illustrate well the processes involved:

- **Healing by first intention.** This is the way in which a clean incised wound will heal if the edges are held together. Surgical wounds heal in this fashion. The sequence of events is:
  - There is bleeding into the tissues (haematoma formation) at the site of incision. The blood clots. Surface clot dehydrates to form a protective scab.
  - An acute inflammatory reaction with polymorphs forms around the clot and more fibrin is released to bind the wound together.
  - Within twenty four hours epithelial cells grow from the basal layer at the edges between viable tissue of the dermis and the clot until they meet up with other epithelial cells which seems to inhibit further growth.
  - Soon the inflammatory exudate begins to be broken up and resorbed by macrophages which have migrated to the site.
  - By about the third day the wound area is filled with granulation tissue which usually replaces the deeper epithelium leaving just the surface epithelium.
  - Collagen fibres are laid down, at first type-III and later type-I, filling the gap.
  - By the end of the first week the inflammatory infiltrate has disappeared and the epidermis has regained its normal thickness. Collagen fibres become organised parallel to the surface in the lines of stress. Blood vessels diminish in number and the scar becomes paler.
  - Contraction of the scar with increasing tensile strength takes place over the next four or five weeks

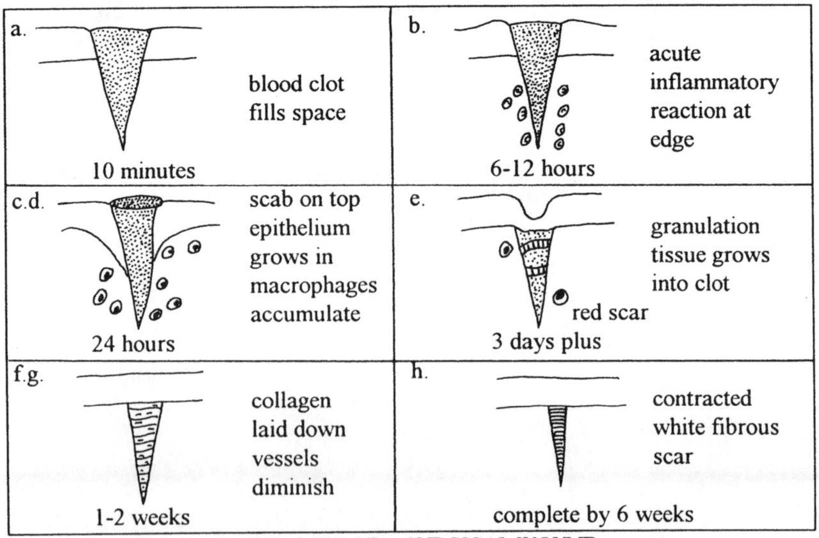

HEALING OF A SURGICAL WOUND

- **Healing by secondary intention** is the name given to the healing of a wound with separated edges. This is also known as healing by granulation.

  The pathological changes are similar but more granulation tissue has to be formed and the process is slower and leaves a larger scar.

  - The wound is filled with a clot, which dries on the surface to form a scab.
  - Contraction of the skin around the edges takes place, reducing the size of the defect. Myofibroblasts may play a part in this process.
  - Epidermis grows in from the sides between the clot and living cells.
  - The exudate is removed by macrophages. Granulation tissue grows into the clot replacing it and
  - gradually becomes covered by epithelium, leaving the scab on top which is completely undergrown and falls off. Collagen fibres are laid down. Hair follicles and sweat glands which have been totally destroyed are not replaced. A scar which undergoes contraction (cicatrisation) and becomes paler is formed

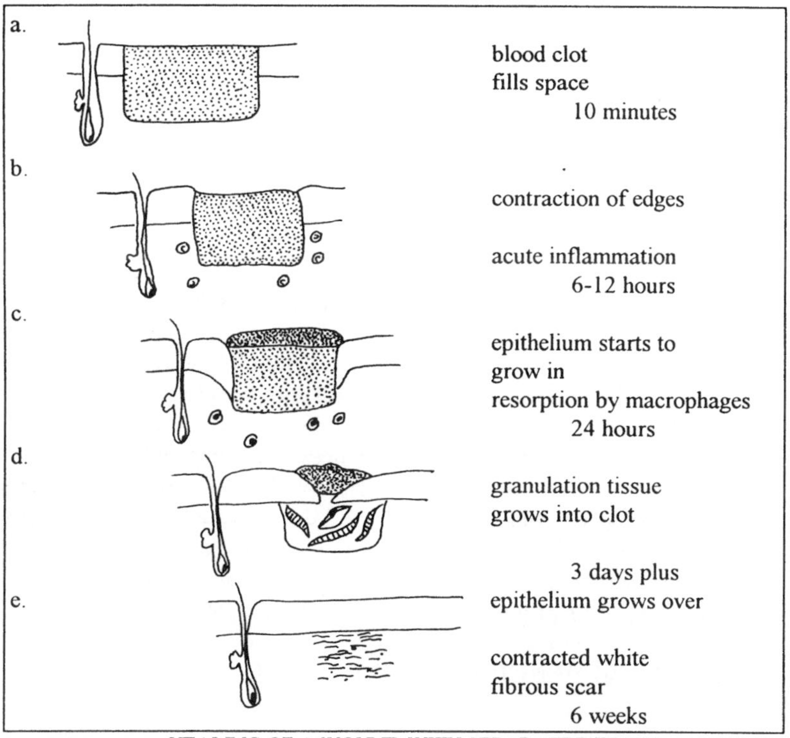

HEALING OF A WOUND WITH SEPARATED EDGES

Similar processes go on in tissues elsewhere without the complication of epithelial overgrowth. Healing in an organ may be associated with parenchymal regeneration.

MECHANISMS OF REPAIR

A number of growth factors have been described which will act on stable cells to bring about proliferation. These include:

- **Epidermal Growth Factor** (EGF)
- **Platelet-derived growth factor** (PDGF)
- **Fibroblast growth factor** (FGF)
- **Transforming growth factors** (TGFα and TGFβ)
- **IL-1 and TNF**

These factors, which seem to be produced by most of the participants in the inflammatory reaction, may stimulate cell division, of both fibroblasts and endothelial cells, but may also have a chemotactic attraction for these cells. Some also stimulate new vessel formation and others promote collagen synthesis.

In the healing process regulatory mechanisms control the course of events. For example, epithelial cells stop growing when the breach in the epithelium has been filled. There seem to be some factors within the cells or on the cell surface which are responsible for this cessation of growth, which has, in the case of epithelial cells, been termed contact inhibition. TGFβ may act as a growth inhibitor.

The extracellular matrix may also influence the proliferation and movement of cells in the tissues by acting on cell surface receptors. Laminin and fibronectin are able to attach to cells, and extracellular substances and are thought to play a role in healing.

WOUND STRENGTH

This depends on a number of factors, but is basically dependent on the formation of collagen and the right type of collagen. Collagen molecules, of which there are many different types, are synthesised in the rough endoplasmic reticulum of fibroblasts and leave the cell via the Golgi apparatus as soluble procollagen. True fibrils form in the extracellular space and these are then joined together by cross-linkages to give additional strength. In certain circumstances, including inflammation, collagen may be split by collagenases and digested, but for the most part it is extremely resistant and strong. Type-III collagen which is laid down early in the healing process, is less strong than type-I collagen, with which it is later replaced. Even so, a healed wound will never fully regain the tensile strength of intact skin, but will have regained some 70-80 per cent of its original strength after six or eight weeks.

FACTORS INFLUENCING WOUND HEALING

Some of the growth factors and controlling mechanisms affecting wound healing have been mentioned very briefly. This area is still the subject of much study and experimental work. Some of the more obvious things which affect healing will now be listed:

General factors influencing healing:

- Age. Healing is said to be faster in the young.
- Protein deficiency
- Vitamin C deficiency (protein and Vitamin C are necessary for adequate collagen strength and synthesis)
- Zinc deficiency will lead to delayed wound healing.
- Glucocorticoid steroid hormones have an anti-inflammatory effect (by inhibiting the arachidonic acid metabolic pathway) and will also delay healing if administered in the early stages.

Local factors influencing healing:

- Blood supply. Well vascularised tissues heal more quickly and efficiently. For example, the face heals more quickly than the leg, and in a leg with a poor circulation ulcers may be almost impossible to heal.
- Infection of a wound, or the presence of foreign material within it, will slow down healing.
- Continuous movement of the injured part will not allow knitting together, as newly formed granulation may be disrupted, there will be additional haematoma formation and the whole process will be slowed down.
- The amount of tissue destroyed, particularly if the supporting framework is destroyed, is important. Big wounds take longer to heal and show more scarring.
- X-ray therapy may impair granulation tissue formation.

Complications of wound healing (see diagram overleaf) include:

- Infection may enter a wound causing additional inflammation with tissue destruction and pus formation.
- Overgrowth of granulation tissue ('proud flesh') may grow through a skin wound, preventing epithelialisation.
- Cicatrisation or excessive contraction of a scar can cause:
    - Skin deformity
    - Obstruction of a hollow viscus, for example, oesophagus, pylorus
    - Contraction of scarred heart valves leading to stenosis.
- Keloid formation can be troublesome, particularly in black Africans. There is excess production of mature collagen which forms a lump in the scar, stretching the overlying skin. These tend to recur after excision and may be difficult to treat.
- Pain can result from a neuroma, formed from proliferating nerve fibres caught up in the wound.
- Weakness. Fibrous scar tissue has reduced tensile strength and can stretch, for example, in a cardiac aneurysm after infarction, or in an incisional hernia.
- Cancer may sometimes develop in a chronic scarred ulcer, for example, chronic leg ulcer.

INFECTION with continued tissue destruction and pus formation

OVERGROWTH of granulation tissue 'proud flesh'

CICATRISATION - i.e. contraction of a scar, for example claw hand, heart valve stenosis

KELOID - raised scar due to excess collagen formation

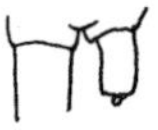

NEUROMA - painful overgrowth of nerve endings

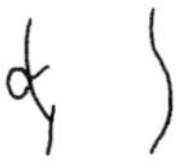

HERNIATION through a weak scar

CANCER may develop in a chronic scarred ulcer

COMPLICATIONS OF WOUND HEALING

GENERAL PATHOLOGY - CHAPTER TWENTY

INFECTION AND SOME INFECTIOUS DISEASES

Some factors determining the body's response to infection, such as those shown in the following table, have already been considered:

ORGANISM	HOST
Virulence (pathogenicity)	**Resistance**
Dose Portal of entry Toxin production and invasiveness	General state of health White blood count Immunity

Of course not every contact with micro organisms results in disease. In fact humans live in constant contact with microbes which inhabit the skin and the gastrointestinal tract. The relationship may be:

- **Commensal,** which literally means sharing a table (*mensa*: is the Latin for table). These are organisms that live together, sharing a food supply. Usually the smaller organisms (for example, bacterial flora of the mouth), are dependent on the larger ones, but cause no harm.
- **Symbiotic,** in which organisms live together to their mutual benefit, such as intestinal bacteria which share a food supply but which manufacture Vitamin K, which is essential for the human host.
- **Parasitic,** where an organism lives parasitically on a host for its own benefit and to the detriment of the host, which suffers from disease. The microorganisms which cause infectious disease are thus true parasites.
- **Saprophytic** organisms live on dead tissues (for example, fungi which live on the keratinous material on the surface of the skin).

Parasitic organisms which cause disease are known as pathogenic. Pathogens may gain access to the body by:

- **Congenital infection**
- **Inhalation**
- **Ingestion**
- **Implantation**
- **Inoculation.**

Spread of infection may take place by:

- **Direct spread** from another infected person.
- **Spread from the environment** from food, water and soil which may have been contaminated by an infected person, or which may contain pathogenic free-living organisms.
- **Foetal infection** from mother either by transplacental transmission or with organisms acquired during the birth process (vertical transmission of infection).
- **Zoonotic infections** which are acquired from animals.

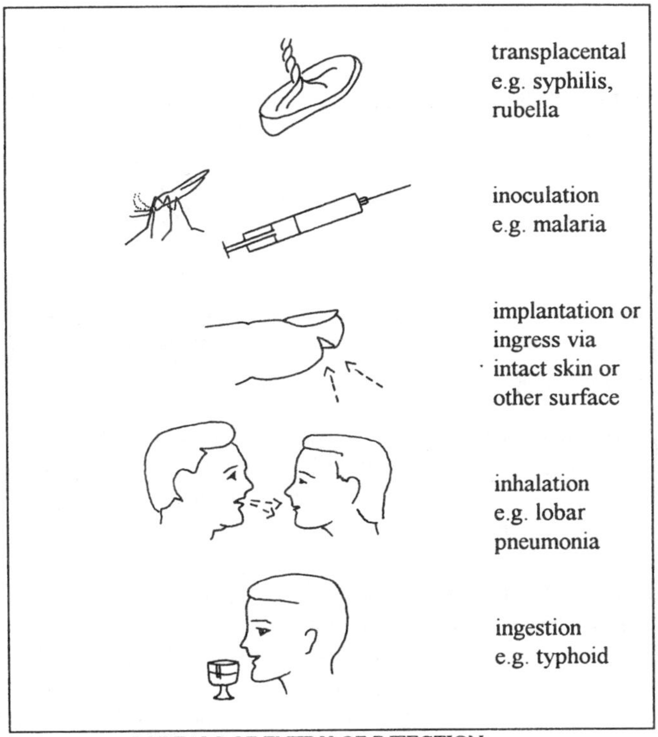

PORTALS OF ENTRY OF INFECTION

Infection with an organism does not always result in infectious disease. For the most part, the tissues of the body are sterile and organism free, but, for example, organisms may gain access to the blood stream and cause a transient bacteraemia without causing disease. To cause disease organisms have to overcome the body's defence mechanisms which include:

- The **intact skin** and **mucous membranes** and **secretions** on body surfaces which may afford both mechanical and bactericidal protection. Normal commensal or symbiotic bacterial populations in the gut can discourage or prevent the establishment of potentially harmful organisms.

- **Phagocytic white cells** which can engulf invading organisms as a part of the inflammatory reaction.
- **The mononuclear phagocytic system** can mop up organisms which have penetrated further than the local site of entry and reached for example, lymph nodes, liver, spleen.
- **The immune system** can make antibodies to invasive organisms and can rapidly dispose of infectious agents to which preformed antibodies exist.

Obviously defects in any of these mechanisms can predispose to infection and make an individual more susceptible to the development of infectious disease.

Thus for infection to take place there has to be a **portal of entry** and there has to be a breach of the **body's defence mechanisms**. Once this has occurred the organism must find the appropriate environment either inside a cell (as in the case of a virus) or in the extracellular spaces as in the case of many bacteria, or on a mucosal surface (for example, cholera) to **multiply** or **replicate**. Then there can be **damage to the host tissues** either as a result of the ensuing inflammatory reaction or as a result of the production of toxins by the organisms. The development of **hypersensitivity** along with **immunity** may also alter the picture. A virulent organism which can resist or overcome the body's internal defence mechanisms may then **spread locally** by virtue of spreading factors such as hylauronidases and collagenases or to a distance via **lymphatics, blood vessels, natural spaces** (for example, serous cavities) or **natural passages** such as the bronchi.

As has been noted in the chapter on inflammation the local effects may be accompanied by systemic effects such as fever and malaise, decreased appetite, loss of weight etc.

The ultimate result is that:

- Either the patient gets better, that is, **recovery**
- Or gets worse with an ultimately fatal result, that is, **death**
- Or enters a more or less steady state, that is, **persistent chronic infection.**

It is not always easy to prove that a disease is due to an infective agent. Robert Koch, a German pathologist who discovered *Mycobacterium tuberculosis*, laid down some rules at the end of the nineteenth century which are known as **Koch's postulates.** These state that for a disease to be caused by a specific organism:

- The organism must be found in all patients with the disease
- The organism must be recovered from the lesions of the disease in pure culture
- The organism must produce similar lesions when inoculated into susceptible animals and
- Be recovered in pure culture from the animal lesions.

Histopathologists can demonstrate organisms in tissues using special stains, and if organisms are consistently seen in the lesions of a particular disease, even if those organisms cannot be cultured (such as *Mycobacterium leprae),* it can be inferred that this is an infectious disease due to a specific organism.

Some examples of infectious diseases due to a variety of organisms follow. The diseases have been chosen to represent the various types of infective agent, such as virus, bacterium or fungus, and the various types of pathological tissue response. Where possible, diseases which affect more than one system (and hence are unlikely to be covered in full detail in systematic organ pathology) are used as examples.

VIRAL DISEASE

Viruses are intracellular parasites which may be of either DNA or RNA. They replicate within host cells and spread. Viral DNA may be integrated into the genome of the host cell. Viruses may either cause cell death or may cause less damage as they multiply. In some instances they induce cell transformation. Some viruses (such as measles and herpes viruses) produce aggregates of virions which may be seen within the nucleus or the cytoplasm of the cell as **inclusion bodies**. Cell fusion may occur as a result of viral proteins inserted into the cell membrane, resulting in the formation of multinucleate giant cells (measles and herpes). The inflammatory infiltrate caused by viral infection is usually composed of lymphocytes and macrophages, polymorphs are rarely seen. Specific diagnosis of virus diseases is by tissue culture, or by immunofluorescent, serological or molecular biological means.

MEASLES

The measles virus is an RNA paramyxovirus which is transmitted by inhalation (or, rarely, passed through the placenta. Unlike German measles this virus does not cause foetal congenital abnormalities).

The virus enters lymphoid tissue in, for example, the tonsils and adenoids, and then enters the blood. Clinical signs appear after an incubation period of about fourteen days during which viral multiplication and spread takes place. The there is fever, conjunctivitis, cough and a cold. Early on in the prodromal phase Koplik's spots appear in the mouth. These are small red areas about three millimetres in diameter with a central white ulcer. After three or four days the characteristic red blotchy skin rash appears, starting behind the ears and spreading over the neck and trunk to the limbs. This lasts a few days and fades as the other symptoms clear. Although generally a fairly benign disease in young children it can be severe in the very young or the very old, the immunosuppressed and the malnourished, and can be a fatal disease in developing countries.

A characteristic feature histologically is follicular hyperplasia of lymphoid tissue together with giant cells (Warthin-Finkeldey giant cells). These giant cells may contain scores of nuclei together with eosinophilic intranuclear and intracytoplasmic inclusions. They are found in lymph nodes and in the lymphoid tissue of the pharynx, bronchi and gastrointestinal tract, including the appendix. In addition, giant cells may be formed from the alveolar lining cells of the lungs. Bronchial epithelium and alveolar lining cells may also contain inclusion bodies and there may be a viral interstitial pneumonia.

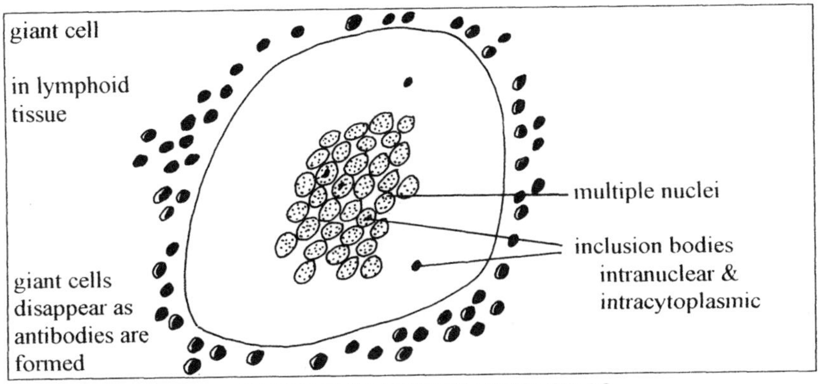

WARTHIN-FINKELDEY GIANT CELLS

Complications include middle ear disease and bronchopneumonia, which may be due to secondary bacterial infection. Viral encephalitis may occur, and at a later stage there may be a neurological disease called subacute sclerosing panencephalitis, with inclusion bodies and destruction of neurones resulting in progressive mental and motor deterioration which is fatal after a number of years. Measles has also been implicated in the aetiology of insulin-dependent diabetes mellitus and thrombocytopenic purpura.

An effective live attenuated vaccine is available to protect against the disease. Gamma globulin from a person who has had the disease, will prevent the disease if given to a susceptible individual within five days of exposure .

CHLAMYDIAL DISEASE

Chlamydiae are intracellular organisms somewhat larger than viruses; they contain both DNA and RNA and have cell walls. They multiply within vacuoles in cells forming elementary bodies which aggregate as cytoplasmic inclusions. These can be demonstrated with immunofluorescent stains or by Giemsa staining. Chlamydiae are sensitive to wide spectrum antibiotics.

LYMPHOGRANULOMA VENEREUM

This is a type of sexually transmitted (venereal) disease caused by certain serotypes (L_1, L_2 and L_3) of *Chlamydia trachomatis* (other types of which cause trachoma of the eye and non-specific genital infections with cervicitis and urethritis).

The primary lesion is usually of the penis, vagina or cervix, and is a blister which soon bursts to become an ulcer. At this stage inclusion bodies are present in the cells of the lesion.

The regional lymph nodes swell and become soft as an abscess (sometimes known as a bubo) forms. This has a characteristic histological appearance, being composed of a central irregular or star-shaped abscess containing pus, which is surrounded by epithelioid macrophages and a few multinucleate giant cells. Peripherally there are lymphocytes and plasma cells and fibrous tissue. Healing is by fibrosis which in untreated cases can be severe and deforming and, for instance, cause rectal stricture in females. Lymphatic obstruction may cause oedema.

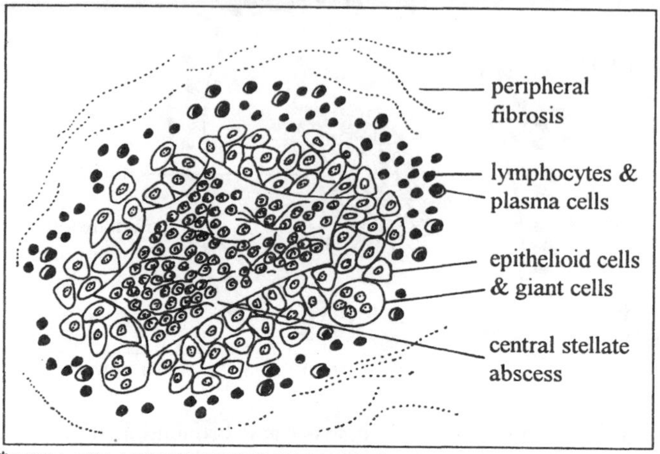

*STELLATE ABSCESS WITH SURROUNDING GRANULOMATOUS INFLAMMATION IN LYMPHOGRANULOMA VENEREUM

*Note that this histological picture is characteristic but not absolutely diagnostic of lymphogranuloma venereum as a similar appearance may be seen in cat-scratch disease and also in some fungal diseases.

RICKETTSIAL DISEASE

These are small gram negative bacteria which have to live within cells and are transmitted to humans from fleas, lice, ticks and mites. These organisms multiply within the endothelial cells of small blood vessels resulting in focal vascular inflammation, small haemorrhages and a rash or with microthrombi causing focal ischaemia. Typhus can be a calamitous epidemic disease in times of upheaval, war or famine. The spotted fevers caused by Rickettsiae are transmitted to man by a chance encounter, usually in the bush, with an infected arthropod. Rickettsiae are sensitive to wide spectrum antibiotics.

TICK-BITE FEVER

In South Africa this disease is due to infection with *Rickettsia conorii.* Following on a tick bite there is a blackish swollen, crusted local lesion called an eschar. Five to seven days later there is severe headache with fever, and some muscle pain. An inconspicuous maculopapular rash may develop a few days later and this may become petechial. Early treatment will prevent the rash appearing.

BACTERIAL DISEASES

These are a major cause of infective disease in humans. A multiplicity of organisms including cocci and bacilli, mycobacteria and spirochaetes can cause a number of pathological responses such as acute inflammation, chronic non-specific inflammation, necrotising inflammation and granulomatous inflammation.

In fact all the patterns that have been described under the inflammatory response may be represented.

A BOIL

A boil or furuncle is a simple form of suppurative inflammation which is due to infection with *Staphylococcus aureus,* which enters a hair follicle and causes acute inflammation with pus production. An abscess which shows the classical cardinal signs of inflammation forms in the dermis and subcutaneous tissue; this is very painful until the boil is lanced or breaks through the overlying skin, releasing the tension and allowing healing to take place.

Multiple subcutaneous abscesses which may drain to the skin through a number of sinuses form in similar fashion and are called carbuncles.

LOBAR PNEUMONIA

This has already been mentioned as an example of acute inflammation due to bacterial infection which can heal by resolution. The causative organism is usually *Streptococcus pneumoniae,* although other organisms can cause this disease. A large dose of a virulent organism when inhaled leads to widespread exudation into the alveolar spaces of one whole lobe of a lung.

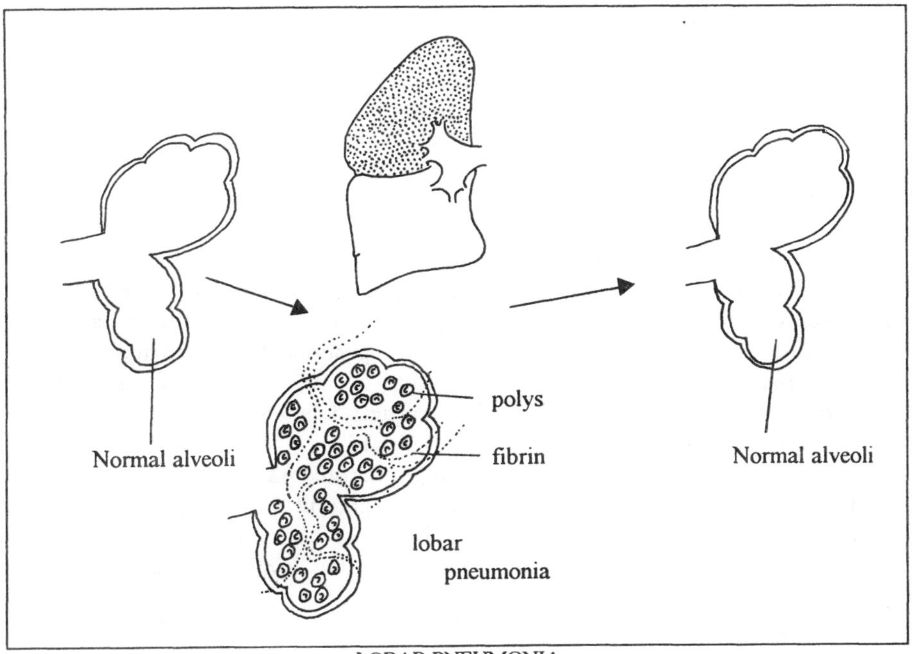

LOBAR PNEUMONIA

Four stages of the disease are recognised pathologically:

- **Congestion** in which there is marked congestion of the alveolar vessels and the fluid exudate moves into the alveolar spaces which may contain many organisms.

- **Red hepatisation** in which there is emigration of polymorphs together with loss of red cells from the vessels into the alveolar space. Fibrin precipitates as strands.

  Macroscopically the lung is red and solid and airless (consistency like liver, hence the term hepatisation). On the overlying pleural surface there is a fibrinopurulent pleurisy.
- **Grey hepatisation** follows as the red cells and polymorphs break down. If there is pus in the pleural space a condition of empyema has developed.
- **Resolution** or healing occurs when the exudate is digested by enzymes in the pus and is then coughed up, cleared up by macrophages or resorbed into the circulation leaving the lung in its normal state.

Clinically the patient may have severe fever with rigors or shivering attacks. There is a cough with 'rusty' sputum being produced in the stage of red hepatisation. There may be difficulty in breathing. Involvement of the pleura gives rise to pleuritic pain, a sharp pain on coughing or on deep inspiration and a pleural rub on listening to the lungs with a stethoscope. The solid lobe or lobes are airless and present the appropriate clinical and radiological signs. Prior to antibiotics the disease used to last for some seven, nine or eleven days, when resolution set in suddenly and the temperature dropped to normal by crisis. In present circumstances antibiotic therapy brings about a rapid cure at any stage.

TYPHOID FEVER

This is due to infection with *Salmonella typhi,* a Gram-negative coliform bacillus. It is the most pathogenic member of a genus which consists of more than 1000 serotypes which may cause, for example, infection of the gastrointestinal tract or septicaemia. Disease is acquired through the ingestion of infected food or water. Typhoid bacilli are contained in the excreta of infected persons, and may be excreted in the bile of otherwise healthy carriers where organisms are contained within the gall bladder.

The ingested organisms penetrate through the mucosa into the lymphoid tissue of the small intestine and thence into the mesenteric nodes and the circulation. Organisms are phagocytosed by macrophages of the mononuclear phagocytic system (MPS) and multiply within them during the incubation period of ten to fourteen days.

Bacteria are then released from the cells and enter the blood causing a bacteraemia. This stage lasts for about one week during which the organisms may be isolated by blood culture. The patients are febrile with a temperature which increases daily, and are seriously ill. Organisms are distributed throughout the body but show a preference for the MPS, the Peyer's patches of the terminal ileum and the gall bladder.

After the bacteraemic phase, organisms multiply again at different sites and there is hepatosplenomegaly. There may be ulceration (longitudinal) of Peyer's patches and perforation of the ileum is a dangerous complication. Bleeding from the ulcers is generally not severe. Clinically there may be diarrhoea and abdominal pain.

The histological picture at this stage is of nodules or collections of macrophages that phagocytose red cells (as well as bacteria). Lymphocytes and plasma cells may be seen, but polymorphs are generally absent from the lesions of typhoid. These lesions may be seen in the intestine, mesenteric lymph nodes, spleen, liver and bone marrow. At this stage organisms can be cultured from the faeces (and also the urine in a number of cases). Antibodies appear about the second or third week and may help to confirm the diagnosis (Widal test). Dying bacilli produce toxins which can damage the heart and kidneys. Pneumonia can occur and there may be small ring haemorrhages in the brain.

Colonisation of the gall bladder may result in the prolonged excretion of bacilli, even after the disappearance of the clinical disease causing a chronic carrier state.

Chloramphenicol and other antibiotics such as ampicillin, amoxicillin and trimethoprim-sulphamethoxazole are effective in the treatment of typhoid.

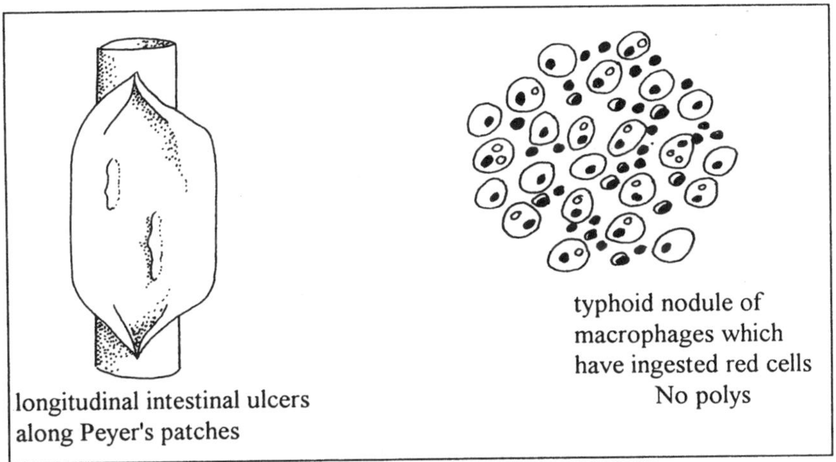

SOME PATHOLOGICAL FEATURES OF TYPHOID

GENERAL PATHOLOGY - CHAPTER TWENTY ONE

TUBERCULOSIS

Tuberculosis is a very important example of a severe bacterial infectious disease characterised by granulomatous inflammation. In Britain and the United States the incidence of active cases of tuberculosis is about 10 per 100 000 population per annum, with a mortality rate of less than 10 per cent. In underdeveloped countries in Africa and Asia the incidence may be many times this. In fact, Swaziland is quoted as having an incidence of 450 per 100 000 population per annum. In Africa and Asia mortality rates may be as high as 50 per cent. Tuberculosis is an important cause of death in areas of the world where poverty, overcrowding and malnutrition are rife.

CAUSATIVE ORGANISM

Tuberculosis results from infection with *Mycobacterium tuberculosis*. There are a number of strains of this organism affecting birds, rodents, reptiles and mammals, but the only two which commonly infect man are the human strain and the bovine strain. The latter is becoming less important in areas where milk, which spreads the infection, is pasteurised and where tuberculosis free herds are kept.

The organism is rod-shaped, that is, a bacillus and about 3μm long. It can be identified in three different ways:

- **Directly** by staining with Ziehl Neelsen stain. The stain may be used on smears from sputum or on tissue sections. The high lipid content of the wall of the organism makes it difficult to stain, and the stain (carbol fuchsin) is applied with heat.

  Once having taken up the stain, it is retained even after treatment with strong acid and alcohol, hence it is called an acid-fast bacillus.

  A fairly high concentration of organisms (10^5-10^6 organisms per ml of tissue) is needed before direct demonstration is possible.
- **By culture**. The organisms grow slowly on special culture media and it may take six weeks to obtain a positive result. Newer methods involve measuring the release of ^{14}C from a liquid substrate as $^{14}CO_2$ when tubercle bacilli are growing and gas chromatographic methods. Techniques using DNA hybridisation are also available in special centres.
- **By animal inoculation** Guinea pigs are susceptible to infection with tuberculosis, and if inoculated with material containing the tubercle bacillus they will develop disseminated tuberculosis over a period of time. The animals are usually killed and examined after six weeks.

As there are a number of conditions which give rise to the picture of granulomatous inflammation it is important to confirm the diagnosis, not only by the demonstration of acid-fast bacilli, but also by culture. This will confirm the nature of the organism and will also allow antibiotic sensitivity testing.

Mycobacterium tuberculosis is oxygen dependent and tends to multiply in areas where the oxygen tension is high, for example, the lungs and especially the apices of the lungs.

The organism is resistant to drying and can exist for a long time in sputum which has been coughed up by an infected patient. Infection with the human strain is acquired from an 'open' case. Infection with the bovine strain is via infected milk.

Infection may be acquired by:

- Inhalation of organisms into the lungs
- Ingestion of organisms into the gastro-intestinal tract (bovine strain)
- Rarely, by inoculation through the skin

FACTORS DETERMINING THE VIRULENCE OF THE DISEASE

- **Virulence** of tubercle bacilli may be related to different lipid compounds synthesised by some strains. These may prevent the action of lysosomes on phagocytosed bacilli within macrophages.
- **Age.** Tuberculosis can occur at any age, but is commonest at the extremes of life in the young and old. It occurs relatively rarely between the ages of five and fifteen years.
- **Host resistance.** There may be some degree of racial variation in that certain groups including North American Indians, black Africans and Australian Aborigines are reputed to be more susceptible to the disease. It may be that their social circumstances also predispose them to tuberculosis.

The effect of previous infection in the host is important. If guinea pigs are inoculated with tubercle bacilli a nodule will develop at the site, the regional lymph nodes will become involved and eventually the animal will die of disseminated disease. If another inoculation at another site is carried out four to six weeks after the first, there is a different response with rapid tissue necrosis and ulceration followed by healing. The regional lymph nodes will not be involved. This is described as the Koch phenomenon (after the investigator who first described it) and is an example of a cell-mediated delayed hypersensitivity response.

Hypersensitivity develops in humans and this accounts for the different responses to infection when it is encountered for the first time and when there is re-infection. It is likely that some degree of immunity accompanies the development of hypersensitivity Immunity may be shown by the improved ability of macrophages to phagocytose and destroy the organisms.

THE DEVELOPMENT OF THE LESIONS OF TUBERCULOSIS

The granulomatous lesion characteristic of tuberculosis is known as the **tubercle.**

- Initially macrophages phagocytose the bacteria, which at this stage they cannot kill. Indeed the bacteria may even multiply within macrophages. The macrophages may transport the live organisms to the regional lymph nodes via the lymphatics.
- Macrophages at the site of infection become transformed into epithelioid cells with ample eosinophilic cytoplasm, and multinucleate giant cells are formed. An outer ring of lymphocytes and fibroblasts accumulates. This is the so-called 'hard' tubercle.
- After seven to ten days central caseous necrosis occurs. The lesion is now a 'soft' tubercle with the characteristic appearance of the lesion of tuberculosis.
- Later still, areas of caseation may coalesce to form a larger granulomatous lesion. At this stage organisms may be difficult to see, but peripheral smaller tubercles may still be found.

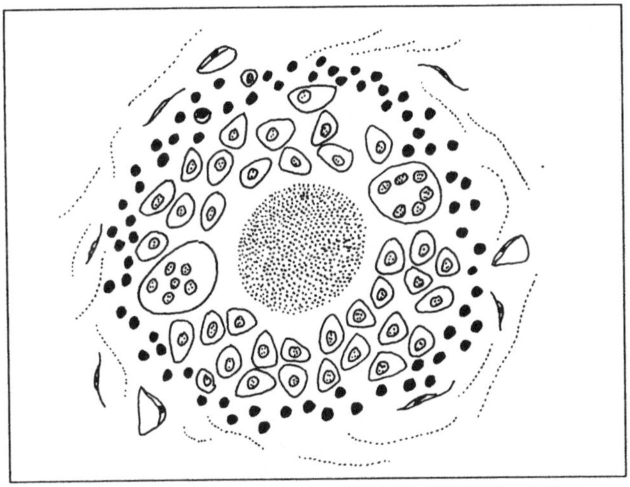

SOFT TUBERCLE - TUBERCULOUS GRANULOMA

The indigestible tubercle bacilli may themselves be responsible for the transformation of macrophages into epithelioid cells. Macrophages transport the bacilli to lymph nodes and present antigen to T cells. Sensitised T cells then activate macrophages; the products of T cells (such as γ-interferon) may also play a role in the transformation of macrophages to epithelioid cells.

The development of caseous necrosis, which occurs in the centre of granulomas is related to cell-mediated hypersensitivity. The mechanism by which caseous necrosis is brought about is still obscure, although the products of activated macrophages such as proteases, toxic oxygen radicals and tumour necrosis factor (TNFα) may play a part.

With the development of hypersensitivity about two weeks or so after infection has occurred, the intradermal injection of a standardised dose of purified protein derivative (PPD), which is obtained from culture medium in which the organisms have grown, will cause an indurated inflammatory reaction of more than 5mm within forty-eight hours. This indicates delayed hypersensitivity reaction and is known as the **Mantoux reaction**. The reaction persists throughout life after the initial contact with the organism and does not necessarily indicate active infection.

Once delayed hypersensitivity of this type has developed, further contacts with the tubercle bacillus will result in rapid caseous necrosis, localisation of the organisms within macrophages at the site of entry, and destruction of organisms within macrophages or epithelioid cells, in a manner resembling the Koch phenomenon in guinea pigs.

INFECTION WITH TUBERCULOSIS

When the tubercle bacillus infects a human being for the first time it results in what is called **primary or childhood type tuberculosis**.

The human strain of the tubercle bacillus usually settles in the lungs in the middle zone (that is the apex of the lower lobe or the lower part of the upper lobe). A small area of inflammation, about 1-1.5cm in diameter, develops sub pleurally. This may be known as a Ghon focus. After about fourteen days a granulomatous reaction with central caseation is established and spread has taken place to the regional lymph nodes, which also develop caseating granulomatous inflammation. The Ghon focus in the lung plus the involved lymph nodes is known as the primary complex or the Ghon complex. In the vast majority of cases healing by fibrosis takes place at this stage. Later calcification may take place. Even so, a few viable tubercle bacilli may persist in the quiescent, apparently healed lesions.

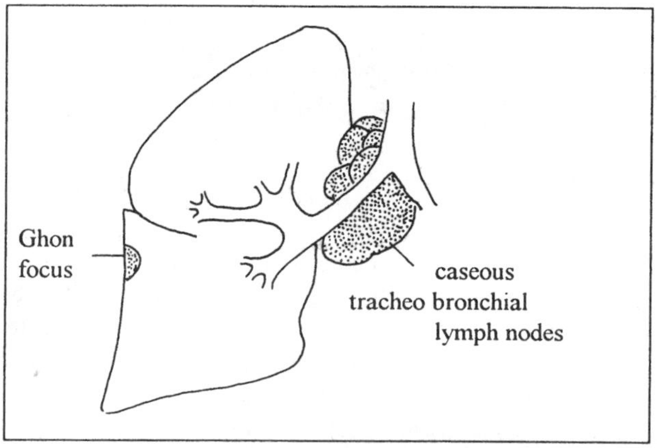

GHON COMPLEX - PRIMARY TUBERCULOSIS

The majority of primary infections in the lungs probably pass unnoticed clinically.

Primary tuberculosis may also affect the tonsils and cervical lymph nodes as well as the intestine and mesenteric nodes. In these cases there has usually been ingestion of tubercle bacilli from drinking infected milk. The bacilli are of the bovine strain.

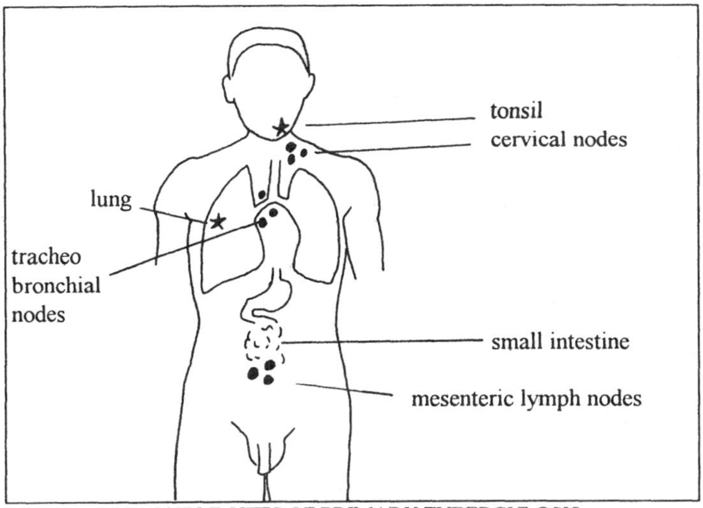

POSSIBLE SITES OF PRIMARY TUBERCULOSIS

While most cases heal, in a few cases infection spreads to give rise to overt disease, either at the time of infection or later.

The local lung lesion may increase in size directly, giving rise to **tuberculous pneumonia**. A bronchus in the area may be eroded allowing infected material to spread via the air passages in the locality, giving rise to **tuberculous bronchopneumonia**. There may be involvement of the overlying pleura with the formation of a serous **pleural effusion**. Less commonly still, rupture of a lesion into the pleura will cause a more serious cellular exudative inflammation of the pleura known as **tuberculous empyema**.

From the enlarged lymph nodes there may be pressure on a major bronchus causing obstruction and collapse of the distal lung. Erosion of a bronchus with spread of infection in the lung via the airways causes **tuberculous bronchopneumonia**. Rarely there may be direct involvement of the pericardium by tuberculous hilar nodes, giving rise to **tuberculous pericarditis**.

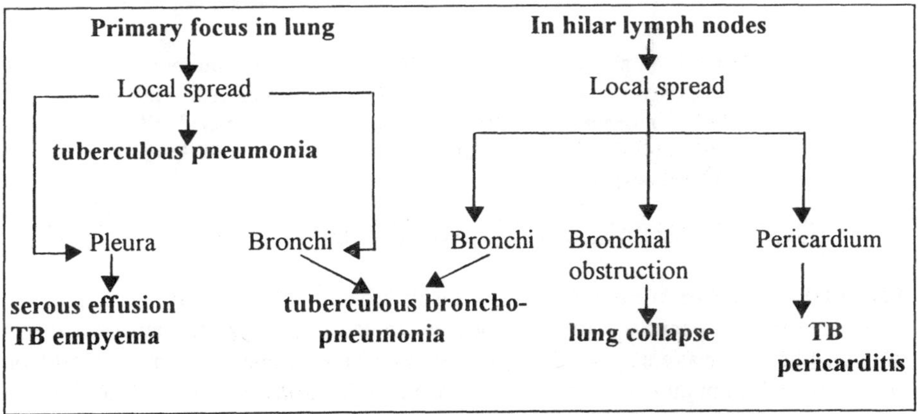

LOCAL SPREAD OF PRIMARY TUBERCULOSIS

There may also be spread of organisms through the blood stream.

The pulmonary vein may be invaded either by the local lung lesion, or by the enlarged tracheobronchial glands. Organisms may then be carried to the left side of the heart and spread systemically via the aorta. When the blood is invaded probably in most cases all the organisms are killed. A few organisms may lodge in certain organs and remain quiescent but capable of being activated at a later stage, possibly to give rise to **isolated organ tuberculosis**. If a large number of virulent organisms enter the blood stream in this way in a patient with little resistance, there may be **generalised miliary tuberculosis** with the development of large numbers of small tuberculous foci, the size of millet seeds, in many organs of the body, with the lungs being relatively spared. Spleen, liver and kidney are the major organs affected. There may also be involvement of the brain with **tuberculous meningitis** resulting from direct invasion via the choroid plexus, or from the rupture of a small focus in the brain into the meninges.

The pulmonary artery A branch of the pulmonary artery may be eroded by enlarged hilar glands. A heavy dose of organisms may be carried to a lobe or segment of one lung giving rise to **localised miliary tuberculosis of the lung**.

The thoracic duct. Organisms from the lymph nodes may enter the circulation via the **thoracic duct** which enters the innominate vein. Organisms can then be carried to the right side of the heart and enter the pulmonary artery circulation and be carried to the lungs giving rise to **generalised miliary tuberculosis of the lungs**.

It is important to be clear about the mechanisms by which tubercle bacilli may be carried around the body to understand the ways in which the disease can spread and can also occur long after the initial infection.

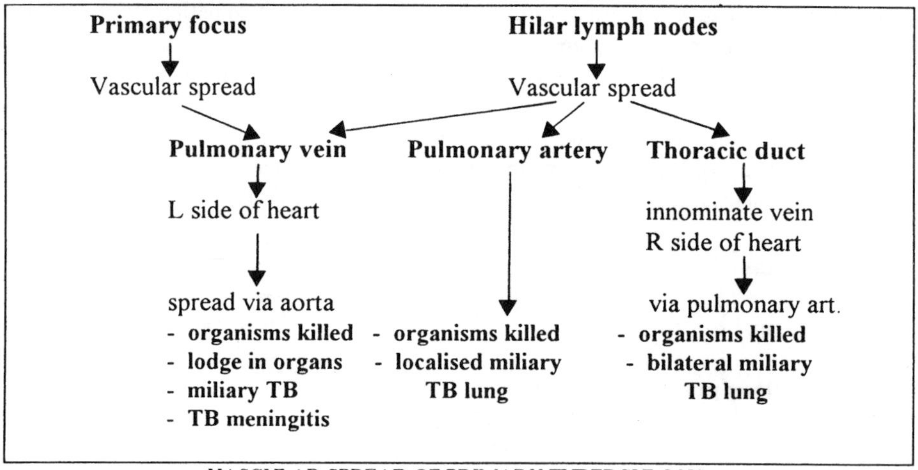

VASCULAR SPREAD OF PRIMARY TUBERCULOSIS

Secondary or adult tuberculosis may result from the reactivation of a primary lesion in which organisms had lain quiescent in the lung (or elsewhere), to cause disease when the patient's resistance is lowered, for example, from malnutrition or from lowered immunity for some other reason such as AIDS. This is probably the most common cause of this type of tuberculosis in the developed countries.

Secondary tuberculosis may also result from reinfection, and this is probably more common in the underdeveloped countries.
It should be noted that in Western countries primary tuberculosis can occur in adults, although the course of the disease may be modified and it may behave more like the adult type. Essentially this is disease arising in a previously sensitised individual.

Classically, the lesions which may be 1-3cm in diameter develop at the apices of the lungs where there is most oxygen. Because of the hypersensitivity and immunity organisms tend to be localised to this area of lung and spread via lymphatics is uncommon. Caseation appears rapidly and the histological appearances are those of a granulomatous reaction with marked caseation. There is also a fibrous reaction to the inflammation and the subsequent course of the disease depends on whether fibrosis predominates confining the lesion, or whether caseation predominates with enlargement and spread of the lesion.

The course of the disease may vary:

- The initial caseous area may be rapidly walled off by fibrous tissue leaving a caseous residue surrounded by **fibrosis**, or a **fibro-calcific scar.**
- Caseation may progress and coalesce. If a bronchus or bronchiole is eroded necrotic material may be drained through the air passages leaving a **cavity**. Draining infected caseous material is usually coughed up as sputum; the patient is highly infectious and is known as an 'open' case.
- Local enlargement of the tuberculous focus can involve more and more of the lungs in cavitating tuberculosis. Arteries traversing cavities may be eroded to give rise to severe haemorrhage with coughing up of blood (haemoptysis).
- Direct local spread may take place:
  - Via the bronchi to give tuberculous bronchopneumonia locally or in some cases with marked hypersensitivity and/or diminished immunity widespread pneumonia can result.
  - The pleura may be involved with effusion, empyema or dense fibrosis.
  - Bacilli in the sputum can cause tuberculous ulceration of the larynx or mouth, or, if swallowed, of the intestine.
  - *Rarely,* lymphatic and blood borne dissemination as described for primary tuberculosis may take place.

Thus there are a host of ways in which tuberculosis can behave, depending upon the circumstances and resistance of the individual patient.

Isolated organ tuberculosis

Tuberculosis of organs other than the lung is thought to be the result of haematogenous spread, usually at the primary stage, with later reactivation.

- **Tuberculous meningitis** tends to occur in younger children and is a common accompaniment of miliary tuberculosis. This is a basal meningitis with a gelatinous or fibrinous exudate. Small tubercles may sometimes be seen within the meninges. The exudate obstructs the flow of cerebrospinal fluid giving rise to hydrocephalus.

There may be marked obliterative endarteritis of vessels in the subarachnoid space, giving rise to ischaemic effects with small infarcts in the brain or cranial nerve roots. Within the cerebrospinal space as in other body cavities the histological lesions may be less characteristic with epitheloid cell granulomas and giant cells being less prominent and a more non-specific type of chronic inflammatory infiltrate.

- **Renal tuberculosis** may start in the cortex of the kidney. Once caseous tuberculosis is established in the kidney it can spread locally to the pelvis and later to the ureter and bladder.
- **Bone tuberculosis** commences in the metaphysis and causes marked local destruction. A distinctive feature is the ability of the tuberculous process to destroy epiphyses and involve joint cavities. In the spine intervertebral discs may be involved.
- **Genital tuberculosis** may involve testes in the male and Fallopian tubes in the female. Genital tuberculosis is an important cause of female infertility in certain parts of the world, including South Africa.
- **Adrenal glands** may be affected with massive destruction of adrenal tissue leading to hypoadrenalism known as Addison's disease.

TREATMENT AND PREVENTION

Tuberculosis can be treated with a wide variety of antibiotics. Long term treatment is indicated. 'Open' cases should be isolated until they are no longer excreting viable bacilli, if possible. Organisms should be cultured and antibiotic sensitivity test performed to identify drug-resistant strains.

BCG (Bacille-Calmette-Guérin) is an attenuated strain of bovine bacillus which is not pathogenic in itself, but which will bring about the development of hypersensitivity and immunity in man and induce a positive delayed hypersensitivity tuberculin test (Mantoux test). BCG vaccination is thought by some to reduce the risk of developing clinical tuberculosis, particularly in Western countries.

It should be remembered that amyloidosis is a possible complication of tuberculosis of long standing.

HIV AND TUBERCULOSIS

Patients with reduced immunity are susceptible to tuberculosis. In HIV reduced numbers of CD4 (helper) T cells reduce cell-mediated immunity. The lesions that develop often resemble aggressive primary tuberculosis with lymph node involvement and often systemic extrapulmonary involvement even if there is a previous history of exposure to tuberculosis. Large numbers of bacilli may be found in the tissues, yet caseous necrosis is not a prominent feature. Tuberculin skin tests are negative. The diagnosis thus may be very difficult to make. A similar pattern of disease may be seen in the elderly, in patients with malignant disease or in immuno-compromised patients.

HIV patients are also particularly susceptible to infection with atypical mycobacteria.

ATYPICAL MYCOBACTERIA

Cases of tuberculosis-like disease may occur in which the organisms are atypical in that they may not be pathogenic to guinea pigs or susceptible to anti-tuberculous drugs. Generally the disease caused is not severe. Often lymph node involvement may be the presenting feature.

These bacilli can be classified according to pigment production on culture and also according to speed of growth and the temperature at which they will grow.

Fast growing organisms such as *M. phlei* and *M. smegmatis* can cause lymph node infection in children. *M. fortuitum* and *M. chelonei* may cause wound infections and have been reported following heart surgery.

M. avium-intracellulare, which is slow growing and does not produce pigment can also produce disease in humans. It is not uncommon to find a positive skin reaction to an extract of this bacillus in the general population.

M. ulcerans is an organism which causes chronic ulceration of the skin. An undermined ulcer overlies necrotic fat. There is minimal inflammatory reaction, but positively staining organisms may be found at the edge of the necrotic fat. Healing is often slow. The resultant indolent ulcer is known as Buruli ulcer in Uganda.

M. marinum will infect skin wounds obtained in swimming pools or from aquaria. There is a non-caseating granulomatous inflammation.

GENERAL PATHOLOGY - CHAPTER TWENTY TWO

OTHER INFECTIOUS AND GRANULOMATOUS DISEASES

In this chapter another mycobacterial disease, namely leprosy will be discussed. Other bacterial, fungal and parasitic diseases will also be mentioned in order to demonstrate the wide range of pathogens and to describe some of the characteristic pathological appearances in chronic inflammation and to deal with some of the generalised diseases which affect more than one system of the body.

LEPROSY

It has been estimated by the World Health Organisation that worldwide over fifteen million people are infected with leprosy. Most of these live in tropical countries, mainly in the southern hemisphere.

Leprosy is caused by *Mycobacterium leprae.* This organism is acid fast, but less so than the tubercle bacillus; for example, it is decolourised by strong acids after staining but can resist 1% H_2SO_4. The organism cannot be cultured but will grow in the footpads of mice. It can also infect the nine-banded armadillo.

The organism appears to be of low infectivity. Probably only about 1 in 200 persons infected will go on to develop overt disease. Prolonged contact with a case is necessary before infection is acquired in the skin or mucous membranes, and the incubation period (that is the time from infection until the manifestations of the disease appear) may be many years. Infection is probably transmitted by inhalation of infected material from the nasal secretions of a case actively excreting bacilli whilst in the lepromatous phase. The organism multiplies best at low temperatures and, after an initial bacteraemia, may affect the peripheral, colder parts of the body such as the nose, ears and limbs.

The tissues affected are the skin, mucous membranes and the peripheral nerves predominantly, with the anterior eye, upper respiratory tract and testes being singled out as likely sites in addition to the lesions of the skin.

There are two main variants of the disease which reflect differences in the cell-mediated immunity of the host.

	In susceptible individuals **LEPROMATOUS**	In the more resistant **TUBERCULOID**
Clinically	often symmetrical, nodular rash which tends to coalesce; leonine face and on limbs; diminished sensation in these areas	irregular macules on skin; marked thickening of nerves. Patchy areas of anaesthesia skin in areas supplied by affected nerves
Histologically	aggregates of foamy macrophages (lepra cells) containing lipid and masses of bacteria (globi) in skin and m.m. nerves filled with bacilli but less thickened or inflamed	tuberculoid granulomas in skin up to the base of the epidermis and around nerves, destroying them. Bacteria rarely seen.
Immunologically	low resistance due to defective T-cell immunity Fernandez - negative Mitsuda - negative. Excessive B-cell response with hyperglobulinaemia	good T-cell response Fernandez - positive .

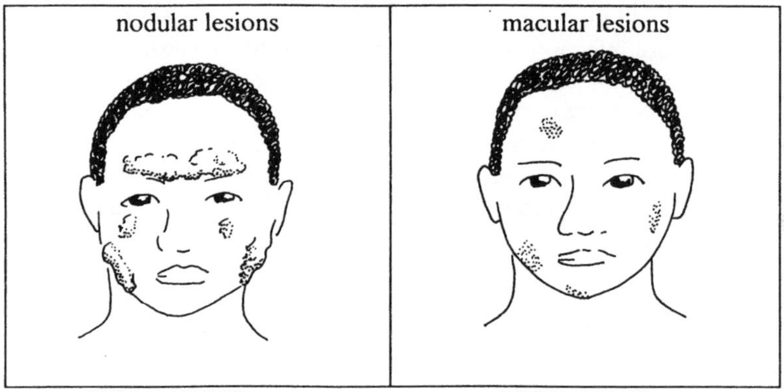

LEPROMATOUS	TUBERCULOID

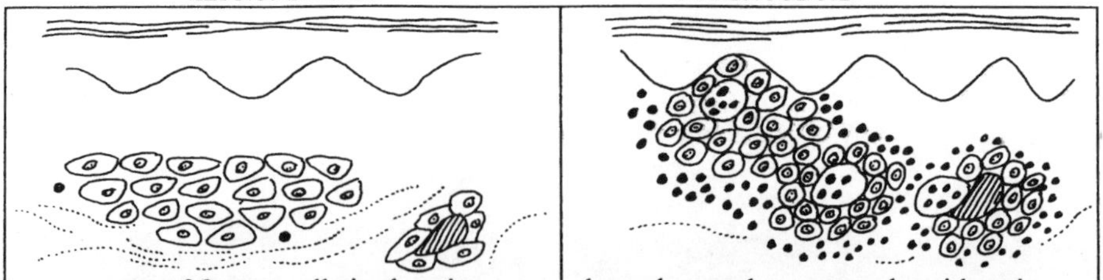

aggregates of foamy cells in dermis many bacilli bacilli in nerves; little inflammation	dermal granulomas touch epidermis few or no bacilli granulomatous inflammation round nerves

In both types of leprosy, nerve involvement and anaesthesia can lead to trophic changes, atrophy, ulceration and fibrous scarring and contractures.

The polar lepromatous (LL) and tuberculoid (TT) forms are at either end of a spectrum of changes, and intermediate forms known as borderline lepromatous (BL) borderline (BB) and borderline tuberculoid (BT) are described. The appearances can change as cell-mediated immunity develops with progress from BL to BT being accompanied by the development of epithelioid cells and giant cells and loss of bacilli in lesions.

The diagnosis is clinical and pathological. If bacteria can be seen in tissues or smears this is good evidence of leprosy, but a tuberculoid infiltrate around nerves in the dermis or subcutis is also fairly diagnostic.

A skin test, the **lepromin test,** consists of injecting an extract of bacilli (from human tissue) into the dermis.

- **Fernandez reaction.** A delayed hypersensitivity reaction (like the Mantoux) occurring after forty-eight hours denotes good hypersensitivity and is positive in tuberculoid leprosy.
- **Mitsuda reaction.** Epithelioid cell granulomas developing in three to four weeks denote innate immunity. This reaction is present in 90 per cent of normal individuals. If the result is negative, the patient is likely to develop lepromatous leprosy if infected, as innate immunity is low.

All forms of the disease are extremely chronic and may go on for years if untreated. Reactive (AA) amyloidosis may be a complication of lepromatous leprosy which is characterised by a polyclonal hyperglobulinaemia with formation of humoral antibodies which cannot react against intracellular bacteria.

Drug therapy is available, with sulphones for long term treatment and rifampicin for rapid reduction of infectivity.

SYPHILIS

This is a disease caused by a corkscrew-shaped organism, *Treponema pallidum* which is about 10-15µm long. Other treponemes (also known as spirochaetes) may cause different diseases such as yaws. Syphilis is a fairly common disease, having undergone a resurgence in the Western world with the advent of sexual permissiveness. However, the widespread use of antibiotics has made the late and serious complications rare. Routine testing in pregnancy has also reduced the incidence of congenital syphilis.

The organism may be difficult to detect, but can be seen in fresh exudates from lesions when these are examined microscopically using dark field illumination. The organism in tissues will take up silver stains.

Infection is transmitted venereally, that is, by sexual intercourse, or transplacentally. Treponemes from an infected lesion on one partner enter the body of the new host through mucous membrane or abraded skin. On entering the body the organisms multiply and are transmitted to regional lymph nodes. Within twenty-four hours they are circulating in the blood stream.

During the incubation period immunoglobulin antibodies are formed and can be detected within a few weeks of infection. The antibodies are of two types:

- Nonspecific, reacting with cardiolipin from beef heart. The Venereal Disease Research Laboratory (VDRL) test and the Rapid Plasma Reagin (RPR) test are two that are commonly used for screening. However false positive results occur in a number of conditions such as infectious mononucleosis, viral hepatitis, leprosy and autoimmune diseases.
- Specific treponemal antibodies can be demonstrated by the Fluorescent Treponemal Antibody (FTA) technique or by treponemal immobilisation techniques.

There are three stages of the disease:

- **Primary syphilis** may occur anywhere from two to ten weeks after infection and is characterised by the development of a solitary, hard, painless ulcer known as a **chancre** at the site of entry of the treponemes. This is accompanied by painless enlargement of regional lymph nodes. Lesions on the penis of males are obvious, but those in the vagina or on the cervix in females may go unnoticed. The chancre will heal in several months, even without treatment. At this stage the patient is immune and cannot develop another chancre.
- **Secondary syphilis** may occur from two weeks to six months after the chancre has healed and is characterised by varied lesions of many parts of the body such as:

  - Skin: Rashes and wart like plaques (condylomata lata).
  - Mucous membrane: Shallow ulcers (snail-track ulcers).
  - Lymph nodes: Follicular hyperplasia with generalised enlargement of lymph nodes.

  Again the lesions heal spontaneously. Antibody tests are always positive by the secondary stage.

- **Tertiary syphilis** In a proportion of untreated cases (about 30 per cent) tertiary manifestations may occur years or decades after the first two stages, both of which can sometimes pass unnoticed by the patient. These are of two types:

  a) **Localised gummas**. These are granulomatous lesions with central necrosis (which looks something like a cross between coagulative necrosis and caseous necrosis). The histological changes of syphilis with obliterative endarteritis and plasma cell cuffing of vessels may be seen in the periphery. Gummas can occur in any tissue but liver, testes, bones and skin are the most common sites. They heal with marked scarring.

b) **Diffuse inflammatory lesions** can occur in
- The **cardiovascular system** causing aortitis, weakening of the wall with aneurysm formation and aortic incompetence due to stretching of the valve ring. There is loss of elastic in the aorta, probably due to endarteritis and periarteritis of the vasa vasorum.
- The **nervous system** can be involved in three ways:
  (i) *Meningo-vascular syphilis* is focal meningitis in which there may be vascular occlusion due to endarteritis. Local paralysis (for example of cranial nerves) can occur.
  (ii) Another example is parenchymatous syphilis which involves the loss of neurones and gives rise to so-called *general paralysis of the insane.*
  (iii) In *tabes dorsalis* there is degeneration of the posterior columns of the spinal cord and posterior nerve roots leading to widespread sensory loss.

It is not usually possible to demonstrate treponemes in tertiary syphilis.

The histological changes involved in syphilis may be seen in all these stages to a greater or a lesser degree.

In the vessels there is endothelial swelling and thickening of the wall of small vessels, and obliterative endarteritis due to concentric proliferation of endothelial cells in small arteries and arterioles.

There is also perivascular cuffing with plasma cells.

Granuloma formation may rarely be seen in the skin lesions of secondary syphilis but is seen in gummas together with gummatous necrosis.

Treponemes are usually plentiful in primary and secondary lesions, often being seen in the walls of vessels.

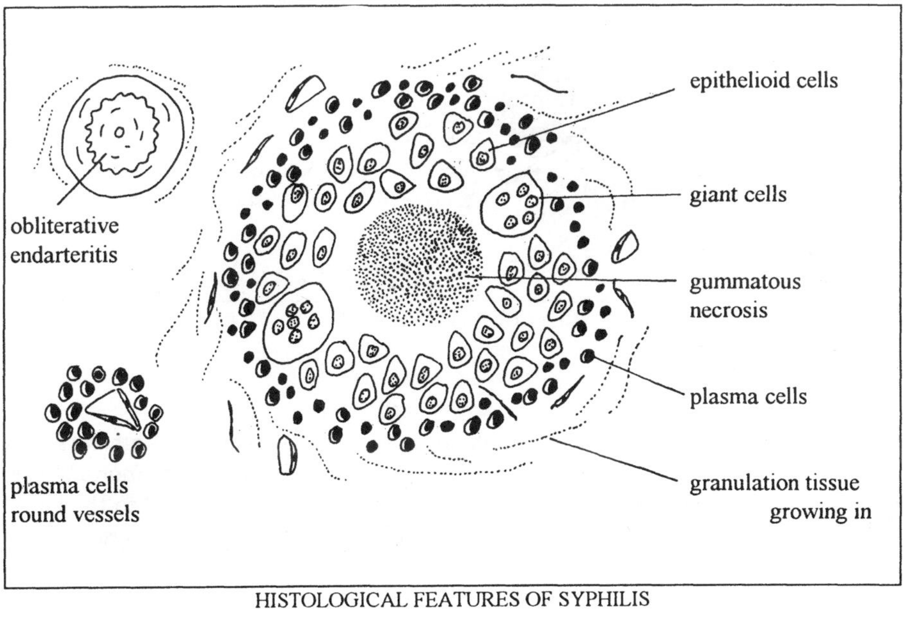

HISTOLOGICAL FEATURES OF SYPHILIS

Congenital syphilis is acquired in utero and may be a cause of stillbirth. It is now fortunately rare. Affected children show a variety of abnormalities of skin, mucous membranes, bone, teeth, liver, lungs and central nervous system.

Treatment. Syphilis responds to treatment with penicillin which will arrest the disease at any stage (a therapeutic test).

Of people who develop syphilis and are not treated, one third will cure spontaneously, one third remain seropositive without tertiary disease and one third develop tertiary syphilis.

ACTINOMYCOSIS

This is a disease caused by organisms called actinomycetes, which are bacteria which have some similarities to fungi. The most common of these organisms which causes human infection is *Actinomyces israeli*. This is a Gram positive anaerobic organism which can live as a commensal in the oral cavity and gut but can cause infection in dead or injured tissue.

Masses of organisms form yellow colonies, which are contained in the pus that discharges from the lesions. These are large enough to be seen with the naked eye as 'sulphur granules'.

Abscesses which discharge through multiple sinuses are characteristic of the disease. These can occur in the neck, the abdomen and the thorax. Histologically, within the abscesses the colonies of the organism consist of clusters of blue staining filamentous bacteria surrounded by eosinophilic material. This eosinophilic material is an immunoglobulin response to the infection.

The disease responds to heavy doses of penicillin given over a long period.

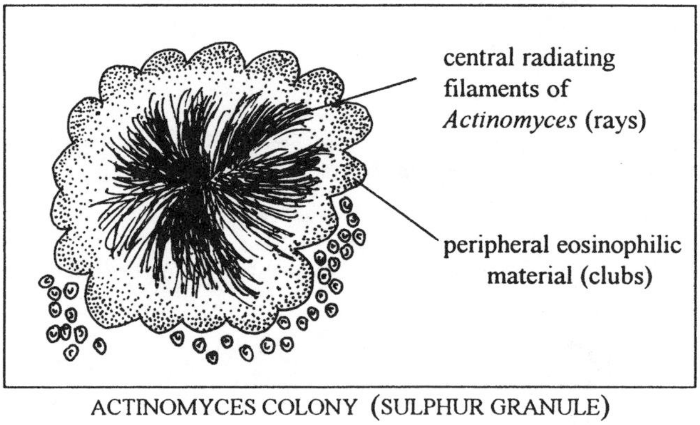

ACTINOMYCES COLONY (SULPHUR GRANULE)

FUNGAL DISEASES

Superficial skin infections or dermatophytoses such as ringworm and athlete's foot are generally unimportant. Dermal or subcutaneous infection with *Sporothrix* in sporotrichosis or chromoblastomycosis usually tend to remain localised to the limbs. Other fungi such as *Candida albicans* can cause superficial, deep or systematised disease.

Although some fungi that cause deep infection are pathogenic in their own right, many are opportunistic and only cause disease in patients with lowered resistance due to immunosuppression, chemotherapy or X-ray therapy for cancer, or intensive antibiotic therapy. Patients with AIDS are particularly prone to develop serious fungal infection, for example, cryptococcosis.

Histologically there is often a mixed pyogenic and granulomatous response with a central abscess surrounded by epithelioid and giant cells, which are in turn surrounded by lymphocytes and plasma cells.

MIXED INFLAMMATION AROUND FUNGUS

Fungi capable of causing deep infection include:

- *Aspergillus spp.*
- *Blastomyces dermatitidis*
- *Candida albicans*
- *Coccidioides immitis*
- *Cryptococcus neoformans*
- *Histoplasma capsulatum*
- *Mucor spp*
- *Sporothrix schenkii*

Aspergillus and *Mucor* species tend to cause opportunistic infections whereas the others can cause infections without known predisposing factors.

SCHISTOSOMIASIS

This is a disease caused by trematodes or flukes which infect humans in the tropics. The adult worms develop in the venous system and lay eggs which are discharged via the bowel or bladder. The parasites are relatively well adapted and cause few symptoms, but problems may be caused by the eggs that are laid in tissues. This is predominantly a tropical disease and it is estimated that some 250 million people throughout the world are infected with these organisms.

Two species are important in Africa, *Schistosoma haematobium* and *Schistosoma mansoni*. Humans develop the infection from bathing or paddling in infected waters which allows the cercariae, released by infected snails, to penetrate the skin. These enter the blood stream as schistosomula or tiny worms and travel through the circulation to the lungs. Migration continues until finally the cercariae reach the liver, where they enter the intrahepatic branches of the portal vein to mature.

Adult worms then migrate to the pelvic veins (*S. haematobium)* or the mesenteric veins *(S. mansoni)* where they mate, the female lying within the gynaecophoric canal of the male. The adult worms are about 15mm in length. Eggs are laid which are about 150µm long, too big to pass through the capillaries. The eggs are discharged through the bladder (*S. haematobium)* or the bowel *(S. mansoni)*. When an egg in urine or faeces reaches water it hatches and a miracidium is released, which then enters the intermediate snail host to complete its life cycle. An adult female worm can lay some 300 eggs daily and may live for as long as twenty years (more often three to five years).

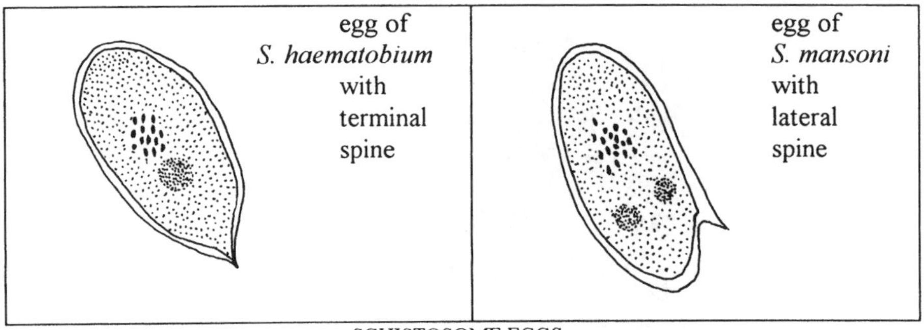

SCHISTOSOME EGGS

Clinically there may be symptoms at the time of penetration of the skin by cercariae (swimmer's itch). Occasionally there are symptoms, fever, eosinophilia, hepatosplenomegaly or allergic manifestations in the early stages of infection, but this sort of reaction (acute schistosomiasis) is usually only found when there is an exceptionally heavy infection in those who have not been exposed previously.

Most symptoms are due to eggs in the small vessels or tissues arousing an inflammatory reaction. Initially this is acute inflammation, but eosinophils predominate and there may almost be an eosinophilic micro-abscess. Later, modified macrophages accumulate and a giant cell granuloma forms.

In *S. haematobium* infections rough inflamed patches may be noted in the bladder wall where eggs have been deposited. Eggs live for about three weeks in the body. The egg shell may undergo calcification after death. Granulomas heal by fibrosis. Fibrosis of the ureters may lead to obstruction and chronic pyelonephritis, which may eventually lead to renal failure. There is haematuria and eggs may be found in the urine on microscopic examination. A late complication may be the development of bladder cancer (squamous cell carcinoma).

In *S. mansoni* infection, there is inflammation of the colon with pseudopolyp formation. Granulomas occur in the liver and there is also an increased amount of fibrous tissue in portal tracts, giving rise to the so-called 'pipe-stem fibrosis', which can cause severe portal hypertension. The worms feed on red cells and excrete a dark haematin pigment which may be found in the liver, which often appears macroscopically to be dark in colour. In some cases a glomerulonephritis (mesangioproliferative or membranous) may develop.

Schistosome eggs may be found in the lungs where they may cause pulmonary granulomatous arteritis resulting in pulmonary hypertension and right sided heart failure. Eggs may also be found in many other sites throughout the body including the central nervous system.

Effective drug treatment is now available for severe cases, but reinfection is a constant threat in endemic areas.

SARCOIDOSIS

This is not an infectious disease, but presents with disseminated granulomatous inflammation in many parts of the body. Sarcoidosis is a disease of unknown aetiology characterised pathologically by non-caseating epithelioid granulomas occurring in any organ or tissue. Other known causes of granulomatous inflammation must be excluded before the diagnosis can be made.

There is a peculiar geographic and racial distribution and the disease appears to be more common in Scandinavians, the Irish and in Blacks in the UK and the USA. the disease seems to be rare in Asians.

Clinically most cases have involvement of the lymph nodes in the hilar regions of the lungs. The lungs themselves may be infiltrated bilaterally. In the lungs lesions may heal by fibrosis giving rise to interstitial pulmonary fibrosis.

The skin is another common site for lesions, often at the site of previous injury.

The eye, the lachrymal glands and the salivary glands may be affected in up to one half of cases.

The liver, spleen and bone marrow may also be the sites of sarcoid lesions.

Histologically the appearances are the same wherever the lesions occur, that is, there is the formation of small epithelioid cell granulomas with multinucleate giant cells. The giant cells may contain inclusions, either laminated calcific bodies known as Schaumann bodies or star-shaped inclusions (asteroid bodies). The granulomas tend to remain separate rather than coalescing as in tuberculosis. There is no central necrosis. There is usually little in the way of peripheral lymphocytic infiltrate, giving the appearance of a 'naked' granuloma.

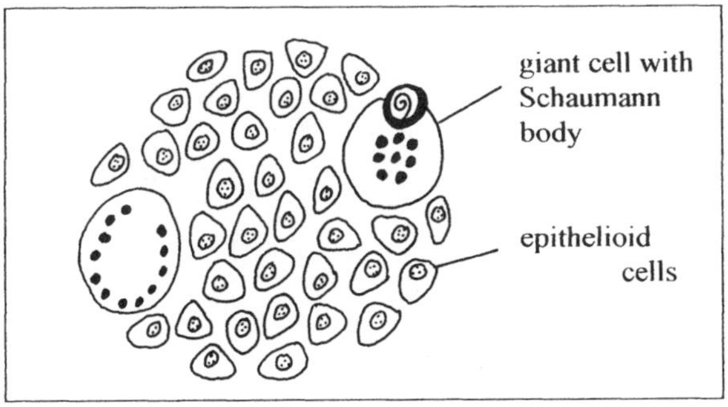

'NAKED' GRANULOMA OF SARCOID

Patients have deficient T-cell responses in the peripheral blood (relative T-helper cell deficiency) with an increase in the proportion of these cells in the lesions. There is a lymphopenia in the blood. The patients also do not show cutaneous delayed hypersensitivity response to tuberculin or other allergens.

Patients may be recognised incidentally, for example, on chest X-ray, or may present with respiratory symptoms or nonspecific complaints such as of fever or weight loss.

There may also be osteolytic lesions in the phalangeal bones. Hypercalcaemia may be a complication.

The diagnosis is made by exclusion. Biopsy of the characteristic sarcoid granuloma may be of help. The disease usually heals itself over a few years, but severe pulmonary infiltration can lead to pulmonary fibrosis and right sided heart failure.

Those cases that need treatment seem to respond to therapy with corticosteroids.

GENERAL PATHOLOGY - CHAPTER TWENTY THREE

DISORDERS OF GROWTH

A CLASSIFICATION OF DISORDERS OF GROWTH			
Failure of growth and development	complete	=	**agenesis**
	partial	=	**hypoplasia**
Acquired reduction in size			**atrophy**
Overgrowth			**hypertrophy** and **hyperplasia**
Abnormal growth or differentiation			**metaplasia, dysplasia** and **neoplasia** (and **anaplasia**)

The above table lists some of the words that will be defined in more detail during the course of this chapter and gives some indication of their meaning or the context in which they are used.

ATROPHY

Atrophy is an acquired decrease in size, of a part of the body, a tissue or an organ, or of individual cells. Atrophy of a part may be due to a reduction in the size of individual cells or to a reduction in the number of cells.

The reduction in the number of cells may be due to physiological death (**apoptosis** or programmed cell death) of cells, which are then not replaced. Reduction in cell size is due to the cell containing fewer mitochondria and myofilaments and less endoplasmic reticulum.

Heterophagy is the name given to phagocytosis of solid matter or pinocytosis of soluble material from outside the cells. **Autophagy** is a process whereby residual material within the cell is membrane bound (from the endoplasmic reticulum) into an autophagic vacuole which may then fuse with a lysosome to form an autophagolysosome. This is a method of removing damaged, worn out or unwanted parts of the cell and is particularly frequently seen in atrophy. Autophagolysosomes containing material which cannot be digested may remain within the cell as residual bodies. When the undigested material is lipid in nature, such as the products of intracellular lipid peroxidation, it may accumulate as **lipofuscin** pigment granules. These may be seen in ageing cells and accumulate in the condition of 'brown atrophy' which may occur in the hearts of elderly people.

Atrophy must be distinguished from:

- **agenesis,** the complete absence of a part dating from embryonic development, for example, renal agenesis.
- **hypoplasia,** the lack of development to full mature size.

Note that the term aplasia which could be taken to be synonymous with agenesis is used by haematologists to describe a condition in which the bone marrow ceases to produce blood cells when it has been damaged by drugs or X-rays.

Thus atrophy is distinguished because it implies an *acquired* decrease to less than normal size.

Physiological atrophy can take place in the developing foetus, or in early infancy when the ductus arteriosus and the foetal adrenal disappear soon after birth. In later life the thymus atrophies, and there may also be atrophy of lymphoid tissue or of muscle.

It should be noted that atrophy may be masked by the ingrowth of connective tissue or fatty connective tissue (stromal fatty infiltration) into the atrophic part.

Pathological atrophy occurs in the more specialised tissues, and can be recognised by the smaller size of the cells, which may seem more closely packed together, for example, in atrophic muscle. There may be lipofuscin deposition. There may be replacement of specialised cells by fibrous or fatty connective tissue.

CAUSES OF ATROPHY:

- **Malnutrition** may cause generalised atrophy as in starvation brought about from lack of food or malabsorption. In patients with inflammatory or malignant disease, generalised atrophy or wasting may be due to tumour necrosis factor (TNF-α) which is released from activated macrophages.

Most other factors are likely to cause local atrophy.

- **Diminished functional activity** or disuse atrophy is exemplified by a limb which has been immobilised in a plaster cast and shows marked muscular atrophy.
- **Interference with nerve supply** can lead to neuropathic atrophy of muscles (which are paralysed and not used).
- **Interference with blood supply** due to arteriosclerosis may account for shrinkage of the brain or kidney in later life. Partial ischaemia can lead to atrophy of a part with focal cell loss and fibrosis, which falls short of actual infarction.
- **Reduced endocrine stimulation,** for example, lack of ACTH may lead to adrenal atrophy.
- **Pressure atrophy** is probably related to ischaemic atrophy in that continuous pressure, such as by an aortic aneurysm on vertebrae, causes atrophy, presumably by interfering with the blood supply
- **Ageing**. Wear and tear with the accumulation of lipofuscin pigment has already been cited as an example of atrophy at cellular level.

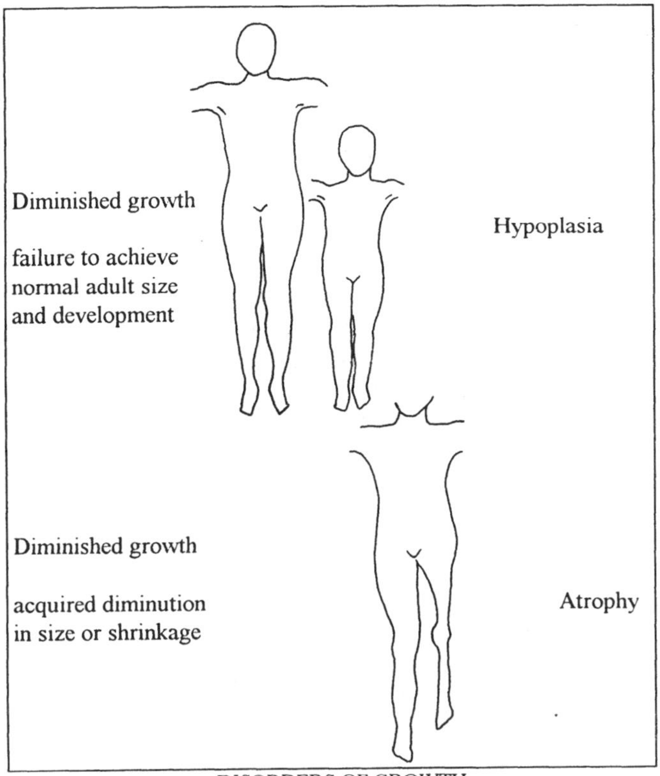

DISORDERS OF GROWTH

HYPERPLASIA AND HYPERTROPHY

As has been mentioned previously, labile cells and stable cells have the power to proliferate and regenerate. The permanent cells, neurones and cardiac and skeletal muscle cells cannot divide (or in the case of muscle, have only limited powers of regeneration) and these can only respond to increased demands by growing in size. Thus if there is loss of muscle tissue for any reason, surviving muscle cells will increase in size to meet the functional demands placed on them. This is known as:

Hypertrophy which is the increase in size of a tissue or organ due to an increase in the *size* of its individual component cells.

Hyperplasia is the increase in size of a tissue or organ due to an increase in the *number* of its component cells.
Hyperplasia should be differentiated from:

- Regeneration in which there is replacement of lost cells without an increase in size.
- Neoplasia may simulate hyperplasia, but differs in that the new growth continues regardless of stimulus - whereas hyperplasia is a response to a stimulus and ceases or regresses when the stimulus is removed.

Due to the fact that in some cases of hyperplasia the nature of the stimulus is not known, these distinctions are not always easy. Also there may be overlap between them. For example in the normal endometrium there is regeneration after menstrual shedding. Excess of oestrogen may lead to hyperplasia and if this state is prolonged the hyperplastic endometrium may proliferate autonomously and become cancerous.

REGENERATION $\neq$ HYPERPLASIA $\neq$ NEOPLASIA

CAUSES OF HYPERPLASIA

- **Endocrine:** Hyperplasia may occur in endocrine glands themselves as in parathyroid hyperplasia accompanying renal failure or in thyroid hyperplasia due to the action of thyroid stimulating immunoglobulins.

  Hyperplasia may also occur in the target organs of endocrine glands. Oestrogens for example may cause enlargement of the breast preparatory to lactation in pregnant females; gynaecomastia or enlargement of the breast occurs in males on oestrogen therapy or with liver disease; hyperplasia of the endometrium in women may result from excess oestrogens and this may be a premalignant condition. Hyperplasia of the prostate in middle aged males is thought to be due to relative oestrogen excess.
- **Increased functional demand:** This is best illustrated in the bone marrow, where there is an increase in erythropoiesis following blood loss for any reason. The red marrow expands in size and extends into the shafts of long bones, which are usually filled with fatty marrow in adults. There may be hyperplasia of granulocytic precursors in infection.

  One kidney may become larger after the other has been removed; this is due to hyperplasia of cells constituting the nephrons, even though the actual number of nephrons does not increase.
- **Chronic irritation:** Chronic irritation or repetitive trauma of an epithelial surface may bring about hyperplasia of the epithelial cells. For example, a corn on the feet due to a badly fitting shoe or hyperplasia of the oral mucosa in relationship to ill-fitting dentures.

  Chronic inflammation may also lead to hyperplasia as in chronic bronchitis, where there is hyperplasia of mucus secreting glands within the walls of bronchi. In many forms of dermatitis there is hyperplasia of the cells of the prickle cell layer of the squamous epithelium (known as acanthosis).

  Epithelium at the edges of chronic ulcers, for example, ulcers of the skin or stomach, may undergo regenerative hyperplasia of such degree that it may be confused with neoplastic proliferation.

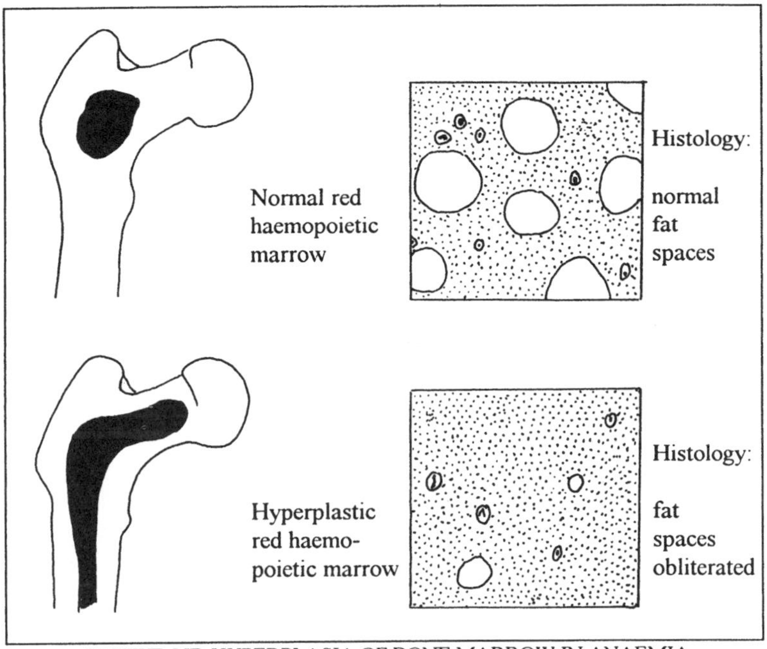

ERYTHROID HYPERPLASIA OF BONE MARROW IN ANAEMIA

CAUSES OF HYPERTROPHY

These can be summed up as being due to **increased functional demand.** Even when there is a hormonal element, as in the muscular hypertrophy that occurs in the pregnant uterus, it is in response to the extra demands that are being placed upon it.

It should also be pointed out that hyperplasia and hypertrophy are not changes that necessarily occur in isolation. Very often they occur together. In most organs which enlarge in size, in fact, there is a combination of the two. For example, in the single kidney which enlarges in size following the removal of the other kidney, both hyperplasia (increase in number of renal tubular cells) and hypertrophy (increase in size of individual renal tubular cells) may take place. Similarly in the enlarged prostate although there may be hyperplasia of glandular epithelial cells in response to oestrogenic stimulation, there is also hypertrophy of smooth muscle cells which enlarge in size. So too in the pregnant uterus, in the early stages there may be some hyperplasia of smooth muscle cells, and this is followed by hypertrophy. Much seems to depend on the regenerative ability of the cells of the organ.

Pure hypertrophy, without accompanying hyperplasia occurs almost always in muscle in response to increased functional demand. For example:

- **Physiological hypertrophy** occurs in skeletal muscles which are subjected to strenuous use. Skeletal muscular hypertrophy can be induced by training, 'pumping iron', as the bulging muscles of athletes and body-builders would testify. Once the stimulus to hypertrophy is removed, that is training ceases, the muscles tend to revert to normal size.

  In the pregnant uterus the changes are mainly those of hypertrophy and muscle fibres may increase up to ten times in length and five times in width.

- **Pathological hypertrophy** may be seen in cardiac muscle and in smooth muscle within the walls of hollow organs.

  The heart hypertrophies when its work load is increased, provided that the blood supply is adequate.

  - The left ventricle will undergo concentric hypertrophy of the muscle of the myocardium in high blood pressure (hypertension).
  - The right ventricle will undergo hypertrophy if there is pulmonary hypertension, or pulmonary valvular disease.

Heart muscle fibres can double in width (from 10-12μm to 20-25μm). The nuclei also increase in size becoming squarer and then distorted in shape.

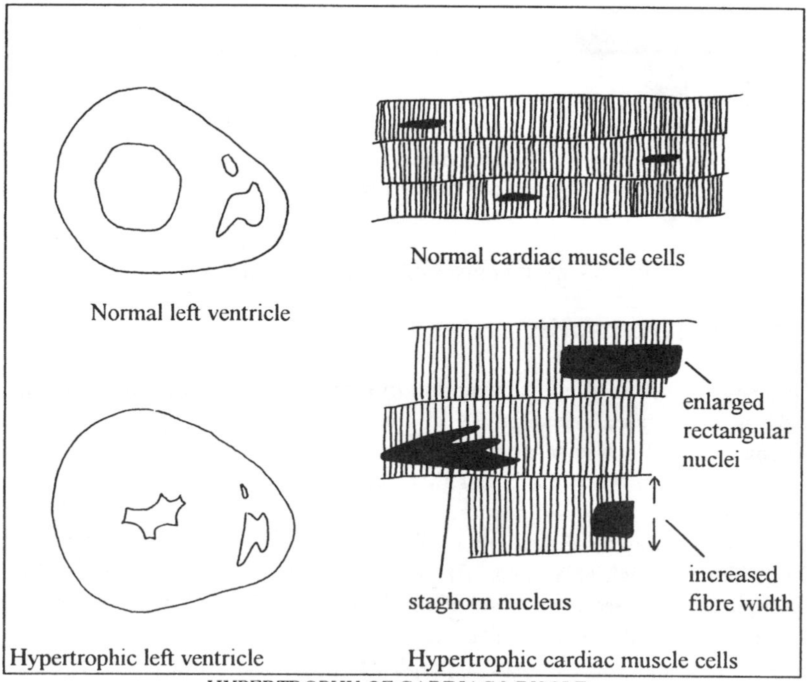

HYPERTROPHY OF CARDIAC MUSCLE

Smooth muscle, especially that lining a hollow organ, will also undergo hypertrophy if it has to work harder for any reason.

- Muscle in the wall of the bladder will undergo hypertrophy if there is an obstruction to urinary outflow, such as may be caused by prostatic enlargement or stricture of the urethra.
- The muscle of the gut wall will hypertrophy if there is intestinal obstruction.
- The media of muscular arteries becomes hypertrophic if the blood pressure is raised.

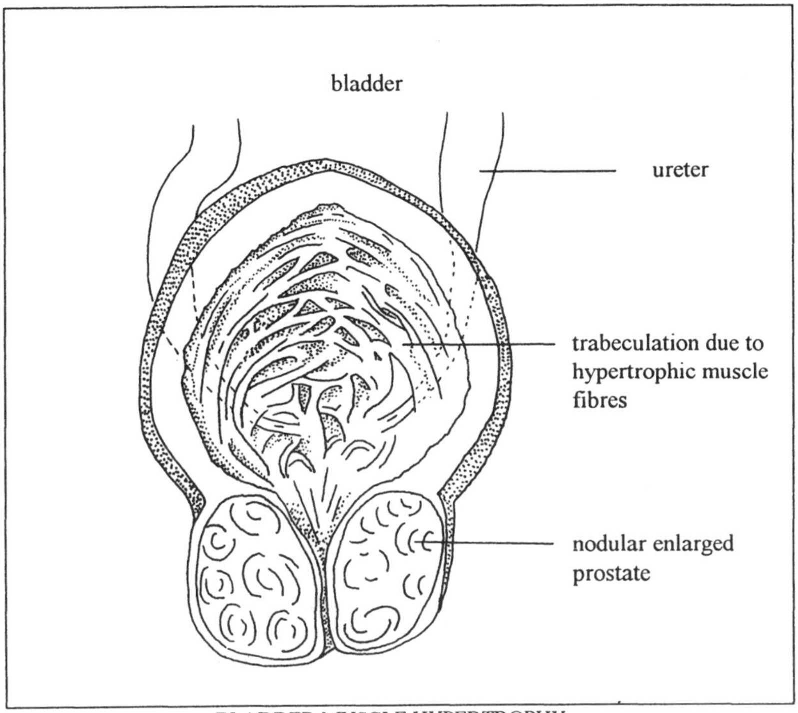

BLADDER MUSCLE HYPERTROPHY

MECHANISMS OF HYPERPLASIA AND HYPERTROPHY

In hypertrophy, for example in muscle, there is an increase in the number of organelles, (including mitochondria and endoplasmic reticulum) and increased synthesis of proteins and filaments to allow the cell to cope with the increased demand.

In hyperplasia polypeptide growth factors and hormones are important. Transforming growth factors (TGFα and TGFβ) have been implicated. The latter may have an inhibitory or switch off effect. There may also be an increase in proto-oncogene expression. An increase in nuclear DNA is necessary for cell division.

In non-dividing cells such as cardiac muscle, although there is only hypertrophy, there is an increase in nuclear DNA content; the growth cycle seems to be interrupted at this stage because of the inability of these cells to undergo mitosis.

ABNORMALITIES OF CELLULAR DIFFERENTIATION

METAPLASIA

Metaplasia is a change from one type of differentiated cell to another type of differentiated cell which is usually less specialised. It is a reversible change.

Examples of metaplasia include:

- **Squamous metaplasia**: specialised and more sensitive epithelia will undergo metaplastic change into a more resistant epithelium, such as squamous epithelium, when subjected to chronic irritation or trauma.
  For example:
  - The pseudostratified ciliated columnar epithelium of trachea and bronchi may be replaced by squamous epithelium in heavy cigarette smokers.
  - The columnar epithelium of the uterine cervix may be replaced by metaplastic squamous epithelium in chronic inflammation or infection.
  - The transitional cell epithelium of the renal pelvis, ureter or bladder may undergo squamous metaplasia in chronic infection, especially if stones are present.
- **Columnar metaplasia:**
  - Pseudostratified ciliated columnar respiratory epithelium may also change to simpler mucus-secreting columnar epithelium as a response to mild irritation.
  - Specialised acid secreting epithelium around gastric ulcers may change to a simpler, pyloric type of mucus-secreting epithelium.
- **Connective tissue metaplasia:** Fibrous tissue can undergo metaplasia to cartilage or bone; fibroblasts seem to retain the ability to differentiate in differing directions, and osseous metaplasia may not uncommonly be found in scars and areas of dystrophic calcification.

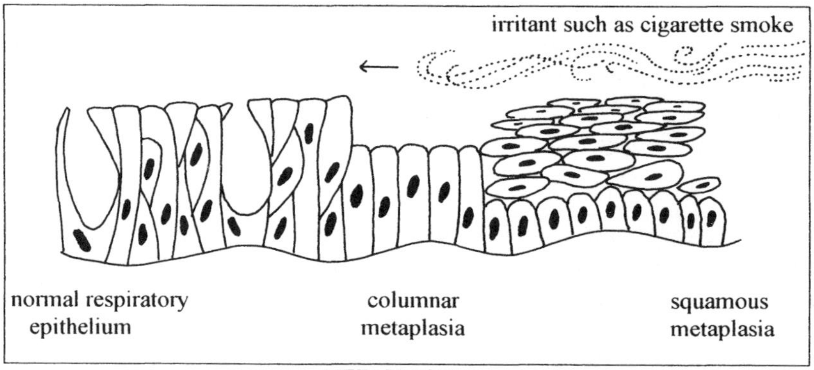

METAPLASIA

DYSPLASIA

Dysplasia is a term used to describe abnormalities in the number, size, shape and orientation of epithelial cells which fall short of true malignant change.

Dysplasia is sometimes referred to as atypical hyperplasia or as intraepithelial neoplasia. It may occur in response to chronic inflammation or to viral infection (in the cervix of the uterus infection with human papilloma virus (HPV) of types 16 and 18 is almost invariably associated with the histological appearances of dysplasia).

In the skin dysplastic changes known as solar keratoses may be found in patients with less pigmented skins who have been exposed to ultra-violet irradiation.

Some cells within an epithelial tissue show nuclear changes similar to those seen in malignant cells, but cytoplasmic differentiation may be better preserved. There may be focal loss of orientation of the cells within the tissue (loss of polarity).

Dysplasia is thought to be a reversible change, but may be a precursor of irreversible true malignancy.

Dysplastic change which involves the full thickness of an epithelium is known as carcinoma-in-situ and has progressed further along the pathway to malignant transformation.

NEOPLASIA AND ANAPLASIA

Neoplasia literally means new growth or tumour. New growths may be benign or malignant. Malignant neoplasms are known as cancer.

No more will be said about neoplasia at this stage although it will be covered in detail in subsequent chapters.

Anaplasia means lack of differentiation and is a term sometimes applied to the cells of malignant neoplasms which lack differentiation to such an extent that their tissue of origin is in doubt.

The cells in a malignant tumour are undifferentiated because they have developed from a stem cell or a reserve cell in the tissues and failed to mature. It does not seem likely that previously differentiated cells lose their specialised characteristics as they undergo malignant proliferation.

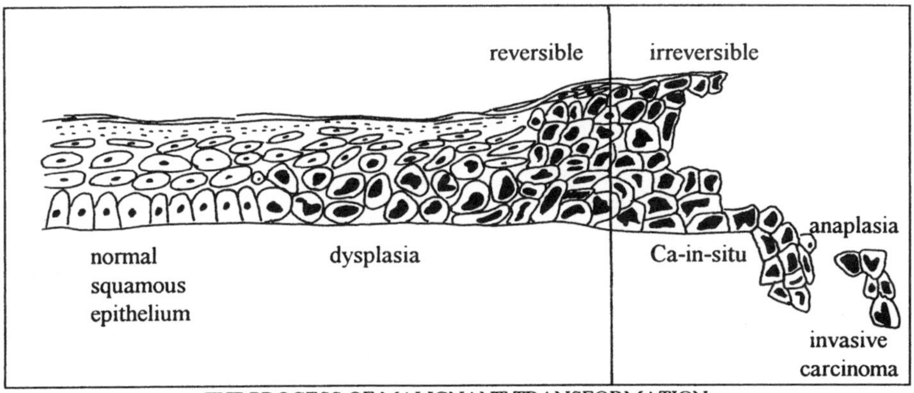

THE PROCESS OF MALIGNANT TRANSFORMATION

GENERAL PATHOLOGY - CHAPTER TWENTY FOUR

TUMOURS AND THEIR CLASSIFICATION

The word tumour literally means swelling and that is all, for example, an abscess is a tumour, although tumour is most often used synonymously with new growth or neoplasm. There are many definitions, all of which are unsatisfactory in some respect or another.

- A **neoplasm** is a semi-autonomous parasite. It is not completely autonomous as it needs, for example, a blood supply. This definition does not distinguish between a foetus and a cancer.
- A **tumour** proper is a mass of cells, tissues or organs resembling those normally present, but arranged atypically. It grows at the expense of the organism without at the same time subserving any useful function. (Powell White).
- A **tumour** is an abnormal mass of tissue, the growth of which exceeds and is unco-ordinated with that of normal tissues, and persists in the same excessive manner after cessation of the stimuli which evoked the change (Willis).

Both the latter two definitions are helpful and include features common to most neoplasms, particularly the lack of responsiveness to normal growth controls.

However, there are such widely differing sorts of neoplasm that it is difficult to find one completely acceptable overall definition and as a result it is difficult to classify tumours.

THE CLASSIFICATION OF TUMOURS

There are five possible methods of classifying tumours:

- **Gross macroscopic and anatomical classification**, for example, at the autopsy level.

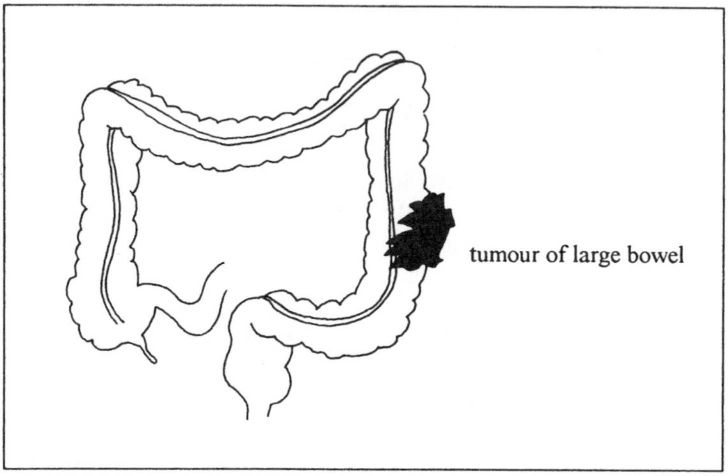

CLASSIFICATION OF TUMOURS: SITE OF ORIGIN

This classification would include both the **site of origin**, for example, tumour of the large bowel, and the **appearance,** for example, annular (growing in a ring), fungating (cauliflower-like growth), scirrhous (hard, for example, fibrous connective tissue stroma), encephaloid, (soft and brain-like, with little fibrous stroma) and mucoid or colloid.

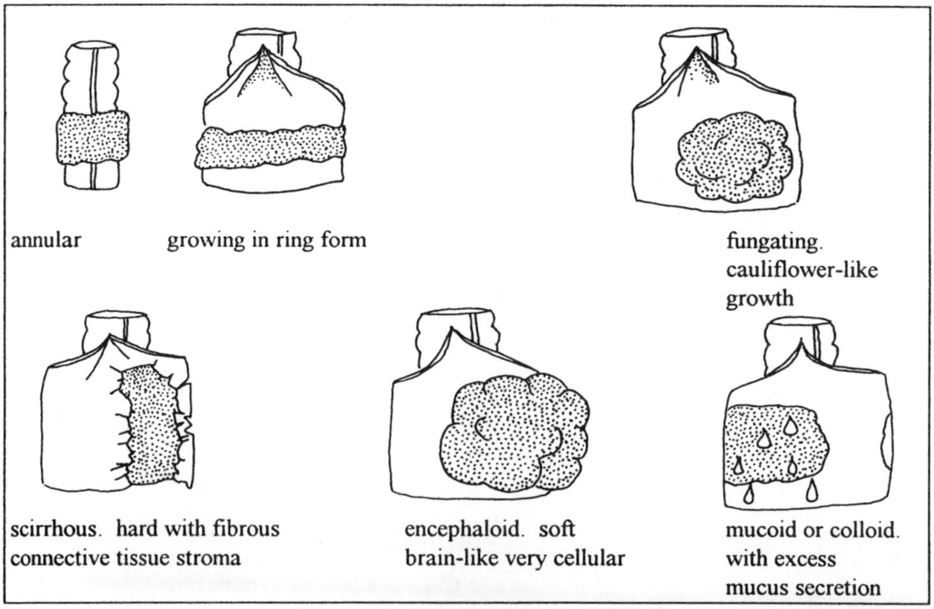

CLASSIFICATION OF TUMOURS: APPEARANCE

- **Histogenetic** classification according to tissue of origin, for example, the tumour derives from epithelium (covering or glandular cells) or from connective tissue.

  This classification is sometimes used in a simple form, but difficulties in ascribing tumours to their tissue or origin are not infrequent, for example, in a very anaplastic growth. There is also confusion as to the classification of the tissue of origin itself, for example, endothelium or mesothelium, brain or lymphoid system.
- **Histological differentiation**: Tumours may be graded as to their degree of differentiation histologically taking into account:

  - Structural and cytological resemblance to the tissue of origin, such as gland formation in tumours of glandular epithelium or keratinisation in squamous cell growths.
  - Degree of variation in the size and shape of the cells and their nuclei, that is, their pleomorphism.
  - The number of mitoses present.

HISTOGENETIC TISSUE OF ORIGIN

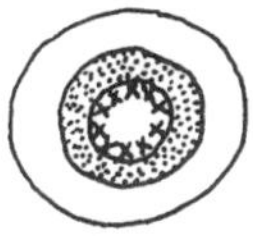

e.g. intestinal epithelium of endodermal origin

HISTOLOGICAL DIFFERENTIATION (resemblance to normal tissue)

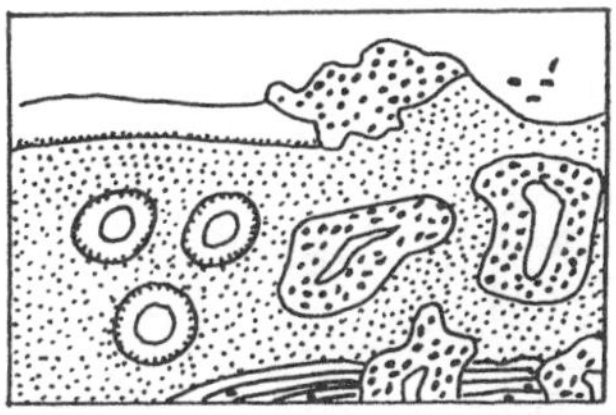

e.g. moderately differentiated adenocarcinoma (derived from intestinal epithelium)

CLASSIFICATION OF TUMOURS: HISTOGENETIC AND HISTOLOGICAL

- **Behavioural** classification - that is benign or malignant. The name implies the likely behaviour of the tumour.

  This type of classification is the one that forms the underlying basis of surgical pathology.

  We shall consider more closely the exact implications of benign and malignant growths and their features later on. Unfortunately, even this distribution is not cut and dried and there are tumours such as basal cell carcinomas of skin which invade locally, but do not metastasise or spread to a distance.

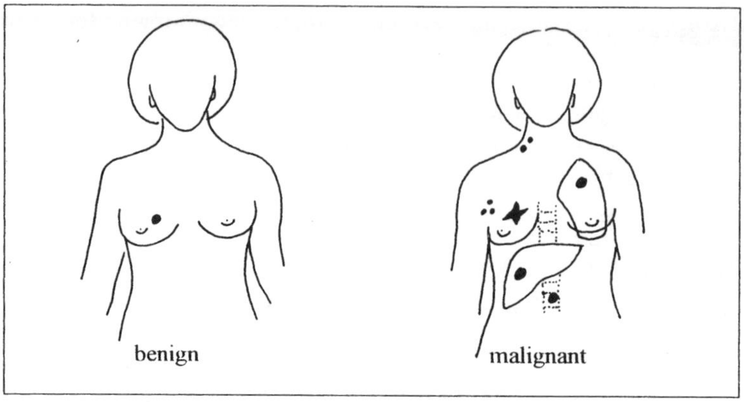

CLASSIFICATION OF TUMOURS: BEHAVIOUR

- **Aetiological** classification: As we do not know the cause of most tumours, and when we do know the cause, it is not related specifically to the type of tumour produced, an aetiological classification is of little use. For example:
  - X-rays or chemical carcinogens can produce similar tumours
  - X-rays alone can produce many different tumours, for example, carcinoma of the skin, osteosarcoma and leukaemia.

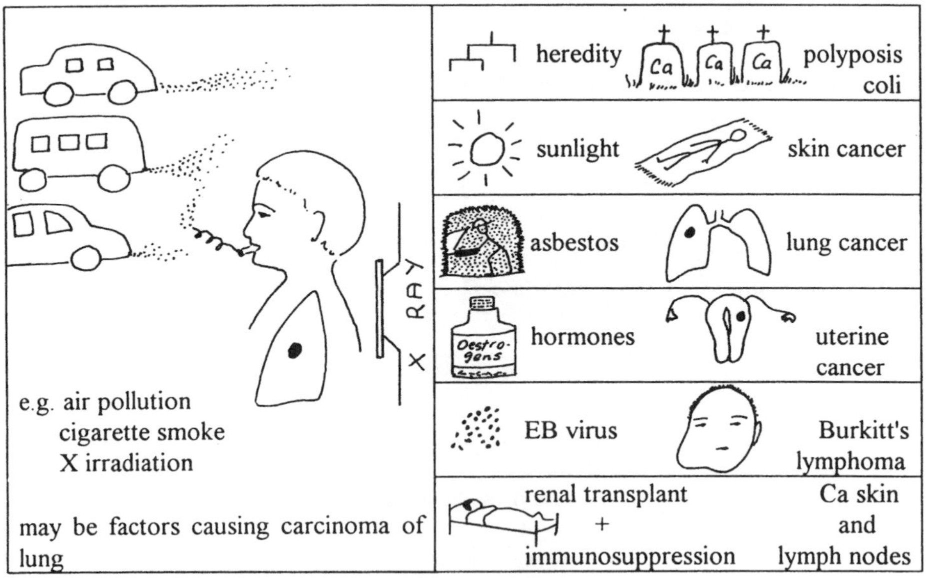

CLASSIFICATION OF TUMOURS: AETIOLOGY

NON-NEOPLASTIC TUMOUR-LIKE CONDITIONS

Neoplastic tumours may be confused with inflammation, hyperplasia or malformation but,

- Tumours differ from inflammatory and reparative proliferations in that the latter fulfil a useful function and are limited in extent according to the stimulus that brought them about.
- Tumours differ from hyperplasia which is a term restricted to the proliferation of cells of a tissue:
  - as a compensatory response to either loss of tissue or increased functional demand
  - as a result of disturbed hormonal control of the activity of the tissue.
- A **hamartoma** is a congenital malformation of tissue normally to be found at a particular site. Tissue arrangement is haphazard, usually with an excess of one or more components. Growth is co-ordinated with that of the patient.

  Such malformations (according to Willis) are disturbances of development or anatomical errors arising in the course of the development of the body to its adult state. The growth of a malformation differs fundamentally from that of a tumour in that it is limited and co-ordinated with that of the rest of the body. However, teratomas and the true embryonic neoplasms are malformations in which the disturbance of development involves neoplastic change, often of a malignant nature.

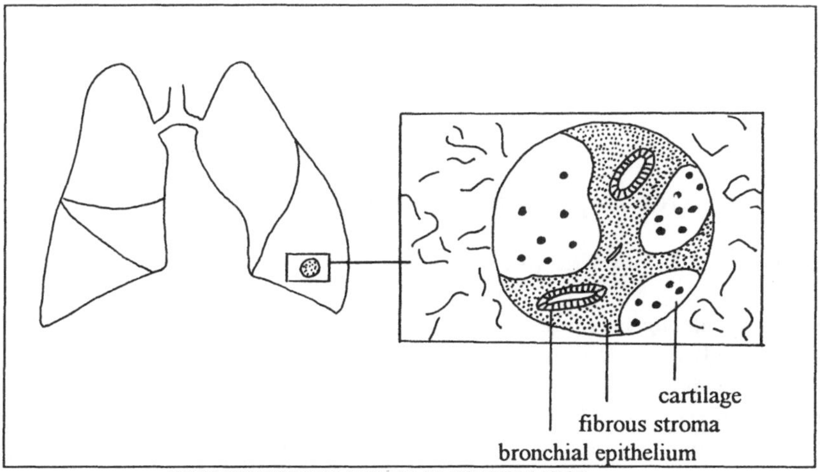

CHONDROID HAMARTOMA OF THE LUNG

- A **choristoma** is an ectopic or heterotopic, that is an abnormally placed, rest of normal cells, for example, an adrenal cell rest in the kidney, or pancreatic tissue in wall of stomach. A rest of cells is a group of cells that are misplaced in embryonic development and are left over to persist into adult life.

A WORKING CLASSIFICATION OF TUMOURS

The following classification based on a mixture of the previous classifications is widely used. Tumours are classified according to the tissue of origin as far as possible, epithelial, connective tissue, haemopoietic, neural and a miscellaneous group and then into benign and malignant within these groups.

TUMOURS OF THE EPITHELIAL TISSUES

If benign, these are **papillomas** or **adenomas** (or glandular epithelium), if malignant, they are **carcinomas**.

Note that names such as 'adenocarcinoma', 'squamous-cell carcinoma' and 'carcinoma simplex' do not denote different species, but merely variations of structure with respect to kind and degree of differentiation.

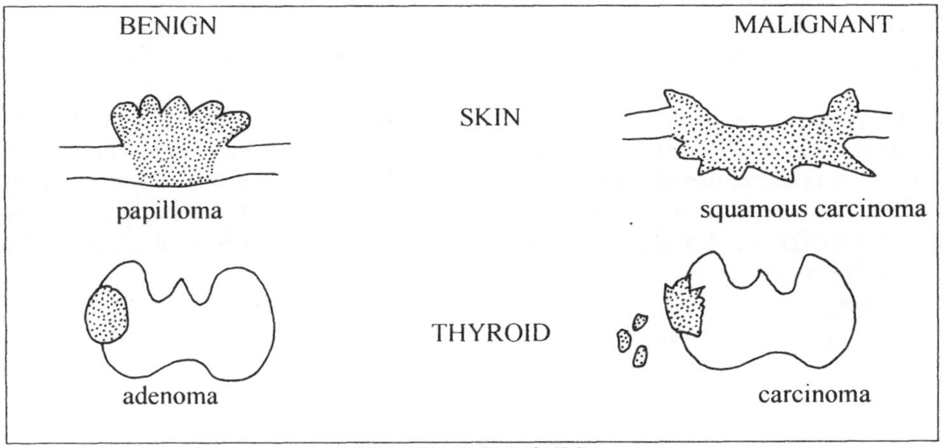

EPITHELIAL TUMOURS

TUMOURS OF NON-HAEMOPOIETIC MESENCHYMAL TISSUES

Benign tumours of connective and skeletal tissue include:

- Adult fibrous tissue - Fibroma
- Embryonic fibrous tissue - Myxoma
- Cartilage - Chondroma
- Bone - Osteoma or benign osteoclastoma
- Synovial membrane - Synovioma
- Fat - Lipoma

Benign tumours of muscle are:

- Smooth muscle - Leiomyoma
- Striated muscle - Rhabdomyoma

Malignant members of the same classes receive the generic name '**sarcoma**' and they are often prefixed according to their benign counterpart. Thus a malignant tumour of fibrous tissue is fibrosarcoma and a malignant tumour of fat tissue is a liposarcoma.

Benign tumours of vascular tissue are '**angiomas**' (such as lymphangioma or haemangioma).

Benign tumours of the meninges are termed **meningiomas**.

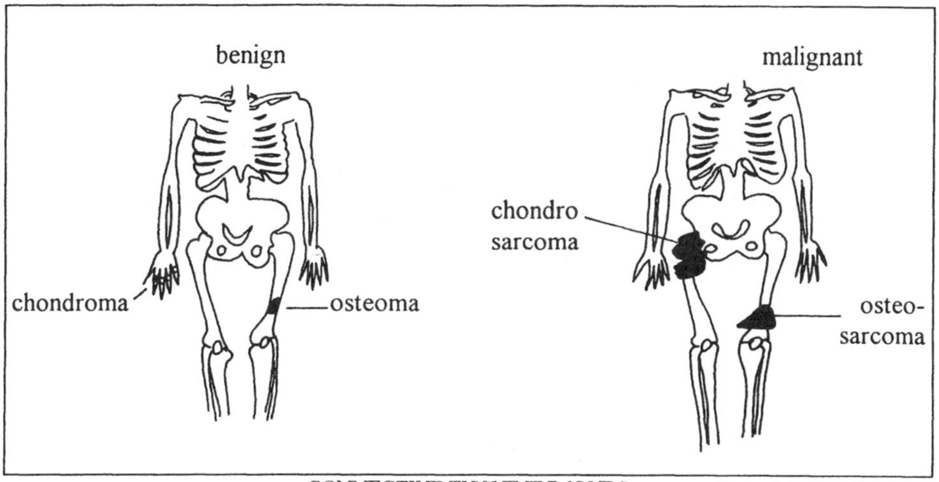

CONNECTIVE TISSUE TUMOURS

TUMOURS OF HAEMOPOIETIC TISSUE (which are nearly all malignant)

- Malignant tumours of lymphoid tissue include Hodgkin's disease and the non-Hodgkin's lymphomas.

  The non-Hodgkin's lymphomas are subdivided into follicular and diffuse forms, and further classified according to cell size and type.
- Leukaemias and myeloproliferative disorders
- Plasma cell dyscrasias and related disorders, for example, multiple myeloma
- Histiocytoses.

TUMOURS OF NEURAL TISSUES

These include gliomas, papillary tumours of the choroid plexus, neurilemmoma, neuroblastomas and ganglioneuroma, chromaffinoma, tumours of the carotid body and allied structures and retinal and ciliary tumours.

SUNDRY SPECIAL CLASSES OF TUMOURS

- Melanoma; chordoma;
- Embryonic visceral tumours, for example nephroblastoma or hepatoblastoma.
- Teratomas; chorion-epithelioma.

There are certain tumours in which the histogenetic classification is difficult as they appear to contain a mixture of tissues. These mixed tumours mostly fall into the group of sundry and special tumours.

Most tumours are 'simple' in that they are composed of one cell type which is neoplastic, plus a non-neoplastic connective tissue stroma.

MIXED TUMOURS

Some tumours, such as fibroadenoma of the breast and pleomorphic adenoma of the salivary gland, are composed of two cell types (believed to take origin from different embryonic germ layers). They are **mixed**. In these tumours there is neoplastic proliferation of both fibrous connective tissue and glandular and ductal epithelium, with the connective tissue most probably being derived from myoepithelial cell proliferation.

A malignant tumour of the uterus may show both carcinomatous elements, derived from endometrial glands and sarcomatous elements, apparently derived from endometrial stroma and sometimes other connective tissue elements, including striated muscle. It is called carcinosarcoma or mixed mesodermal tumour. Presumably the tumour arises from a multipotential stem cell which can differentiate along a number of different lines.

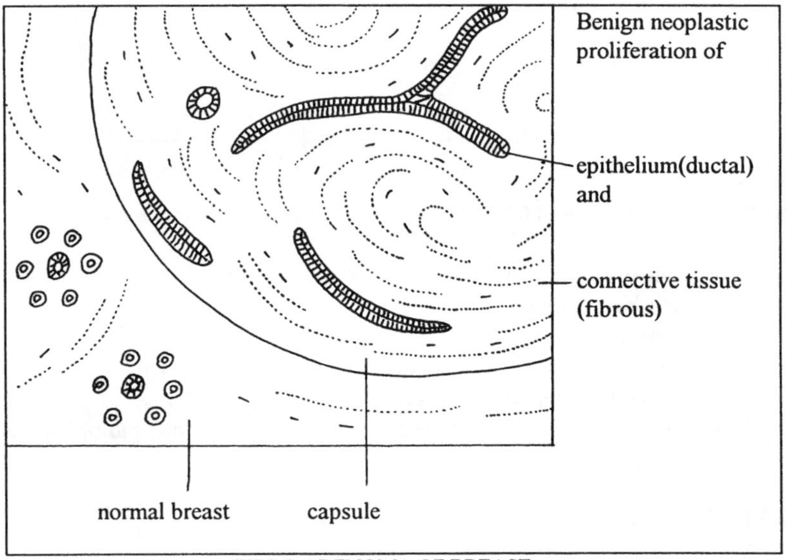

FIBRO ADENOMA OF BREAST

Some embryonic tumours, such as nephroblastoma, can also contain other elements, for example, muscle and cartilage, in addition to the primitive embryonic cells of the primitive renal blastoma. These are of mesodermal origin. The nephroblastoma is thought to be an example of a mixed tumour derived from pluri-potential cells.

The **teratomas** are tumours demonstrating a variety of tissues, and usually containing representatives of the three primitive layers in situations where these tissues do not normally occur.

They are thought to arise from totipotential cells capable of differentiating into any of the tissues of the body.

Teratomas occur in the gonads, (the ovaries and testes) and less commonly in the midline of the body, retroperitoneally in the mediastinum or in the sacro coccygeal region.

It is helpful, though perhaps not strictly accurate to think of the teratomas of the gonads as arising from attempted parthenogenesis, that is, one germ cell trying to go it alone and develop into a complete organism.

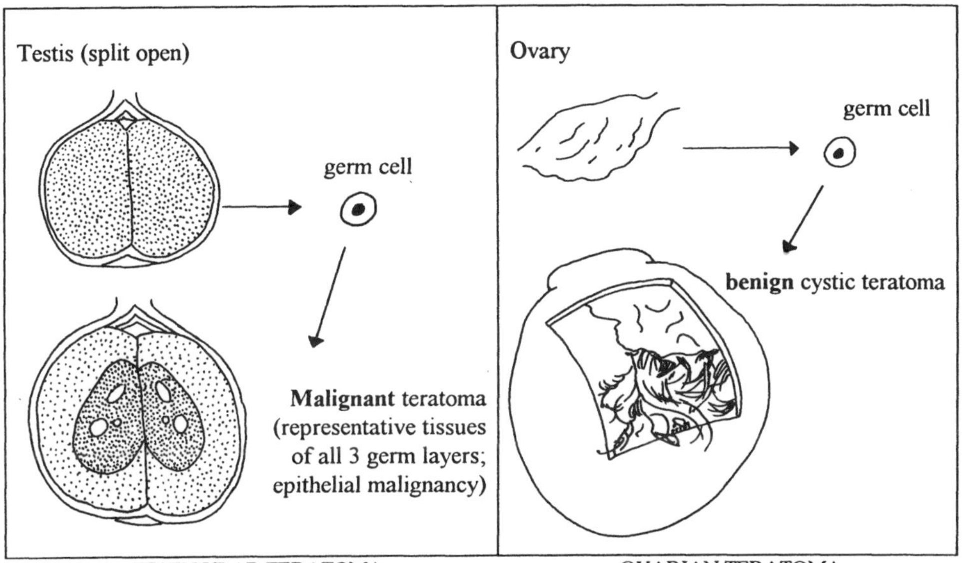

TESTICULAR TERATOMA
(MOST COMMON TESTICULAR TUMOUR
IN YOUNG MEN)

OVARIAN TERATOMA
(MOST COMMON OVARIAN TUMOUR
IN GIRLS AND YOUNG WOMEN)

Teratomas may be well differentiated and benign, for example, the benign cystic teratoma of the ovary shows mainly ectodermal differentiation and consists of a cyst filled with keratinous and sebaceous material, often containing hair and lined by squamous epithelium. Teeth may be found in the wall, as may representatives of other germ layers, for example, brain, muscle, cartilage and gut.

Teratomas of the testis on the other hand are usually malignant. Generally an epithelial component shows the malignant change, whilst other tissues such as muscle and cartilage have a more benign appearance.

Teratomas occurring elsewhere are thought to arise from sequestration of totipotential embryonic cells. Thus, the sacro coccygeal teratoma of infants is sometimes not unlike an abortive Siamese twin.

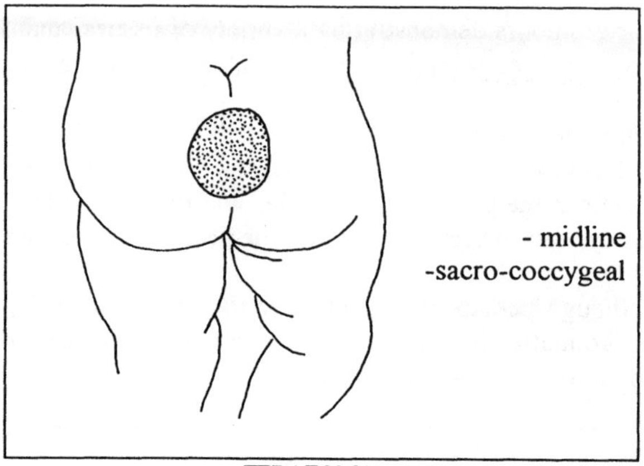

TERATOMA

GENERAL PATHOLOGY - CHAPTER TWENTY FIVE

THE CHARACTERISTICS OF MALIGNANCY

Tumours can be classified, as already mentioned, as benign or malignant. From the point of view of treatment and outlook (or prognosis) this distinction is very important.

This chapter deals with the various ways of distinguishing between benign and malignant tumours and the implications of these diagnoses.

A start can be made at a clinical and anatomical level, mentioning some histological features too, and then proceeding to look under the high power lens at the characteristics of individual malignant cells.

Clinically most tumours or neoplasms present as lumps.

- Benign tumours grow by expansion, compressing the tissues around them and forming a fibrous capsule. They usually feel round, smooth and regular. Benign tumours tend to grow slowly, do not recur after removal and rarely cause severe problems. However, they may cause pressure effects, and tumours of endocrine glands may secrete hormones. Occasionally benign tumours may undergo malignant transformation.
- Malignant tumours grow by local invasion and infiltration with destruction of surrounding tissues. They have no capsule. This method of growth means that the malignant tumour is fixed to surrounding tissues, feels rough and irregular and if it is close to a surface that surface may be ulcerated.

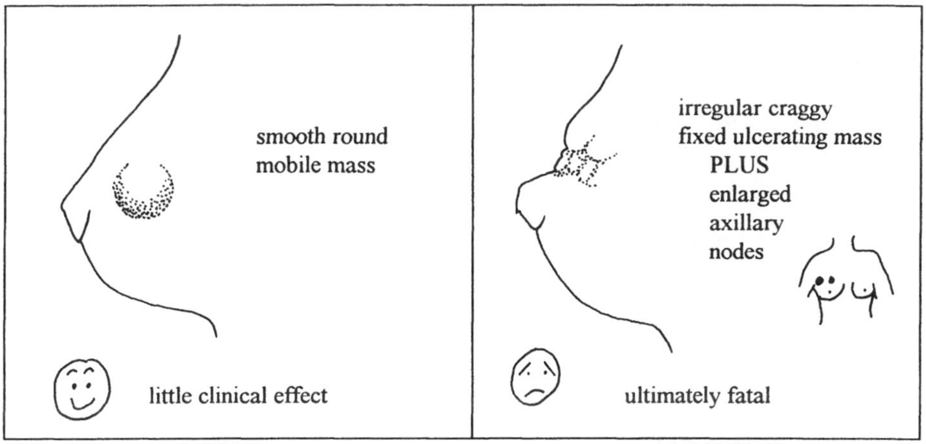

BENIGN MALIGNANT

Malignant tumours grow rapidly, and are difficult to remove surgically in their entirety (because of their infiltrative, tentacular mode of growth). The chief definitive characteristic of malignant tumours is their tendency to spread locally, and by various routes to distant sites. Spread of tumours to distant sites is called **metastasis.**

Histologically benign tumours are well differentiated (that is closely resembling their tissues of origin), localised and encapsulated.

Malignant tumours are poorly differentiated and can be seen to invade and infiltrate the surrounding tissues in tentacular fashion with destruction. Invasion of vessels may also be seen. Rapidly growing tumours show frequent mitoses (which may be atypical).

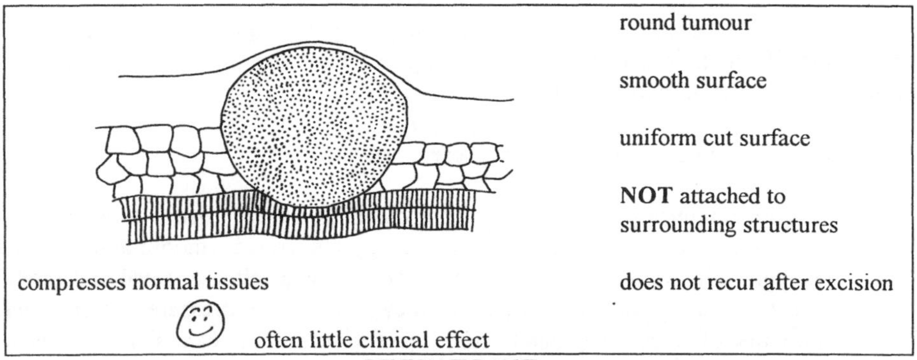

BENIGN TUMOUR

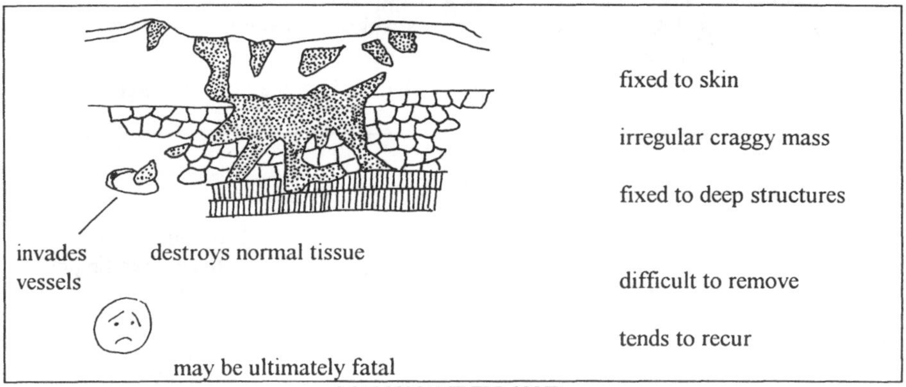

MALIGNANT TUMOUR

Malignant tumour cells show a loss of normal arrangement or a loss of polarity as it is called. The cells are not arranged in rows with the nuclei parallel to each other or perpendicular to the basement membrane, and may present a jumbled-up appearance. This is one way in which their differentiation or resemblance to normal tissue can be lost.

THE CYTOLOGICAL CHARACTERICS OF MALIGNANCY

Many of the characteristic features of malignant tumours occur as a result of the mode of growth of malignant cells. Benign cells multiply regularly and slowly in an even fashion, forming a rounded mass. The growing edge pushes the surrounding tissues away and induces the formation of a fibrous tissue capsule. Malignant cells multiply rapidly and irregularly and spikes and cords of cells are formed which can infiltrate the surrounding tissues.

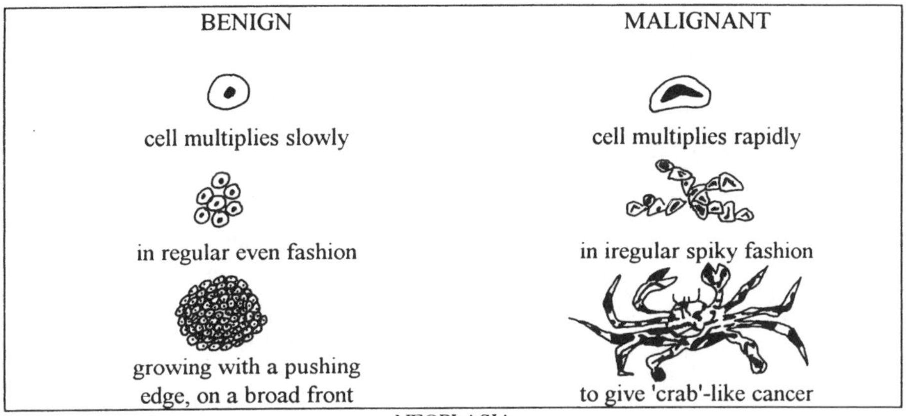

NEOPLASIA

Differentiation, that is resemblance to the tissue of origin, is also applicable at cellular level where it is determined by the resemblance of the tumour cells to the cells of the tissue of origin. This resemblance is assessed by cytoplasmic characteristics, for example, keratin or mucin production.

Cellular differentiation is poor in malignant tumours. Very poorly differentiated, undifferentiated or anaplastic malignant tumour cells may show no resemblance to the tissues of origin. The cytoplasm of malignant cells may show basophilia due to increased RNA content.

The cytological changes in the nuclei may best be summed up as those of **irregularity**.

- **Irregularity of size** or variation in cell size. Malignant nuclei tend to be larger than benign nuclei.
- **Irregularity of shape** or pleomorphism, that is, a variety of shapes and forms. For example, normal nuclei are round or oval and regular in outline, while malignant nuclei may be of varying shape and have irregular wrinkled nuclear membranes.
- **Staining characteristics.** Generally malignant nuclei stain more deeply basophilic, that is, they are hyperchromatic (possibly due to an increased chromatin content). The chromatin tends to show irregular clumping and intervening pale areas are present in the nucleus.
- **Nucleoli** are often prominent, large and irregular in outline in malignant cells.

- **Mitoses** and atypical mitoses are frequent. There may be as many as 20/1000 in malignancy, compared with 1/1000 in benign tumours. An approximation of the mitotic rate may be obtained by counting the number of mitoses in ten high power fields (HPF) of the microscope. Benign tumours generally have less than one mitosis per ten HPF. Malignant tumours usually have more than ten mitoses per ten HPF.
- **The nuclear cytoplasmic ratio** is increased in malignancy. In most normal epithelial cells, the maximal diameter of the nucleus is less than the distance between the nucleus and the cell wall. This may be reversed in malignant cells. The nuclear cytoplasmic ratio may be increased from 1:5 to 1:1.
- Malignant cells tend to **adhere less strongly** to adjacent cells than do normal or benign cells. Hence their tendency to metastasise is enhanced. This property of lack of cohesiveness of malignant cells is used in the diagnostic technique of exfoliative cytology.

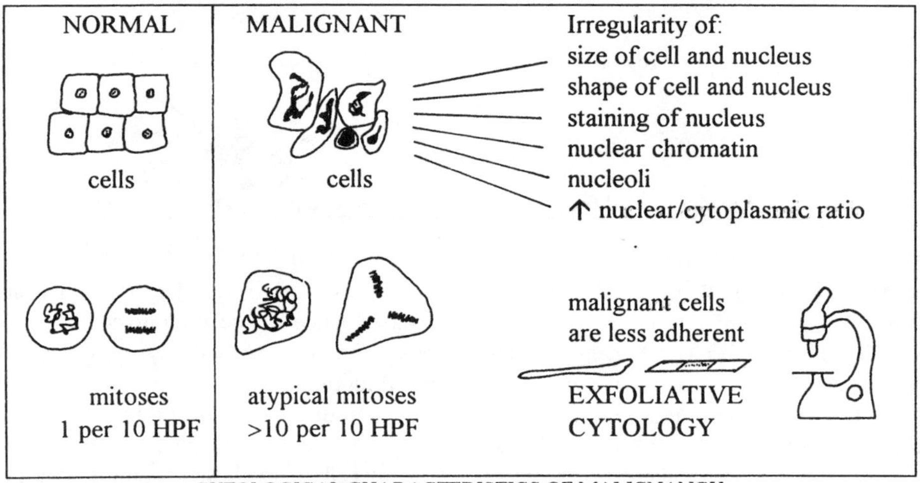

CYTOLOGICAL CHARACTERISTICS OF MALIGNANCY

DIAGNOSTIC PROCEDURES

Clinical diagnosis is made from the patient's history and by physical examination. See table at end of chapter.

Histological diagnosis requires tissue to be taken for pathological examination, that is, a biopsy. Where the tumour is small, excision biopsy is to be preferred with removal of the entire lesion. If the tumour is large, an incision biopsy may be necessary, although this theoretically poses the hazards of spread of the tumour via incised vessels or by implantation in other tissues during removal.

In some cases needle biopsy of deep organs, such as liver, kidney and bone marrow, can now be done. This may be either a core biopsy for histological examination or a fine needle aspiration biopsy for cytological examination.

Before a section suitable for microscopic examination can be cut, the tissue must be made firm. A thin section cannot be cut from fresh, soft, floppy tissue. The usual method involves fixing the tissue in formalin (so that further processing will not affect the cellular patterns), dehydrating it and embedding it in paraffin wax. A section is then cut, put on a slide and stained. This process usually takes up to twenty four hours.

Frozen section technique permits more rapid diagnosis. If tissue is frozen this fixes it temporarily, hardens the tissue and a section can then be cut, put on a slide and stained, all in ten minutes, in or near the operating theatre. This technique has many applications, for example, when there is a lump in the breast, frozen section done in theatre may reveal its nature and if malignant, further operative treatment such as mastectomy can be undertaken straight away with convenience and saving of time.

CYTOLOGICAL DIAGNOSIS

Fine needle aspiration of solid lesions is carried out by inserting a needle attached to a syringe, into a tumour and aspirating cellular material by suction. The aspirate can be smeared on a slide, fixed, stained and then examined.

Otherwise cells are either scraped from the surface of the lesion, obtained from fluids in serous cavities, or found in sputum or urine. They are examined with the microscope. Malignant cells may be recognised by their cytological characteristics as described.

Exfoliative cytology is used most widely in gynaecological practice in the diagnosis of early carcinoma of the cervix of the uterus.

It is used in two ways:

- For screening purposes to detect early unsuspected cases which can be found and treated at a stage when cure is possible.
- As a diagnostic procedure in clinically suspicious cases.

It is not absolutely reliable, and suspicious or positive results should always be confirmed by biopsy when this is possible.

Note: it should be realised that not *all* the characteristics of malignancy are present in *every* malignant tumour. Thus, experience is needed in many cases to evaluate the clinical, histological and cytological features before the diagnosis of benign or malignant can be reached. Also, as mentioned when discussing the classification of tumours, there are some intermediate tumours (for example, basal cell carcinoma of the skin and some low grade connective tissue tumours) which are locally invasive but virtually never metastasise.

BIOCHEMICAL DIAGNOSIS

Occasionally, hormone or enzyme studies can help in the diagnosis of neoplasms. Persistence of abnormally high gonadotrophin levels in blood or urine in women may reveal a placental cancer, that is, chorio-carcinoma.

In prostatic carcinoma there are high levels of prostate-specific antigen in the serum and often high levels of prostatic acid phosphatase also if the tumour is extensive.

Carcinoembryonic antigen (CEA) is normally produced by certain embryonic tissues, such as gut and pancreas. In adults it may be elaborated by carcinomas of colon, pancreas, stomach and breast and less frequently by other cancers.

Raised alpha-foetoprotein levels may be found in cancers of the liver and testis.

Increased levels of substances produced by tumours and found in the blood are somewhat variable and not always reliable for diagnosis. However they may be of use in giving some idea of the extent of the disease and can also be used to indicate response to therapy or recurrence of a tumour after treatment.

This review of the properties of tumours and the diagnostic techniques used has shown how a tumour may be classified by its features and behaviour as benign or malignant. The purpose of this is to enable us to predict its future behaviour and hence decide what treatment is most likely to be effective.

PROGNOSIS

This is the prediction of the future progress of a disease or tumour - the future outlook.

In benign tumours the prognosis is good. The tumour may cause little trouble if it remains with the patient and is unlikely to recur after removal.

In malignant tumours all of the features we have mentioned must be considered.

- **Site.** Carcinomas of the viscera, for example, stomach or caecum, tend to cause symptoms only at a late stage when they are extensive. Hence they are generally diagnosed late and have a bad prognosis.
  Also one sort of tumour, for example squamous carcinoma, will grow slowly and metastasise late in the skin, whereas in the pharynx or lung, growth is more rapid and metastatic spread occurs earlier.
- **Nature.** Malignant melanoma, for example, is a rapidly spreading, highly fatal tumour even though it often occurs on the skin and can be diagnosed early.

Prognosis according to the site and nature of the tumour can only be assessed from past experience with such tumours

The overall prognosis for any tumour of a particular nature at any site also depends upon the rapidity of growth, the clinical staging of the growth and the histological grading (or degree of differentiation).

- **Growth**. A rapidly growing tumour is likely to behave worse than a more slowly growing one.
- **Extent.** An extensive tumour, especially if there are metastases, is more difficult to treat and will have a worse prognosis than a small localised malignant tumour. This forms the basis of clinical staging of tumours. Staging is based upon the size of the primary tumour (T0-4), whether there is lymph node involvement (N0-3) and the presence or absence of more distant metastases (M0-4). This is the basis of the TNM classification.
  [Note: T0 indicates carcinoma-in-situ; N0 and M0 indicate absence of tumour in lymph nodes and no distant metastases; numbers 1-4 indicate increasing size of tumour or number and distance of metastases according to the site of the primary tumour.]
  An American system of staging uses similar criteria to divide tumours into stages I-IV.
- **Differentiation.** Histological grading is made according to tissue and cellular differentiation and the mitotic rate of the tumour. Three grades of tumour, well, moderately and poorly differentiated, are usually recognised.

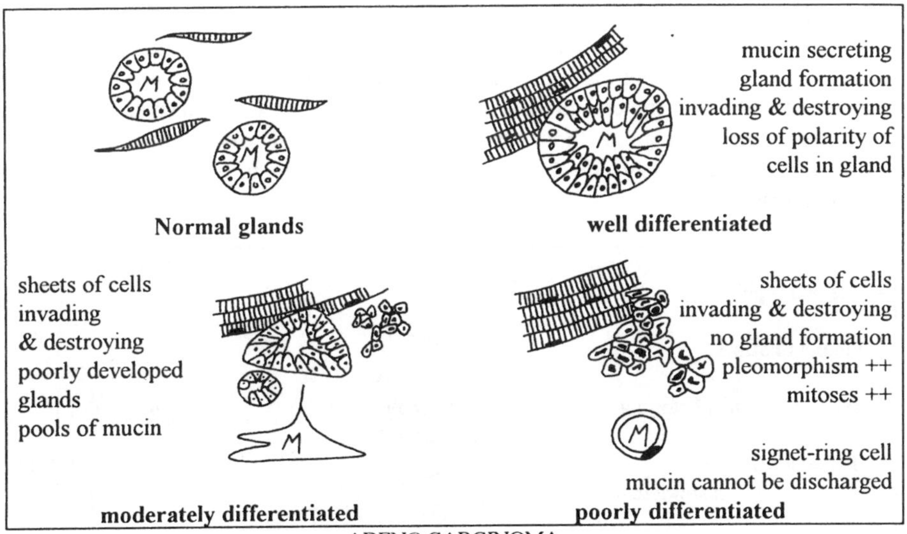

ADENO CARCINOMA

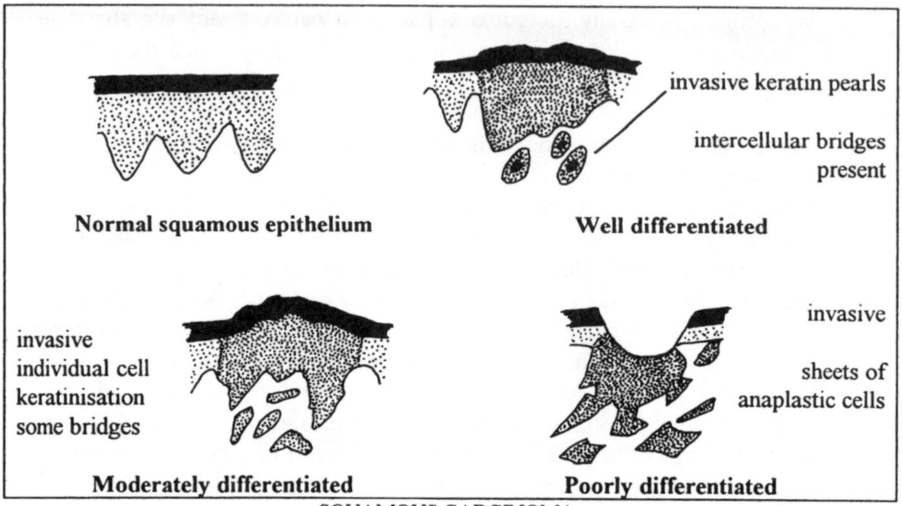

SQUAMOUS CARCINOMA

Generally speaking, poorly differentiated or anaplastic tumours which have a high mitotic rate, indicating rapid growth, behave badly.

Incidentally, those tumours that are growing rapidly and are poorly differentiated may often respond well and quickly to X-ray treatment (radiotherapy), at least in the first instance.

Statistical studies have been performed on the outlook and results of treatment for most common tumours at various stages and grades. Thus, the best treatments can be evolved and a general prognosis in terms of a five year survival rate can be made although of course there are always individual exceptions. Sometimes (unfortunately extremely rarely) a highly malignant tumour can disappear spontaneously.

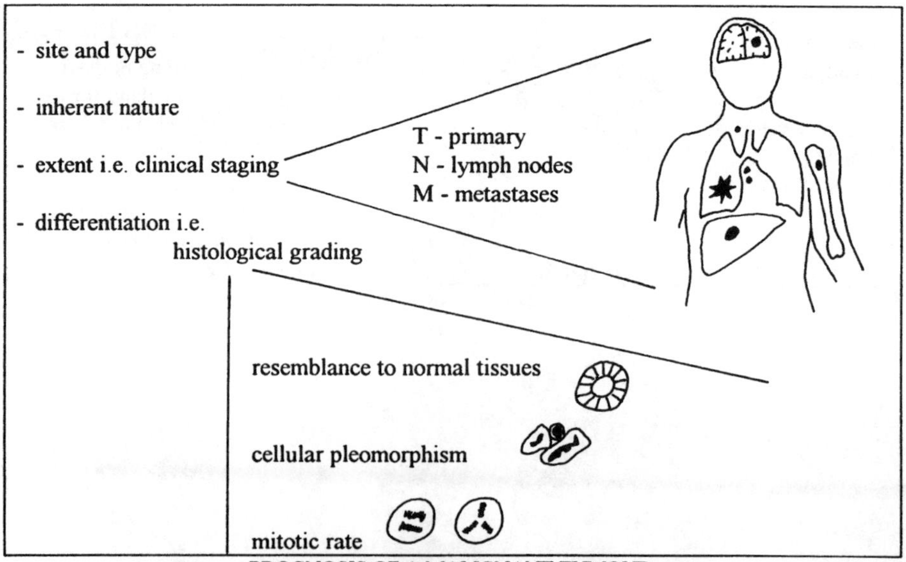

PROGNOSIS OF A MALIGNANT TUMOUR

CHARACTERISTICS OF BENIGN (B) AND MALIGNANT (M) NEOPLASMS

	Clinical/Anatomical	Histological
PRESENTATION	(B) as lump (M) as lump, ulceration common	(B) lump well differentiated - resembling tissue of origin (M) lump poorly differentiated or anaplastic with **loss of polarity** - jumbled arrangement and crowding of cells, etc.
MODE OF GROWTH	(B) grow by expansion compressing adjacent tissues. They may feel round, smooth, regular, mobile (M) grow by invasion - no capsule, feel hard, craggy, irregular, fixed	(B) fibrous capsule can be seen, the surrounding tissues are compressed. (M) tentacular growth. **Destruction** and replacement of surrounding tissue (may be surrounding inflammation)
RATE OF GROWTH	(B) grow slowly - may stop growing (M) grow rapidly and continue to grow if untreated	(B) mitoses infrequent 1/1000 cells or less (<1 per 10 HPF) (M) mistoses frequent 20/1000 cells or so and are atypical - tripolar, multipolar, etc (mitoses >10 per 10HPF). Rapid growth of tumour cells may outstrip rate at which new stromal blood vessels can form and this may cause <u>necrosis</u> of tumour
RECURRENCE	(B) rarely recur after removal (M) often recur after removal possibly because of invasive and rapid nature of growth	
METASTASIS	(B) never metastasise (M) nearly all metastasise in time	(M) tumour cells may be seen invading lympatic or blood vessels or as emboli within them **HISTOLOGICALLY SIMILAR TUMOUR AT DISTANT SITE = METASTASIS**
OUTCOME	(B) usually harmless, but may exert pressure or cause local complication. Secrete hormones May undergo malignant change (M) by virtue of invasion + metastasis will kill patient eventually	

GENERAL PATHOLOGY - CHAPTER TWENTY SIX

THE SPREAD OF MALIGNANT TUMOURS

Cancer (in a general sense meaning any malignant tumour) spreads throughout the body by much the same routes that spread infection.

LOCAL SPREAD

Local spread takes place by direct **invasion** and **infiltration** between tissues. Tumours can grow and extend considerably in this way, infiltrating, destroying and replacing the tissues around them. In fact, to some extent they choose the easiest direction of extension by following tissue planes. Bone, cartilage and fascia are more resistant to direct invasion of this sort, although they are by no means exempt from it. Body surfaces can be invaded and as we have seen, malignant neoplasms may often cause **ulceration**. Another characteristic of malignant tumours is that they may appear **fixed** either to the skin by virtue of the invasion of skin (before the stage of ulceration) or to the underlying tissues by virtue of invasion of these. An example of fixation of a tumour can be seen in carcinoma of the breast, where the tumour mass may be fixed to skin and underlying fascia or muscle by virtue of its invasiveness in contrast to benign growths of the breast which are always mobile within the breast fat.

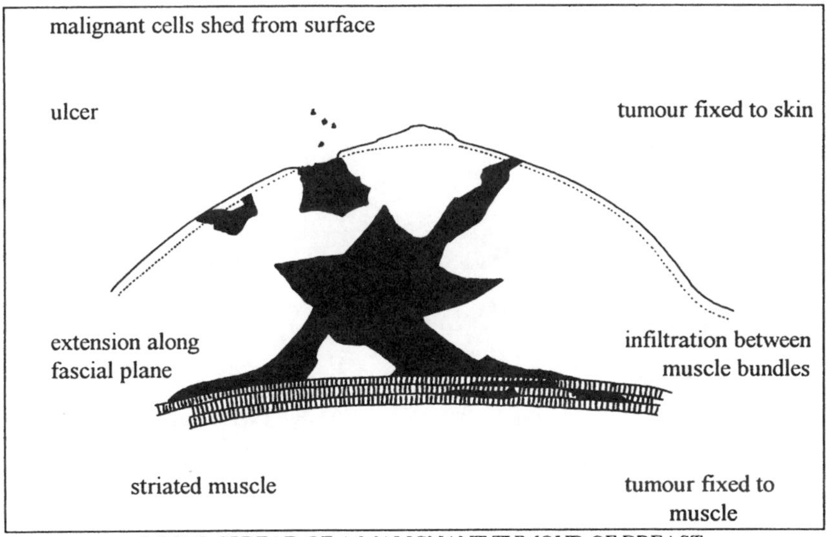

LOCAL SPREAD OF A MALIGNANT TUMOUR OF BREAST

Histological evidence of invasion is important. This is evidence of true malignancy as opposed to carcinoma-in-situ and in any circumstances histological evidence of invasion and destruction of adjacent tissues is one of the most important criteria in determining the malignant nature of a tumour.

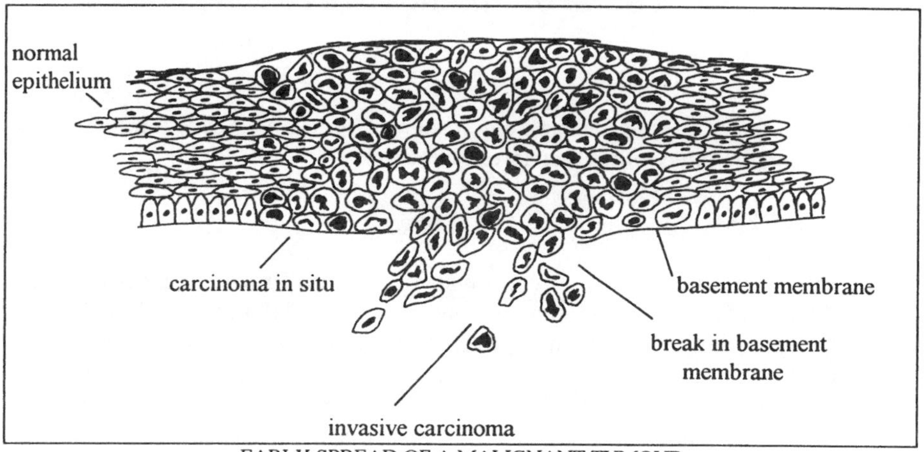

EARLY SPREAD OF A MALIGNANT TUMOUR
BY LOCAL INVASION

DISSEMINATION

Distant spread occurs when tumour cells become detached from the main tumour mass and travel through the body to a distant site where they can grow and multiply (at a distance from the primary site) forming an isolated secondary mass of tumour - a **metastasis**. Any tumour that metastasises is malignant, that is cancer; but not all cancers metastasise, although they invade locally (for example basal cell carcinoma of the skin and gliomas of the central nervous system).

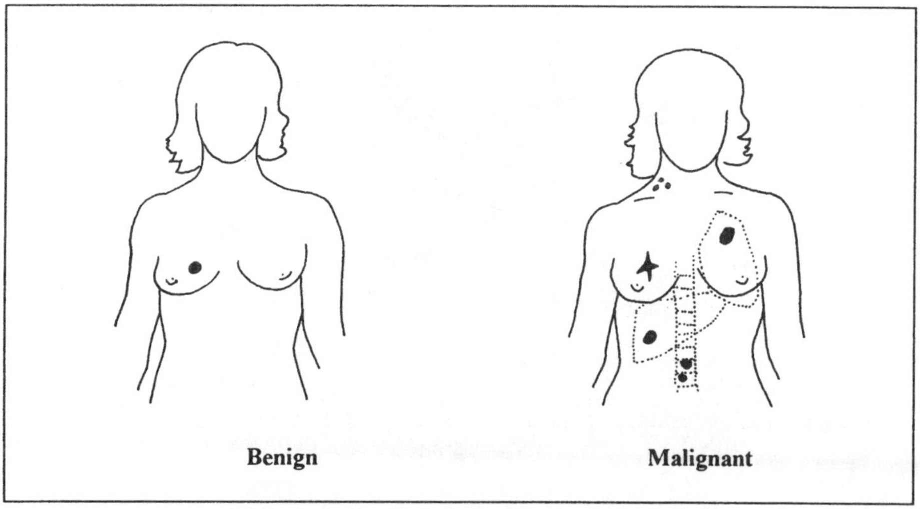

CLASSIFICATION OF TUMOURS : BEHAVIOUR

Malignant tumours are transported :-

- most often via the lymphatic vessels and
- via the blood vessels, or
- through body cavities, or
- by implantation, for example, during surgery

The original tumour is called the primary tumour while metastases are often referred to as secondary deposits.

LYMPHATIC SPREAD OF CARCINOMA

Malignancies of epithelial tissues (carcinoma) tend to invade lymphatics and spread by the lymphatics at an early stage, as opposed to sarcomas of connective tissue which spread by the blood rather than via the lymphatics.

Tumour cells gain access to lymphatics as they invade locally. Once there they may:

- Detach to become emboli or
- Form a continuous growth along the vessel - lymphatic permeation.

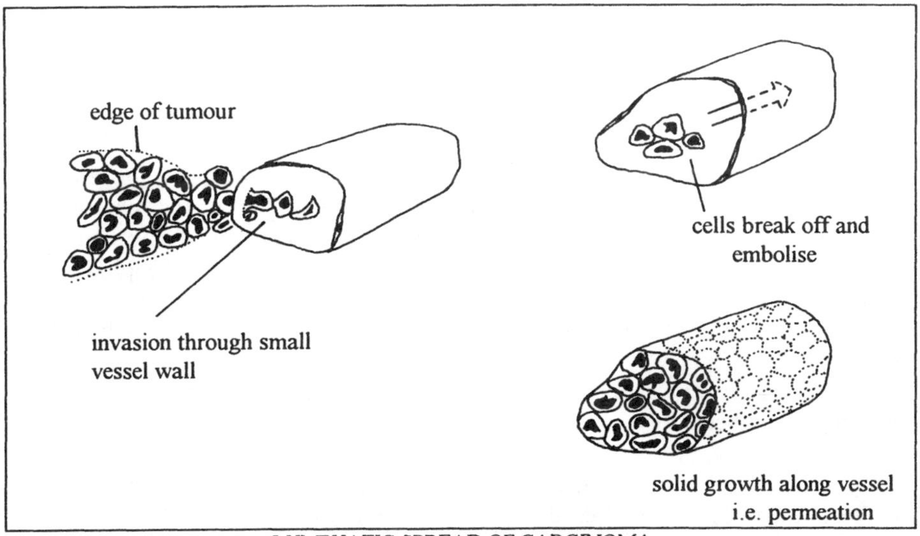

LYMPHATIC SPREAD OF CARCINOMA

Lymphatic emboli usually travel in the normal direction of lymph flow to the regional nodes. There they become lodged in the subcapsular peripheral sinus. If they survive they may grow and multiply until the whole node can be replaced and enlarged by tumour. The capsule of the lymph node usually remains intact until a late stage. Malignant cells can progress from a metastatic lymph node via the efferent lymphatics to more central nodes, and eventually malignant cells may enter the blood stream via the thoracic duct.

Less commonly retrograde embolism may occur. If the onward flowing lymphatics are blocked, emboli may travel against the normal direction of flow.

An example of this is when cervical lymph nodes on the left become enlarged in cancer of the stomach (Troisier's sign). It is thought that the thoracic duct becomes blocked by tumour at its entrance into the left subclavian vein, and thus retrograde spread can take place to the cervical nodes.

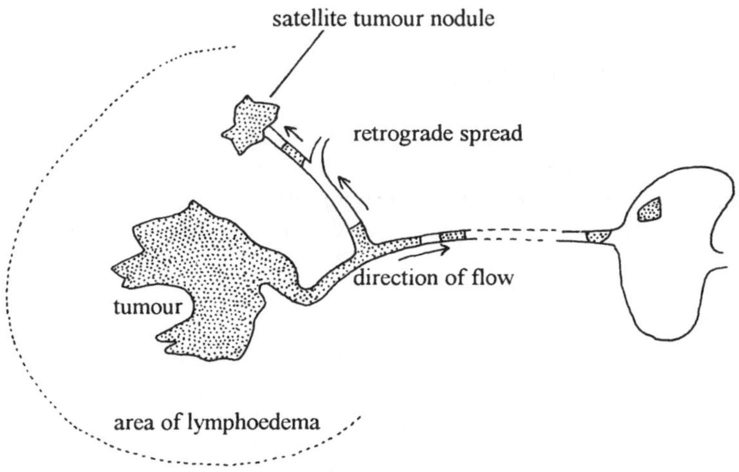

SPREAD VIA LYMPHATICS

Most carcinomas first metastasise to lymph nodes, and looking for enlarged tumorous regional nodes forms an important part of both the clinical and pathological examination of a malignant tumour. The presence of any metastasis worsens the prognosis. Obviously, a knowledge of the anatomy and lymph drainage is important.

Obviously also, once there has been metastasis, local treatment will not be able to cure the disease.

Why some carcinomas metastasise at an early stage, (for example, breast, tongue and lung) and others (such as skin) at a much later stage is not known.

Lymphatic permeation occurs when tumour cells grow and infiltrate along lymphatic channels in a solid cord. The tumour may reach the regional lymph nodes in this way, and retrograde spread may give rise to local satellite tumours. Blockage of the local lymphatics may cause local lymphoedema. This can be seen sometimes in the skin of a breast where underlying carcinoma has permeated the lymphatics widely, giving an orange-peel (*peau d'orange*) appearance.

Tumour-permeated lymphatics may be visible to the naked eye, especially on the pleural surface of the lung or in the subperitoneal lymphatics of the abdomen.

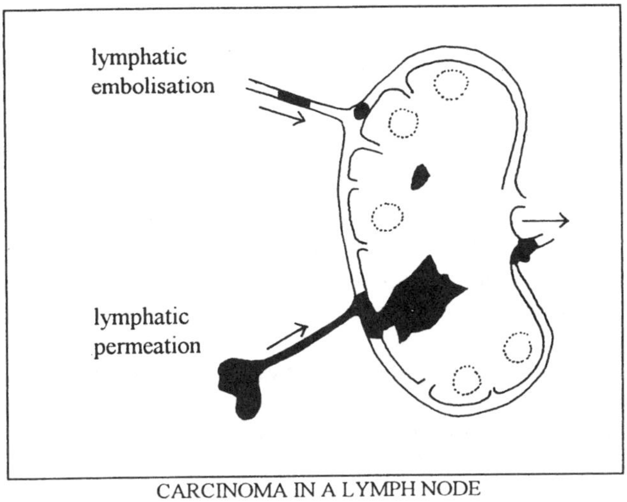

CARCINOMA IN A LYMPH NODE

Tumour cells can sometimes be seen apparently infiltrating along nerve sheaths, for example in prostate and salivary gland. Some people say this is due to the invasion and permeation of nerve sheath lymphatics, whereas others deny the very anatomical existence of such lymphatics, and call this spread via the peri-neural space.

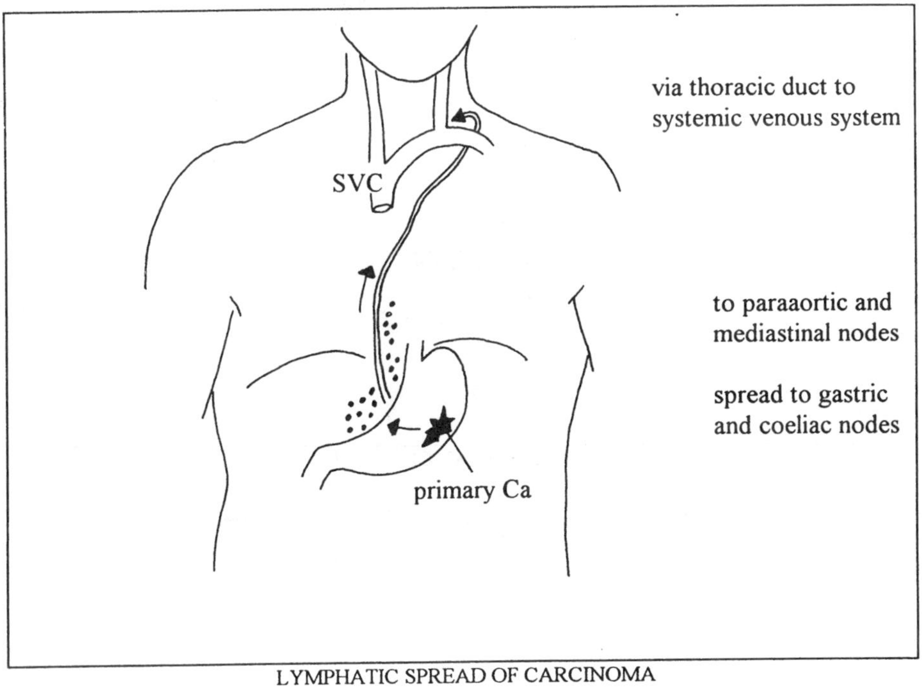

LYMPHATIC SPREAD OF CARCINOMA

HAEMATOGENOUS OR BLOOD SPREAD

This is the main method of spread in sarcomas, but usually occurs after there has been lymphatic spread in carcinomas. A malignant tumour may enter the circulation by invading the blood vessels directly or by tumour cells reaching the blood stream, as we have just seen, via the thoracic duct. Direct invasion is probably the most important, as at operation malignant cells have often been detected in the veins draining the tumour. It is noteworthy that tumours find it easier to invade the relatively thin walled veins than they do the arteries.

Arterial invasion (except perhaps in the lung) is thus not common. Common sites for blood borne metastases are :

- The lungs, from tumour which has invaded the systemic veins and traversed the right side of the heart and
- The liver from tumours which have given rise to emboli in the portal venous system and impacted in the portal hepatic capillary system.

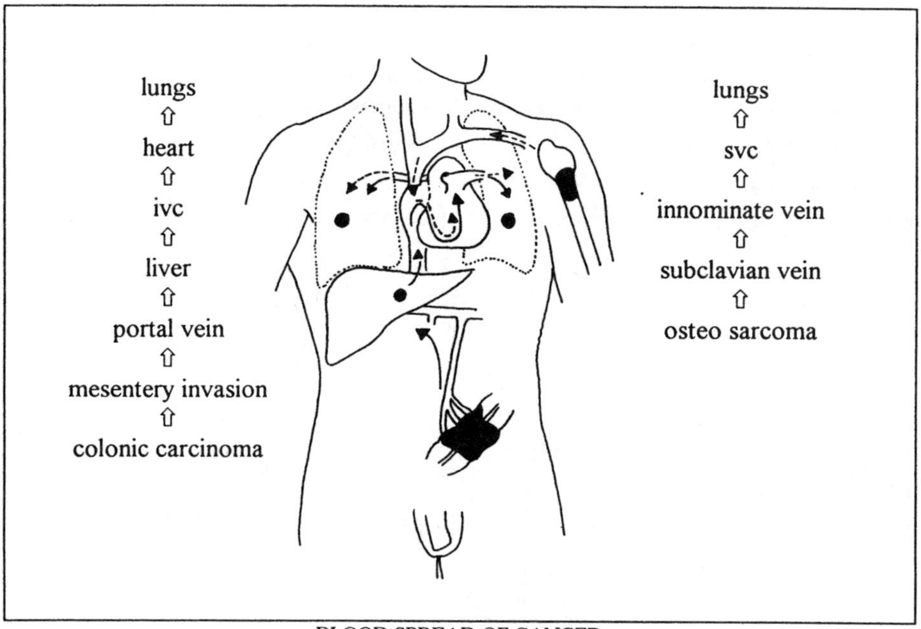

BLOOD SPREAD OF CANCER

However, systemic metastases in bones, brain, adrenals and other organs also occur.

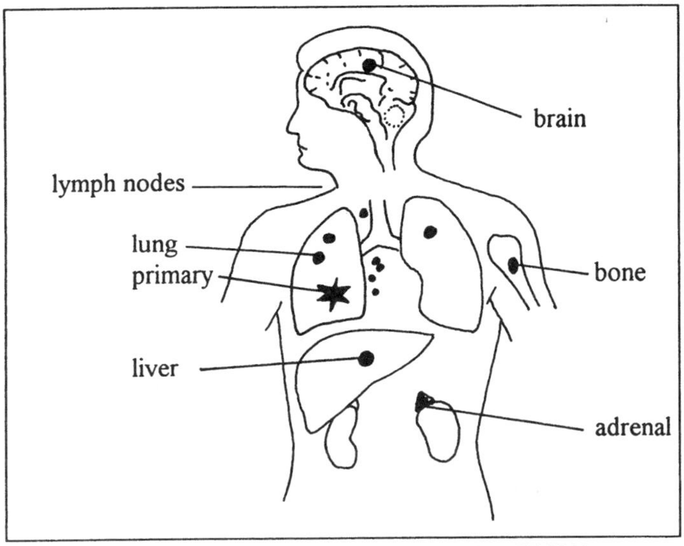

DISSEMINATED SPREAD - CARCINOMA OF LUNG

There are a number of ways in which this can happen.

- Lung secondaries invade the pulmonary veins giving rise to systemic metastases.

  However, systemic metastases may appear in the absence of obvious lung deposits, assuming that there are not in fact small lung secondaries which may have been overlooked.
- It may be that small emboli can pass through the pulmonary circulation, perhaps via a small arterio-venous shunts to reach the systemic circulation.
- There may be retrograde venous spread. Just as tumour growth can obstruct lymphatics, so can it obstruct veins, either directly by growth in the lumen or indirectly by pressing upon the wall. Emboli in obstructed veins may pass backwards against the flow to lodge elsewhere, particularly in anastomotic venous plexuses of low pressure.

  An often cited example is carcinoma of the prostate which frequently metastasises to the bones of the spine. It is thought that this may be due to retrograde venous spread via the vertebral venous plexus.

There are a number of stages necessary for metastasis to take place :

- There must be actual **invasion** and **penetration** of tumour into the vessel. This involves the tumour cells binding to proteins of the basement membrane (for example laminin) and extracellular matrix (for example fibronectin) via cell surface receptors. Tumour cells then secrete proteolytic enzymes such as type IV collagenase and other collagenases. These then destroy the extracellular matrix to allow tumour cells passage.

Cell motility is also involved in the active penetration of tumour cells through the wall of the vessel and the endothelial cells; this may be mediated by cell secretion which affects the cell itself, an autocrine motility factor.

- Within the vessels, tumour cells must survive. They are at risk from natural killer cells, from macrophages and sensitised cytotoxic T lymphocytes. Cells may aggregate to form a tumour cell embolus and there may also be adherence of platelets to these cells. Platelet adherence seems to favour metastasis formation, perhaps by virtue of growth factors (for example PDGF) derived from the platelets.
- Tumour cells must then **settle** at a new site after transport through the circulation. This may be mediated by tumour cell surface receptors or endothelial receptors.
- **Emigration** from the vessel can take place in much the same way as invasion did in the first instance, and depends on many different properties of the tumour cells.
- Tumour cells **proliferate** at the new site and a **metastasis** is established.

It is possible that all malignant tumours invade vessels at some time. There are also many instances in which cells have been detected circulating in the blood without the development of metastases and in fact with apparent subsequent cure of the patient. It may be that the likelihood of metastasis depends upon the number of malignant cells which reach the circulation.

Thus for metastasis to take place the tumour cells must have certain attributes, such as :

- The right sort of cell surface receptors
- The ability to produce proteases such as collagenases
- Motility
- Powers of replication
- Ability to secrete angiogenesis stimulating factors

It may be that none of the cells in a malignant tumour are capable of metastasis, or that only certain sub-populations of cells have the necessary powers.

THE SITE OF METASTASES

Once systemic embolisation and metastasis takes place it might be thought that those areas with the richest blood supply would be the most common sites for secondary deposits to occur. In fact this is not necessarily so. Skeletal muscles, the heart and the spleen have a rich blood supply but are rarely the sites of secondary tumours, whereas the liver and adrenal glands are quite commonly involved.

Certain carcinomas, namely those of lung, breast, prostate, thyroid and kidney appear to have a predilection for producing secondary deposits in bone.

Similarly, the liver is a favoured site for metastases not only from cancers arising in the areas drained by the portal venous system, but also from cancers of the breast and lungs. These latter two cancers have a tendency also to spread to adrenals.

Surface receptors on tumour cells and on the endothelium of vessels in certain tissues may determine the likely site of metastases, with different groups of tumour cells having the ability to 'home in' on specific organ sites.

The ability to metastasise may depend on the genetic properties of the cell. Experimental evidence suggests that the presence of the *ras* oncogene may promote metastasis perhaps by inducing the autocrine secretion of transforming growth factors (TGFs). Levels of a tumour suppressor gene, nm23, have been shown to be low in breast cancers with extensive nodal metastases.

TIMING OF METASTASES.

Another feature of clinical importance is the time at which metastases appear. About one third of newly diagnosed patients with solid tumours of internal organs may have metastases at the time of presentation. Obviously this worsens the prognosis. But even in cases in which the primary has apparently been successfully removed metastases may occur at a later date.

In some tumours the patient may be said to be cured after a certain period of time has elapsed without recurrence after treatment. For example, in children with Wilm's tumour of the kidney (nephroblastoma) three year survival without recurrence following treatment may be equated with cure of the disease. In patients who have had carcinoma of the colon, survival without evidence of recurrent or metastatic tumour for five years can generally be taken as being good evidence of a cure. But certain cancers, particularly those of breast and kidney and malignant melanoma of the skin, may present with metastases many years after apparently successful treatment of the primary tumour.

A favourite examination case in Britain at one time was a patient who presented with a large liver and a glass eye. The latter was the clue. Once it was noticed that an eye had been removed (perhaps many years previously) then the possibility of malignant melanoma of the eye with the subsequent development of liver metastases had to become the prime differential diagnosis.

It would seem that in some instances and in some cancers, tumour cells may lie dormant at distant sites for many years. A number of factors including other illnesses, reduced immunity or some hormonal disturbance may be postulated as precipitating new growth of possibly fatal metastatic disease at this later date.

SPREAD THROUGH BODY CAVITIES

Spread can certainly occur when a tumour has infiltrated, say, a serous membrane. It is not uncommon in the pleural and peritoneal cavities to find a tumour which has broken through the serous membrane, accompanied by numerous seedling tumours on the serous membranes. There is often also a haemorrhagic inflammatory malignant exudate.

Tumour seeding of the peritoneal space will also involve the ovary. Malignant brain tumours involving the meninges or the ventricles can also give rise to secondary deposits which are seeded by the cerebro-spinal fluid.

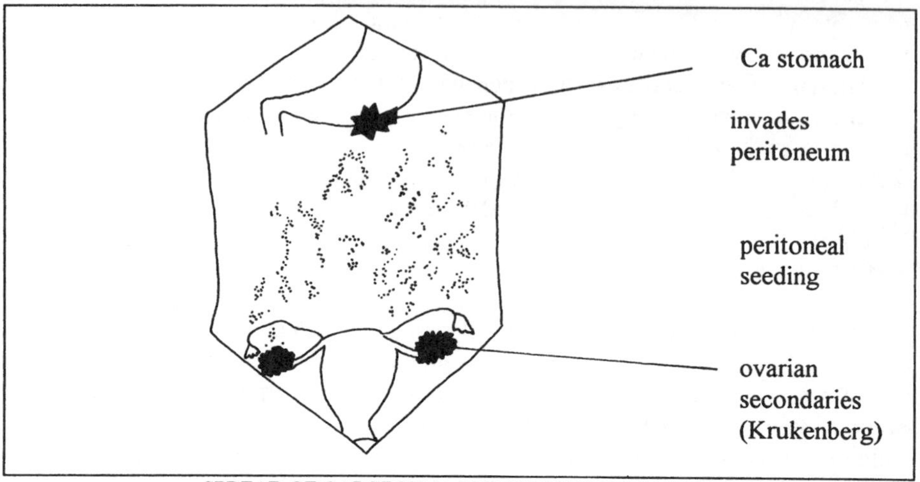

SPREAD OF CARCINOMA THROUGH BODY CAVITY

SPREAD BY IMPLANTATION

Implantation of tumour cells is a hazard of surgical removal of malignant tumours. Obviously efforts are made to avoid cutting into the tumour. Gloves, instruments and towels are changed after taking a biopsy or after removal of a tumour before proceeding to a more radical operation. Nevertheless, nodules of implanted tumour may develop in or around a scar following removal of a cancer. Breast and bladder cancers are particularly prone to implant in this way.

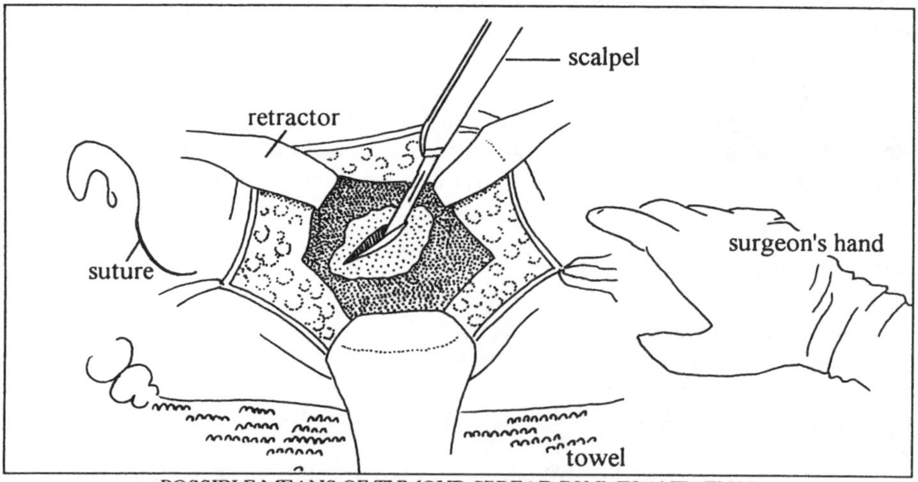

POSSIBLE MEANS OF TUMOUR SPREAD BY IMPLANTATION

Spread of tumours via natural passages is a possible but dubious mode of spread. Carcinoma of the pelvis of the kidney may be associated with similar tumours in the ureter or bladder, but this is now thought to be a multicentric malignant change (perhaps due to a carcinogenic influence in the urine) rather than tumour spread.

GENERAL PATHOLOGY - CHAPTER TWENTY SEVEN

CARCINOGENESIS

Cancer is a common cause of death in the countries of the Western world, second only to vascular disease. It is responsible for almost 25 per cent of all deaths. The table shows a breakdown of estimated cancer deaths by site and sex for 1991 for the United States.

Estimated cancer deaths 1991			
Male		**Female**	
Lung	34%	Lung	21%
Prostate	12%	Breast	18%
Colon and rectum	11%	Colon and rectum	13%
Leukaemia & lymphoma	9%	Leukaemia & lymphoma	9%

It is worth noting that the diagnosis of cancer is not an automatic death sentence and that certain common cancers are not included in this list, for example squamous and basal cell carcinomas of the skin, which can be cured by surgery or other therapies.

There are some changes in the importance of various cancers over time and space. Thus lung cancer in males is levelling off as the most common fatal cancer as the dangers of cigarette smoking become more widely appreciated. Unfortunately this is being accompanied by a marked increase in the rate of lung cancer among females, who seem to have adopted the habit of smoking in more recent years. Also in females, uterine cancer, notably carcinoma of the cervix is becoming less important as a cause of death due to detection of this disease in its early stages, whilst it is curable, by cytology screening programmes. There are also changes in incidence and mortality which are less easy to explain, such as the fall in mortality from carcinomas of the liver and stomach in the Western world.

There is also a marked variation in the geographic incidence of tumours. Liver cancer (hepatocellular carcinoma) is common and a very important cause of death in Africa, including South Africa, but is rare in the United States where it accounts for some two per cent of all cancer deaths. Breast cancer is four or five times more common in Europe and the United States than it is in Japan, whereas gastric carcinoma is seven times more common in Japan than in the United States. Some regional variations are more easily explained. Australia has the highest incidence of malignant melanoma in the world, due to its sunny climate and a mostly fair skinned population of northern European descent with a predilection for exposing their scantily clad bodies to the sun.

It is known that ultra-violet irradiation plays a part in the causation of melanoma, the incidence of which is highest in the tropical areas of Australia.

It is likely that most of the regional differences can be explained in terms of the environment rather than on genetic grounds, but in many instances the explanation is still elusive.

It is from observations such as these made by epidemiologists that our early knowledge of the causes of cancer developed. In 1775 Percival Pott in Britain described the high incidence of cancer of the scrotum in chimney sweeps, who were in intimate contact with soot. This appears to be the first description of an occupational cancer. Since that time further observations have revealed other associations:

AGE: Cancer is more common in older people, and hence as life expectancy increases becomes ever more important as a cause of death. This may be explained in part by the increasing likelihood of unfavourable genetic mutations associated with malignancy with increasing age. However tumours of children are also important, including leukaemia and lymphoma, tumours of the brain and sarcomas of soft tissues and of bone and a group of embryonic 'blastomas' (such as nephroblastoma, neuroblastoma, retinoblastoma and medulloblastoma).

HEREDITY: Certain tumours are hereditary and a study of these has considerably widened our knowledge about the genetics of cancer in general.

Retinoblastoma is a tumour of the eye, which may be bilateral in familial cases. A predisposition to this tumour may be inherited as an autosomal dominant trait, and seems to be due to lack of a tumour suppressor gene.

In familial polyposis coli, which is also transmitted as an autosomal dominant trait, affected individuals develop multiple benign adenomatous polyps of the colon during early life, and one or more of these is destined to become a carcinoma by middle age.

In addition there are certain rare conditions, such as xeroderma pigmentosum, in which there is a defect in the DNA repair mechanism. This is a recessive condition in which affected persons acquire sun-induced cancers of the skin at an early age.

ENVIRONMENT: The multiplicity of agents which have been implicated in the causation of cancer can be divided into three main groups:

- Radiation, including ionising radiation (X-ray) and ultraviolet (sunshine)
- Chemical carcinogens of various sorts
- Viruses

Radiation: Sun exposure can clearly be associated with the later development of skin cancers, and studies on early radiologists and the survivors of atomic bomb explosions have demonstrated the increased incidence of a variety of cancers in these individuals.

Ultraviolet light appears to generate pyrimidine dimers in DNA, which leads to transcriptional errors in replicating cells, some of which may give rise to cancer.

Ionising radiation has a more damaging effect cleaving DNA in the cell. There may be errors in repair of single stranded DNA giving rise to single base mutations, while double stranded DNA damage may result in chromosome breakage with consequent translocations and deletions.

Chemical carcinogens: Some chemical substances (**direct carcinogens**) can act directly on the cells to transform them into cancer cells. Many other substances are not active in themselves (**procarcinogens**) but give rise to active carcinogens (**ultimate carcinogens**) by virtue of their metabolic products within the body.

Direct carcinogens are usually weak carcinogens, possibly because there are many intracellular protective molecules, such as glutathione, which react with them before they can affect the cellular DNA. They are important because they include anti-cancer drugs which are alkylating agents such as cyclophosphamide, chlorambucil and the nitrosoureas. There is a slight, but nevertheless real, risk of treating one malignant condition successfully only to induce another type of cancer later on.

Most of the better known carcinogens are procarcinogens which are active through their metabolic breakdown products.

- **Polycyclic hydrocarbons** such as benzanthracene which will cause cancer in animals if painted on the skin; benzpyrene is found along with other carcinogenic agents in tobacco smoke.
- **Aromatic amines, amides and azo dyes** include β-naphthylamine which used to be used in the aniline dye industry and was also contained in an antioxidant used in the rubber industry. The ultimate carcinogen is produced in the liver, detoxified by the conjugation with glucuronic acid and then excreted in the urine. The bladder mucosa secretes the enzyme glucuronidase which releases the active ultimate carcinogen to cause cancer of the bladder.
  Other members of this group, such as 2-acetylaminofluorene and dimethylaminoazobenzene (once used as a food dye stuff) tend to be metabolised in the liver where the ultimate carcinogen is released to bring about liver cancer.
- **Natural products** such as Aflatoxin B, which is produced from a fungus, *Aspergillus flavus*, which grows on badly stored grains and peanuts has been associated with hepatocellular carcinoma and may act together with Hepatitis B virus to produce this cancer.
- **Other carcinogens** include minerals such as asbestos and elements such as nickel, chromium and arsenic together with more complex organic compounds.

Chemical carcinogens act as electrophiles, that is they have electron-deficient atoms which react with electron-rich atoms in DNA to bring about a number of genetic mutations. Some of these may be associated with neoplastic transformation; that is they are **mutagens**, in that they bring about **mutations.**

In some instances there may be DNA alteration brought about by a chemical carcinogen that acts as an initiator, but which requires the subsequent action of other substances known as promoters before cancerous growth can take place. Promoters act as **mitogens**, stimulating the altered cells to divide by **mitosis.**

Viral carcinogens. It is from investigation of viruses which cause tumours in animals, known as acute transforming retroviruses (RNA viruses) that much of our knowledge of the molecular biology of cancer is derived.

At this stage it is therefore proposed to postpone discussion of viral carcinogens to discuss cell growth and proliferation and its genetic control.

CELL GROWTH

Parenchymal cells are the cells which are associated with the function of a tissue or an organ and are supported by connective tissues, which contain blood vessels, lymphatics and nerves.

Cells arise from a precursor cell or stem cell, which divides to give rise to another stem cell and other cells which can travel along the path of differentiation or specialisation. Cells which are in this process may still divide (for example, haemopoietic cells in the bone marrow) but may enter a terminal differentiated phase in which division can no longer take place (for example, mature neutrophil). Stem cell divisions are infrequent and most division takes place among differentiating cells. However most cell in a particular tissue may be in the terminal differentiated phase and may themselves be incapable of further division, for example, the cells of the epidermis.

Growth is encouraged by growth factors produced by cells or hormones and an elaborate system of regulation or control is thought to exist. Proliferating cells are in the cell cycle which has four phases:

- G_1 In which the cell prepares to undertake DNA synthesis
- S In which synthesis actually takes place
- G_2 In which the cell prepares the mechanisms for the distribution of the newly synthesised chromosomes to the daughter cells which are produced in
- M The stage of mitosis or cell division.

Non-proliferating cells may be in a resting phase, G_0 and can be stimulated to enter the growth cycle by growth factors of the type which were mentioned in the chapter on healing and repair (for example, PDGF, FGF, TGFs) or they may be in the terminal differentiated phase and totally incapable of division in any circumstances.

Growth factors may act by attachment to cell receptors and initiate a series of events in the cells which activates genes to commence transcription of DNA (phases G_1 and S)

TUMOUR GROWTH

It seems as if tumours arise from the multiplication of one cell to form a clone of abnormal cells. Evidence for this is provided by the demonstration that tumours in females fail to show X-chromosome mosaicism, such as may be seen in the normal tissues; they display a monoclonal pattern of X-chromosome inactivation.

When normal cells and tumour cells are cultured certain differences can be noted:

- Normal cells form an even monolayer, whilst tumour cells show less contact inhibition and grow in multilayered fashion. In addition tumour cells are less dependent on added growth factors and will continue to proliferate in a semi-solid medium. These properties are those of a transformed phenotype.
- Transformed cells (tumour cells) will cause tumour formation when inoculated into immunosuppressed mice.
- Whereas most normal cells can only be subcultured and replicated a finite number of times (say around forty or fifty times) tumour cells can continue to divide continuously in culture and are said to be immortal.

The first stage in the development of a tumour then is the transformation of a cell so that it acquires some or all of the properties listed above, and can proliferate in relatively uncontrolled fashion until it becomes clinically obvious.

From what has already been said it can be inferred that non-lethal damage to DNA or mutation is the basis of cancer formation, whether this is brought about by radiation, chemicals or viruses. One cell that mutates to a cancer cell is all that is required for the development of a monoclonal neoplasm.

But a single mutation does not make a cancer. There may be several mutations and genetic alterations required to bring about the development of a tumour. And even during the life of a tumour, as it proliferates, further mutations may take place and be passed on to daughter cells to give rise to different sub-clonal cell populations which may have different properties, such as the ability to metastasise to particular organs.

CANCER GENES

Mutations can affect the two classes of genes that regulate growth, that is the protooncogenes which stimulate growth and the tumour suppressor genes (antioncogenes) which inhibit growth. It should be noted that alteration of one protooncogene may be enough to bring about transformation of cells, whereas both alleles of a tumour suppressor gene must be inactivated before transformation can take place.

ONCOGENES

Two major lines of research led to the discovery of protooncogenes which promote normal growth in the cell but can be changed into cancer producing oncogenes.

- Acute transforming retroviruses which cause animal tumours were shown to contain DNA sequences not found in non-tumour forming retroviruses. These sequences were termed viral oncogenes or v-*oncs*. It was then discovered that very similar, but not identical sequences of DNA could be found in normal human cells.
  The v-*oncs* tended to be named after the tumour viruses in which they were first discovered and were given three letter names such as v-*ras* from the *ra*t *s*arcoma virus or v-*sis* from the *si*mian *s*arcoma virus
  The similar sequences in normal human cells are concerned with normal growth and are known as protooncogenes and are designated by the same letters without the v-prefix, thus *ras* and *sis*.
- It was also found that DNA could be inserted into target cells by direct injection or in particles which were taken into the cell by endocytosis. Some of this DNA could then became integrated into the cell nucleus.
  When DNA from human tumours was introduced into mouse fibroblasts in culture in this way, these cells became transformed into malignant cells.
  It turned out that transforming DNA sequences similar to those found in viruses and protooncogenes were present in these tumours and these actively transforming sequences were named cellular oncogenes or c-*oncs*. Some new oncogenes were also found.

Thus cells may undergo malignant transformation by the incorporation of a retrovirus v-*onc* into the genome (in certain animal tumours) or by the changing of a normal protooncogene into a transforming cellular oncogene. This activation of a protooncogene may take place:

- As a result of a change in the structure of a gene with synthesis of an abnormally functioning gene product, or
- As a result of a changed gene expression, with increased production of normal gene product which stimulates growth.

These changes may be brought about by **point mutations** of a protooncogene (such as *ras*), or by **translocations** which may bring a protooncogene close to an area of intense transcriptional activity, with consequent overexpression of gene product (such as *myc* in Burkitt's lymphoma; this may be moved from chromosome 8, where it is strictly regulated, to chromosome 14 where it is subjected to stimulation from nearby active transcription which is taking place).

There may also be **amplification** of the actual DNA sequences of the gene itself, sometimes to such an extent that these become visible as tiny separate pieces of chromosomal material (double minutes) or as homogeneous staining regions within chromosomes. The multiplicity of oncogenes will give rise to an increased amount of gene product.

Abnormal protein product from an oncogene or excessive amount of normal protein product all may have the same effect in bringing about the transformation of cells.

Viral carcinogens (continued from page 238):

RNA viruses: The discussion of oncogenes has shown how acute transforming retroviruses may cause tumours in animals by incorporating oncogenes (v-*oncs*) into the genome of a cell and causing malignant transformation.

Slow transforming retroviruses insert viral DNA into a nucleus near a protooncogene which is then stimulated into overactivity.

In man only one retrovirus is associated with cancer and that is the Human T-cell Leukaemia Virus, type 1 (HTLV-1). As its name suggests this virus is associated with the development of T-cell leukaemia (in about one per cent of infected persons). Infection seems to be most common in Japan and the West Indies The virus is related to the Human Immunodeficiency Virus (HIV) which causes AIDS, and acts on CD4+ T-cells which undergo malignant transformation. A special gene known as *tax* within the viral genome seems able to activate protooncogenes within the cells and to bring about transcription of viral m-RNA.

DNA oncogenic viruses: Three of these viruses have been shown to be associated with human cancer, that is Human Papilloma Virus (HPV), Epstein-Barr virus (EBV) and Hepatitis B Virus (HBV). They cause neoplastic transformation by integrating viral DNA into the genome of the human cell. Because of this integration the virus is unable to complete its life cycle, but viral DNA is transmitted into the progeny of cells attacked in this way.

- Human Papilloma Virus is important because DNA sequences of types 16 and 18 are found in the great majority of cases of intraepithelial and invasive carcinomas of the cervix of the uterus.
  It appears that viral genes encode for proteins that bind and inactivate tumour suppressor genes *RB* and *p53*. This in itself may not be enough to induce carcinoma and it is thought that other environmental factors may contribute to cancer formation.
- Epstein-Barr Virus is associated with Burkitt's lymphoma and nasopharyngeal carcinoma. The virus infects B cells and stimulates them to proliferate. Malignant transformation only occurs in the presence of some other factor, such as malaria, which will reduce immunity. There is then more marked proliferation and the scene is set for a mutation such as t(8,14) which seems to activate c-*myc*. Other mutations may occur and malignant lymphoma follows on.
- Hepatitis B Virus is associated with hepatocellular carcinoma, but the mechanism of malignant transformation is not yet fully elucidated, as a number of different changes may be observed but these do not appear to be constant. Liver cell injury and active regeneration, co-carcinogenic action of dietary agents such as aflatoxin, and inactivation of *p53* together with other inconstant genetic disturbances are postulated as being responsible for the development of liver cancer.

ACTION OF ONCOPROTEINS

The protein products of oncogenes are called oncoproteins and they act on the various stages of cell proliferation.

- Binding of growth factor to specific receptor on cell membrane. Some oncogenes, for example *sis*, will encode for growth factors such as PDGF. If the tumour cells also possess receptors for the growth factor an autocrine loop may be set up.
- Growth factor receptors may be altered by mutated oncogenes which encode for them. Examples are c-*erbB-1* and c-*erbB-2* (named after viruses giving rise to *er*ythro*b*lastosis in chickens). The altered receptors perpetuate the activity of tyrosine kinases on the interior of the cell membrane and thus deliver a continuous stream of mitogenic messages to the cell. There may be amplification of c-*erbB-2* with increased production of normal protein, leading to increased sensitivity to growth factors, in cancers of breast ovary and stomach.
- A signal is transmitted from the cell membrane to the nucleus. Oncogenes may be associated with tyrosine kinases not linked to growth receptors (c-*src* and c-*abl*) and located in relation to the interior of the cell membrane. As many as 25 per cent of human cancers may contain an altered version of the *ras* gene. The protein products of this gene bind guanosine nucleotides (GTP and GDP). Thus *ras* protein-GTP is active, transmitting signals to the regulators of proliferation. In normal circumstances GTPase soon hydrolyses GTP to GDP, which is inactive, under the influence of a protein GAP (*G*TPase *a*ctivating *p*rotein). Mutations in the gene inhibit the GTPase activity and maintain the *ras* protein-GTP in the active state.
- Many genes produce proteins that act directly on the nucleus including *myc*, *jun* and *fos*, which appear to be regulators of nuclear transcription and the oncogenes from these would increase DNA transcription. *Myc* oncogenes are found in many human tumours.

  Other oncogenes such as c-*rel* and c-*Erb-A* prevent the expression of genes for cell differentiation, and may maintain the cell in a less differentiated, constantly proliferating phase.
- *bcl-2* is located on mitochondria and acts by preventing apoptosis or programmed cell death. This oncogene is activated by translocation and is found in B cell lymphomas

TUMOUR SUPPRESSOR GENES

These act to prevent cell proliferation and hence tumour formation. It is in the case of absence or malfunction of these genes then that tumours may occur.

The *RB* gene. The best known example of tumours due to disturbance of these genes is retinoblastoma. The *RB* gene lies on chromosome 13q14 (the long arm of chromosome 13, region 1, band 4).

In familial cases of retinoblastoma (about 40 per cent of all cases) one allele is lost or defective and this is passed on to 50 per cent of the children. Tumours develop in one or both eyes when spontaneous mutation of retinal cells knocks out the remaining *RB* gene. tumour occurs.

It has since been shown that the *RB* gene is not specific for retinoblastoma but maybe missing (deleted) or mutant on both chromosomes in a number of tumours including osteosarcoma, breast cancer, small cell carcinoma of the lung and some brain tumours.

Other tumour suppressor genes acting in similar fashion include:

- ***p53* gene**. This is located on chromosome 17p13 and appears to be a tumour suppressor gene that is lost in a homozygous manner in many tumours such as carcinoma of the lung, colon, breast and ovary. The tumour suppression effect can be demonstrated *in vitro* by the introduction of a cloned normal gene into transformed cell culture and demonstration of reversion of the transformed phenotype. This is the usual action of *p53*, but mutated forms have also been shown to act as oncogenes which can activate transcription and transform cells.

- ***APC* and *NF-1* genes.** These are involved in the development of *a*denomatous *p*olyposis *c*oli and *n*euro*f*ibromatosis type 1. The loss of one copy of the gene may allow the development of benign tumours; homozygous loss is associated with the development of malignant tumours.

- Other tumour suppressor genes include *DCC* (*d*eleted in *c*olon *c*arcinoma) and *WT-1* which is associated with the development of Wilm's tumour, a childhood tumour of the kidney.

These genes function by preventing entry of cells into the growth cycle. The genes may be lost (deleted), or mutated, or inactivated by binding to the protein products of DNA tumour viruses. Malignant transformation may then take place in an affected cell.

A concatenation (or chain) of genetic events: It has been shown that a single oncogene cannot induce malignant transformation unaided, but that the action of several oncogenes is required. In fact in most human cancers there are several *active oncogenes* and several *inactive tumour suppressor genes*. In other words, carcinogenesis is brought about by an accumulation of genetic events. In the evolution of carcinoma of the colon from normal, through hyperplastic, adenomatous and dysplastic stages to unequivocal malignancy, a number of specific genetic changes have been identified including loss of *APC*, *DCC* and *p53* tumour suppressor genes and mutation of the *ras* gene to an activated cellular oncogene (c-*ras*).

THE GROWTH OF TUMOURS

Lengthy discussion of the genetic control of tumour development has concentrated attention on the first step towards malignancy, that is neoplastic transformation of a cell. This is followed by monoclonal proliferation leading to observable tumour formation. Only then may the distinctive characteristics of malignancy such as invasive growth and metastatic spread to a distant site occur.

The rate of growth of a tumour depends on the doubling time of dividing cells, and how many of the tumour cells are actually in the growth cycle.

It has been estimated that a small cell must undergo some thirty doublings to produce 10^9 cells, which would weigh approximately 1g. A further ten doublings would give rise to a tumour weighing 1,000 times that or 1 kg! But this would only apply if all the cells were multiplying, and this is certainly not the case.

In fact a surprising number of tumour cells may be in the resting or quiescent G_0 growth phase, perhaps more than 90 per cent. In addition many cells may actually be lost from tumours by being shed from the surface, for example. This means that for tumours to reach a clinically detectable stage may take many months or years.

If *all* the cells were doubling with a cell cycle time of three days, it would take 120 days or four months for a tumour to reach 1kg, or in other words to achieve fatal proportions. In reality, clinical observations show that many common tumours (for example, lung and large bowel) have a doubling time of two to three months, although there is wide variation ranging from one month to one year. Thus many malignant tumours may have a natural history of forty multiplied by two to three months, that is 80-120 months or seven to ten years.

A consequence of the growth rate of a tumour is its susceptibility to anticancer agents, which act mainly on cells in the S phase. If most cells are not in the growth cycle only a small proportion of tumour cells can be reached. Rapidly dividing tumours, which may have as many as thirty per cent of cells in the growth cycle phase are obviously more susceptible to this sort of treatment.

If tumour mass is diminished, for example by surgical therapy, then more cells may enter the growth cycle and susceptibility to chemotherapy will be increased, which is why many tumours are treated by a combination of surgery and chemotherapy.

THE STROMA OF TUMOURS

Tumour cells do not grow in isolation, but require a supporting connective tissue framework, which is not in itself tumourous. The tumour cells induce the formation of such a framework which carries in it the blood vessels that are essential for tumour survival and further growth, as well as lymphatics and nerves.

Tumour cells secrete growth factors such as fibroblast growth factors (FGF) which attract and stimulate the proliferation of endothelial cells and fibroblasts, and at the same time produce proteolytic enzymes that allow the newly formed vessels to penetrate the fibrous stroma of the tumour. Transforming (tumour-derived) growth factor (TGF-α) and epidermal growth factor (EGF) may also be secreted by tumour cells and can promote angiogenesis.

It will be recalled that certain oncogenes act by stimulating growth factor production (for example, c-*sis*) so that it can be seen that factors bringing about angiogenesis are closely related to those factors responsible for malignant transformation in the first place.

TUMOUR PROGRESSION

This has already been mentioned in connection with the development of cancer of the colon, where a number of genetic events accompany the hyperplasia-adenoma-dysplasia-carcinoma progression.

In many carcinomas there is an accumulation of mutations and chromosomal abnormalities, with the development of subpopulations of cells which have different properties in terms of invasiveness, proneness to metastasise and susceptibility to therapeutic measures. This accounts for the way in which tumours seem to become more highly malignant with the passage of time, or after recurrence following treatment. It is a case of survival of the fittest of the subclones In the case of the tumour, the fittest can be equated with the most malignant.

GENERAL PATHOLOGY - CHAPTER TWENTY EIGHT

SOME TUMOURS OF EPITHELIAL TISSUE: SKIN TUMOURS

In this chapter on epithelial tumours, emphasis will be placed on tumours of the skin. This is for two reasons:

- They are common and accessible and you may already have seen examples of them. Fifty per cent of Whites are likely to have had at least one skin cancer by the time they reach the age of 65.
- Courses of lectures in systemic pathology do not always include lectures on the pathology of the skin as a separate subject.

Thus, tumours of transitional cell epithelium, columnar cell epithelium and other various glandular epithelia will be left till later. Epithelial tissue in the pathological sense consists of covering cells or lining cells, for example of skin and hollow organs such as gut, urinary tract and respiratory tract (but not blood or lymph vessels or serous cavities) and of glandular or secretory cells of all types regardless of the embryonic germ layer from which they are derived. Thus there are considerable quantities of epithelium and in fact epithelial tumours are the ones most commonly found in man.

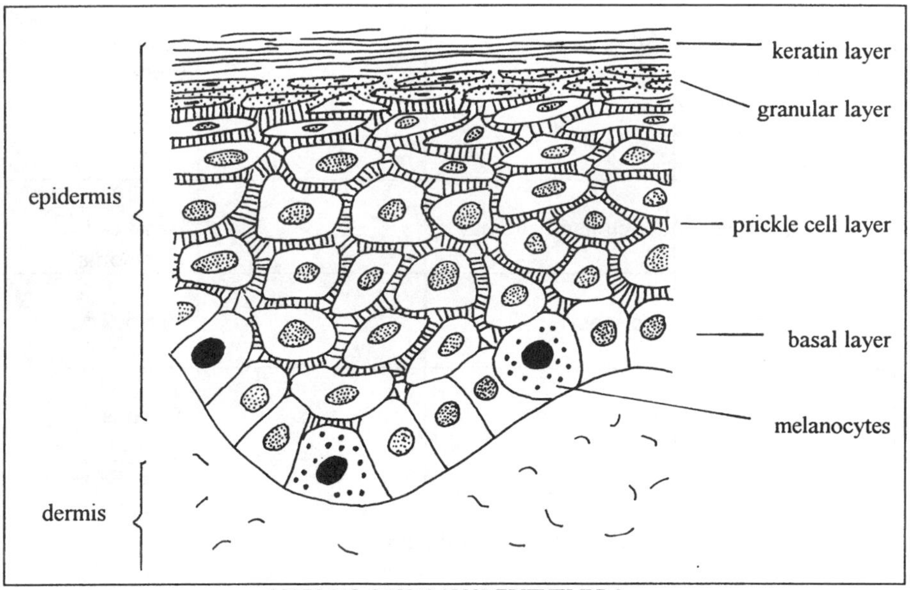

NORMAL SQUAMOUS EPITHELIUM

In the classification of tumours we mentioned that tumours

	Benign	Malignant
of surface epithelium are called	papillomas	carcinomas
of glandular epithelium are called	adenomas	carcinomas

In the skin this classification may be further broken down as shown in the following table.

	BENIGN	PRE-MALIGNANT	MALIGNANT
EPIDERMIS Squamous	Simple squamous papilloma Keratoacanthoma	Keratosis Carcinoma-in-situ	Squamous cell carcinoma
Basal	Basal cell papilloma (seborrheic wart)		Basal cell carcinoma
Pigmented	Cellular naevus	Dysplastic naevus Lentigo maligna Melanoma-in-situ	Malignant melanoma
DERMIS Glandular Epithelium (sweat glands and sebaceous glands)	Adenoma		Carcinoma (very rare)
Hair follicles	Trichoepithelioma		Basal cell carcinoma
Connective tissue	Various *		Various *
Nerves, vessels and smooth muscle	Various *		Various *

* See chapter on 'Classification of Tumours'

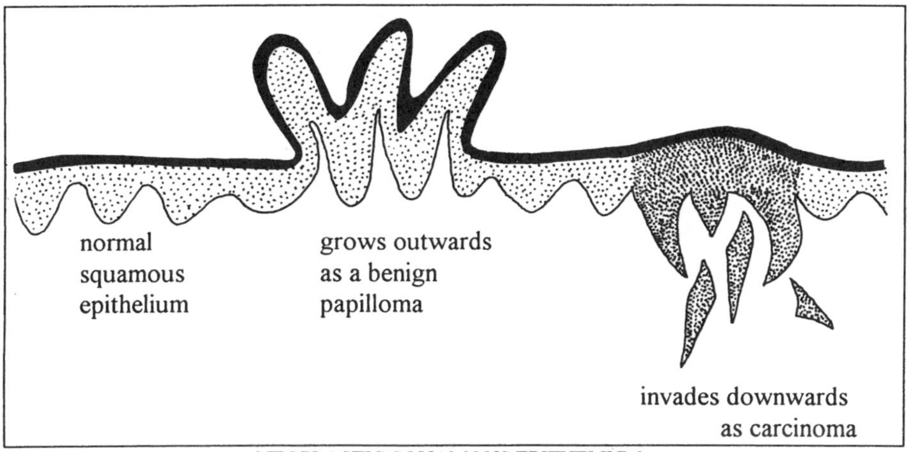

NEOPLASTIC SQUAMOUS EPITHELIUM

The remainder of this chapter will be devoted to the consideration of tumours of the epidermis.

TUMOURS OF SQUAMOUS CELLS

Benign: Simple squamous papilloma: There are a number of different types of which the viral wart is the most common.

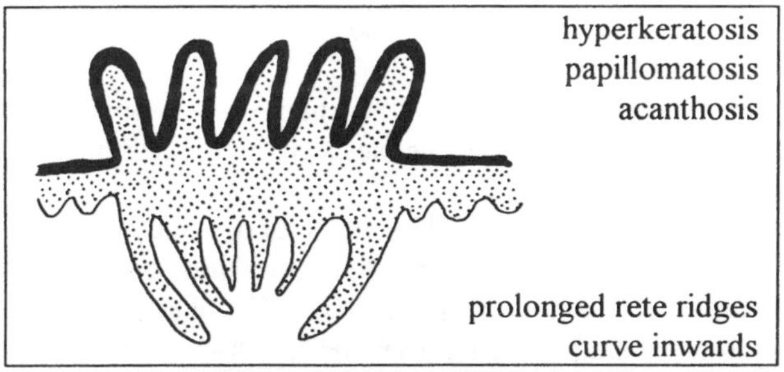

SIMPLE SQUAMOUS PAPILLOMA SUCH AS A VIRAL WART

In addition to the features shown in the diagram, inclusions may be seen in the granular layer, and vacuolated squamous cells may be prominent. The typical viral wart results from epithelial proliferation in response to a human papilloma virus (HPV), particularly types 2, 4 and 7.

Benign - Keratoacanthoma: This is a benign lesion that tends to occur in sun-exposed areas of skin. It presents on a nodule with a central, keratin-filled crater. Histologically it may resemble well differentiated squamous cell carcinomas very closely. However, lesions grow rapidly over six to eight weeks and then proceed to undergo spontaneous involution and heal over the next few months.

Pre-malignant: Keratosis: Ultra-violet light exposure causes these lesions as well as indisputable carcinomas, so they are usually called solar keratoses. They are a form of dysplasia. Similar growths may occur due to arsenic ingestion (arsenical keratosis), pitch or tar products (pitch warts), or due to X-rays.

Pre-malignant: Dysplasia (intraepithelial neoplasia): This occurs on mucous membranes for example vulva and cervix. In some situations there may be superficial hyperkeratosis which has a matt white appearance in moist situations and lesions may then be referred to by the clinical term of leukoplakia.

Pre-malignant: Carcinoma-in-situ: This occurs when the changes are more marked and extend through the whole thickness of the skin. Carcinoma-in-situ of the skin is sometimes called Bowen's disease.

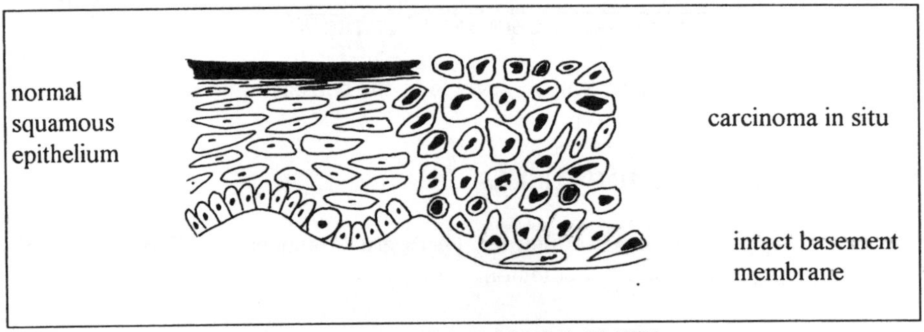

PRE-MALIGNANT CARCINOMA IN SITU

These premalignant lesions all illustrate the changes of dysplasia. When dysplasia increases in severity so that the full thickness of the epithelium is involved it then becomes carcinoma-in-situ.

In addition to occurring in the skin and mucous membrane of the lip such changes are well known in the epithelium of the bronchus, oesophagus and cervix of the uterus. In the cervix, dysplasia is known as Cervical intraepithelial Neoplasia (CIN) and carcinoma-in-situ, along with severe dysplasia, is classified as CIN III.

The hall mark of true carcinoma is invasion. Keratosis, leukoplakia and even carcinoma-in-situ do not necessarily always proceed to invasive carcinoma and may be reversible.

Malignant: Squamous cell carcinoma in the skin can arise *de novo* and need not be preceded by dysplastic change or carcinoma-in-situ, although it often is. Sun exposure is a frequent causative agent. Other predisposing factors include arsenic, tar, X-irradiation, chronic ulceration or scarring. It occurs most commonly in elderly fair skinned people in sunny climates on the exposed parts of the body such as the face and back of hands. Squamous cell carcinoma of the skin is characterised by **invasion** of the underlying dermis. It grows as a mass which is usually ulcerated with raised, rolled, everted edges and is composed of malignant cells which may be well differentiated producing much keratin, or more poorly differentiated. Well differentiated carcinomas may contain epithelial pearls or cell nests (nests of keratinised cells with a central keratin mass deep in the tissue). Metastasis to lymph nodes by the sun-induced carcinoma of skin occurs at a late stage - probably only in a very small proportion of cases (less than 0.5 per cent).

Sun exposure is not a cause of squamous cell carcinoma in Blacks. Nonetheless Blacks may develop squamous carcinomas, often secondary to chronic ulceration or scarring of the legs. Such tumours are more likely to metastasise than those caused by sun exposure.

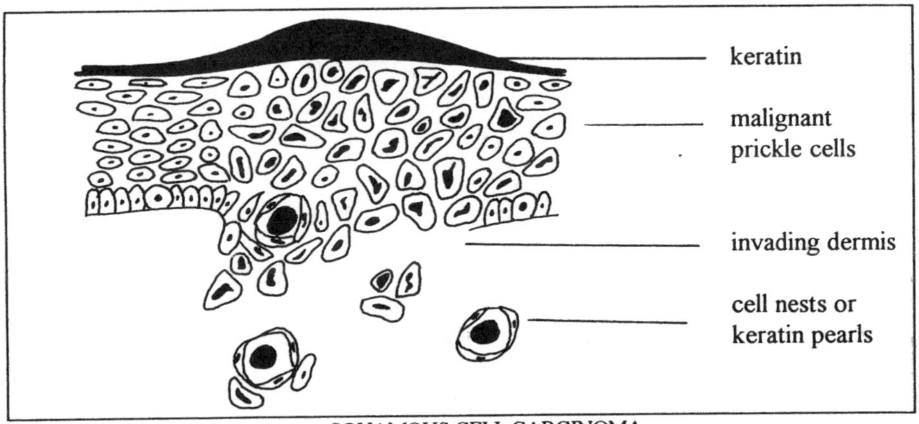

SQUAMOUS CELL CARCINOMA

Squamous cell carcinoma of the skin has a better prognosis than squamous cell carcinoma elsewhere, for example tongue, pharynx, lung and urinary tract.

TUMOURS OF BASAL CELLS

Benign: Basal cell papilloma is a proliferation mainly of superficial basal cells that generally occurs in the elderly (another name is senile wart, yet another name is seborrhoeic keratosis or seborrheic wart). It appears to be stuck on the top of the normal skin. Clinically it is flat, gray and greasy, often brownish-gray due to melanin pigment.

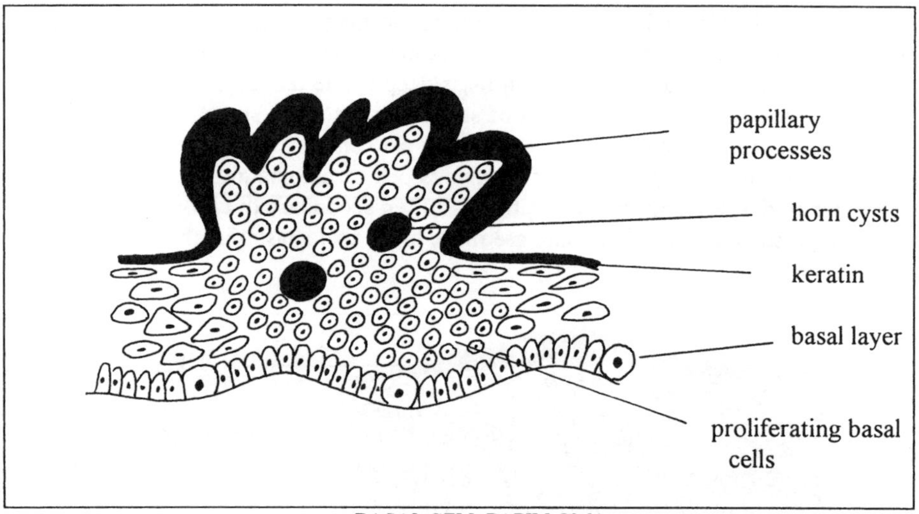

BASAL CELL PAPILLOMA

Basal cell papillomas are hyperkeratotic and contain keratin-filled horn cysts. Melanocytes often seem to proliferate along with the basal cells and the lesion may be quite markedly pigmented, sometimes so deeply that it may be mistaken clinically for a malignant melanoma.

Malignant: Basal cell carcinoma occurs in the same sites and after the same types of predisposing conditions as squamous cell carcinoma. It is a tumour of intermediate malignancy, locally invasive but virtually never metastasising, arising from the basal cells of the skin or hair follicles.

Clinically it presents as a nodule or an ulcer (rodent ulcer is an alternative name for this tumour). Histologically it is composed of islands of small darkly staining cells which may show palisading of their nuclei at the periphery. The overlying epidermis may be intact in the nodular type, or ulcerated. The invading islands of cells often show cystic change. Sometimes (particularly when situated on the back) proliferating basal cells remain contiguous with the overlying epidermis and only invade superficially; these tumours may be multicentric and are known as superficial basal cell carcinomas. There may be metaplastic changes in tumour cells, usually squamous metaplasia, but sometimes there may be sebaceous metaplasia or the formation of sweat gland ducts. On occasion, these tumours may show quite dense melanin pigmentation. They may invade deeply, as far as bone or even through bone, causing gross deformity or even death, but usually if diagnosed and treated at an early stage the prognosis is good.

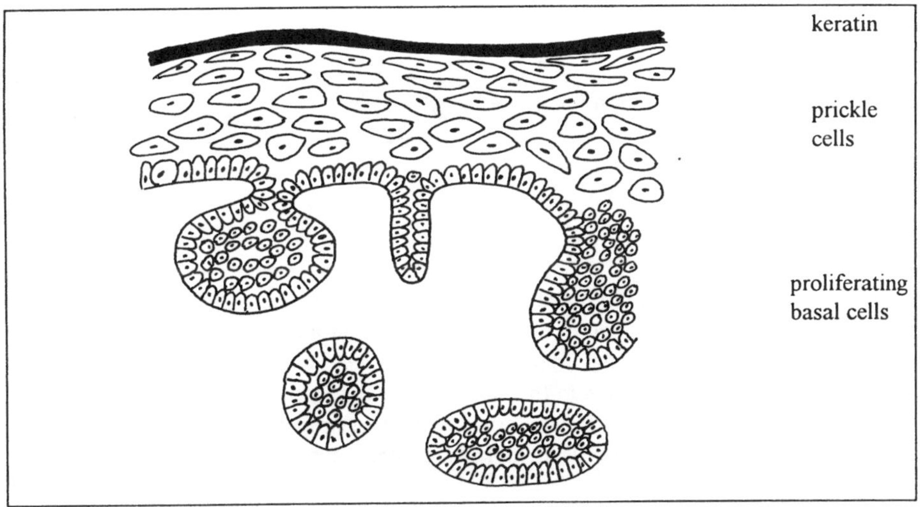

BASAL CELL CARCINOMA

Basal cell carcinoma is the most common malignant growth in white South Africans, but is hardly ever found Blacks. It occurs predominantly on the face, and is particularly common in those who have fair or sandy complexions and who are exposed to sunlight, for example sailors and farmers. It is easily treated in the early stages by surgical means.

TUMOURS OF PIGMENTED CELLS

Benign naevi: The benign tumour of pigmented cells is the common mole. Pathologically it is called a cellular naevus and is derived either from the melanocytes of the basal layer or from the nests of naevus cells.

Naevi may be actual benign neoplasms, but many people think that they are harmartomas, that is congenital developmental malformations.

Melanocytes or naevus cells are thought to be derived from migrating neuro ectoderm.

Naevi are of three main types, junctional, compound or intradermal, and often proceed through these stages sequentially as the individual grows and matures.

A junctional naevus is one in which clusters of naevus cells are present in the epidermis at its junction with the dermis. These are found mainly in children and in adults on the palms of the hands and soles of the feet.

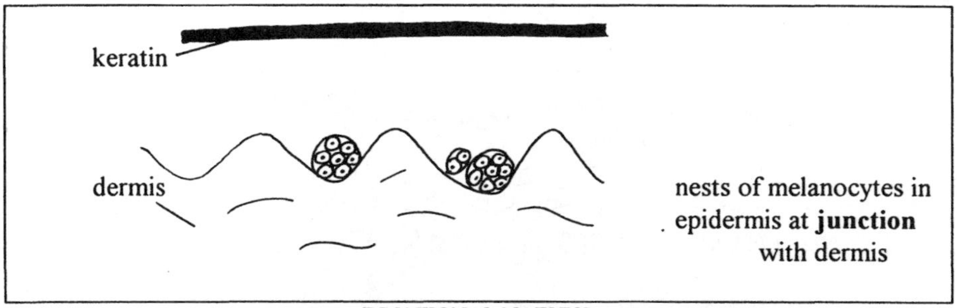

JUNCTIONAL NAEVUS

Junctional naevi appear as flat black spots, usually less than 7mm in diameter and are the least common of the naevi, except in children. The junctional naevus appears to be the first stage in cellular naevus development, with the multiplication of melanocytes in nests or clusters in the basal part of the epidermis.

Often these clusters from the epidermis appear to drop off into the dermis and at this stage with both epidermal and dermal components the lesion is known as a **compound naevus.**

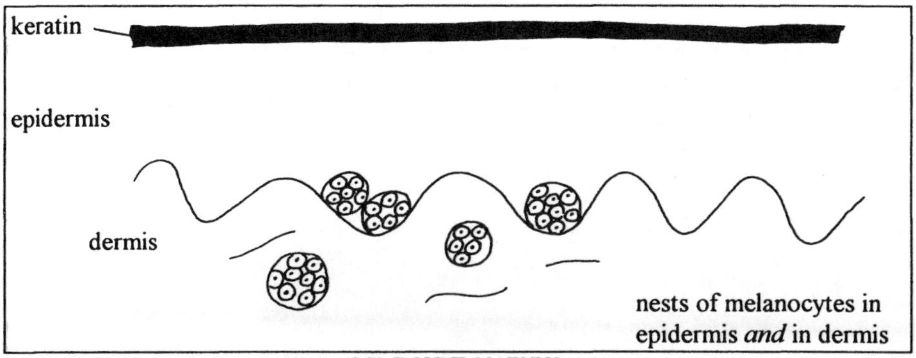

COMPOUND NAEVUS

However most naevi (in adults especially) show no junctional activity, but collections of round, regular, often pigmented cells are present in the dermis. These are **intradermal naevi**.

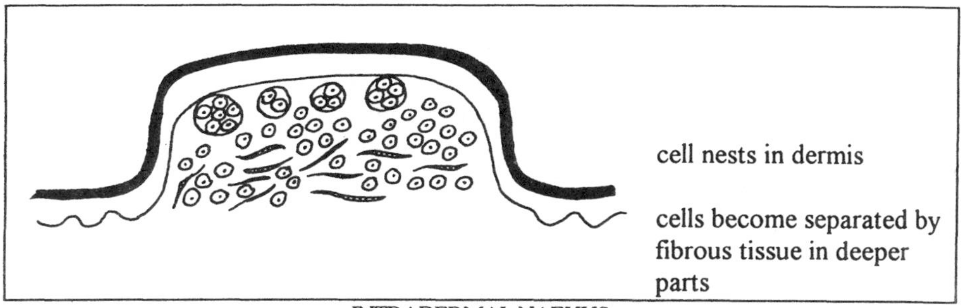

INTRADERMAL NAEVUS

Naevi as you know are common, but they are even more common and more obvious in fair skinned people. Europeans have an average of about twenty moles per person.

Dysplastic naevi may be found in patients who have a familial history of malignant melanoma or they may occur sporadically in otherwise normal persons. They are larger (more than 7mm in diameter) and more irregular than common naevi. Microscopically there is excessive proliferation of somewhat atypical large melanocytes in the basal or junctional portion of the epidermis. The proliferating melanocytes may adopt a spindle shape and may also show architectural abnormalities with fusion of nests of melanocytes in adjacent rete ridges. There may also be a fine dusting of pigmentation in these cells. There is usually lymphocytic infiltration of the dermis.

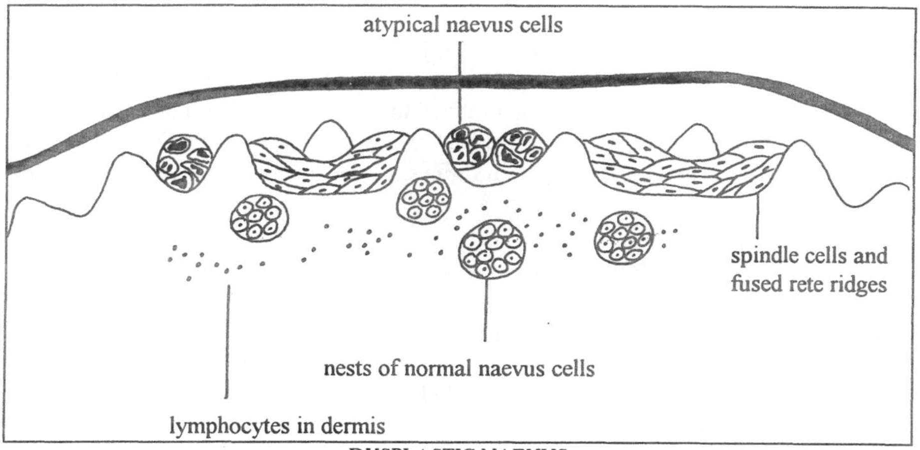

DYSPLASTIC NAEVUS

Malignant melanomas are fortunately much less common than moles. They are malignant tumours of melanocytes or naevus cells. Probably about one third of all melanomas arise from pre-existent naevi.

Malignant melanoma account for about three per cent of all cancers and about 0.5 per cent of all cancer deaths The chances of a mole undergoing malignant change are very slight indeed, about one in several thousand. Malignant melanoma occurs in adults and is extremely rare before fifteen years of age. It occurs on the skin more commonly in Whites than in Blacks. The distribution in Whites is variable, but the lower leg is a common site in females and the trunk and back in males. In Blacks the most common site (about two thirds of melanomas) is the sole of the foot and the palm of the hand. Melanomas can also grow in the eye, and in the mucous membranes of the mouth, the anus and in the meninges.

Grossly a naevus which is larger than normal (more than 7mm, so that it cannot be covered by the end of a pencil) and which has an irregular outline and irregular pigmentation, with areas of dense pigmentation and also areas of depigmentation should be regarded with suspicion. Also changes in a naevus may be ominous. Naevi which are increasing in size or in pigmentation, or which ulcerate and bleed, or which develop satellite lesions round about may be undergoing malignant change.

Histologically malignant melanocytes tend to start multiplying in the basal layer at the junction of epidermis and dermis and then invade upwards through the epidermis to the surface and downwards into the underlying dermis. There is usually some lymphocytic infiltrate around invading melanoma cells in the dermis. Occasionally melanomas may be lacking in pigment. The diagnosis must always be confirmed histologically as there are many other pigmented lesions of the skin that can mimic melanoma.

The prognosis of malignant melanoma depends on a number of factors :

- **The histological pattern of growth:** In the earlier stages of malignant melanoma there may be a radial growth pattern with spread of the tumour laterally within the epidermis. At the stage the lesion may be regarded as melanoma-in-situ as it has not broken through the basement membrane of the epidermis. With the passage of time there may develop a vertical growth phase. This is the stage at which the lesion becomes invasive. The radial growth phase persists longer in different types of melanoma, lasting longest in :

  - **Lentigo maligna** (Hutchinson's melanotic freckle), a flat pigmented lesion in sun exposed areas consisting of a linear proliferation of atypical melanocytes within the epidermis. This may persist for years before becoming invasive; even when there has been invasion the prognosis for this type of lesion seems to be better.

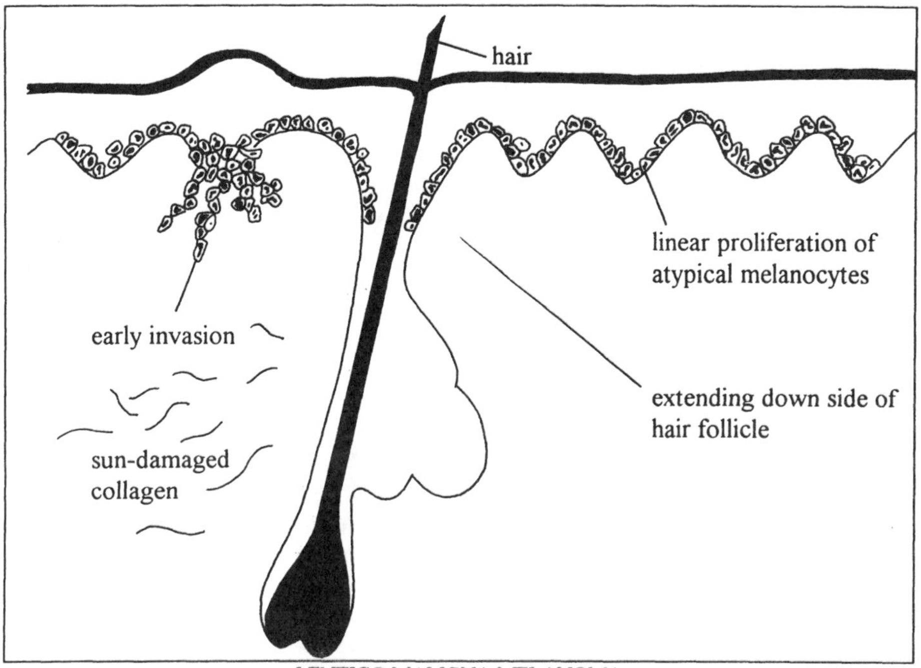

LENTIGO MALIGNA MELANOMA

- **Superficial spreading melanoma** is characterised by nests of atypical melanocytes which grow in the junctional region at the edge of the invasive part of the tumour. Individual melanocytes migrate upwards through the epidermis. Unfortunately this is rarely diagnosed in the radial growth phase because invasion appears to take place at an earlier stage in the development of the tumour.

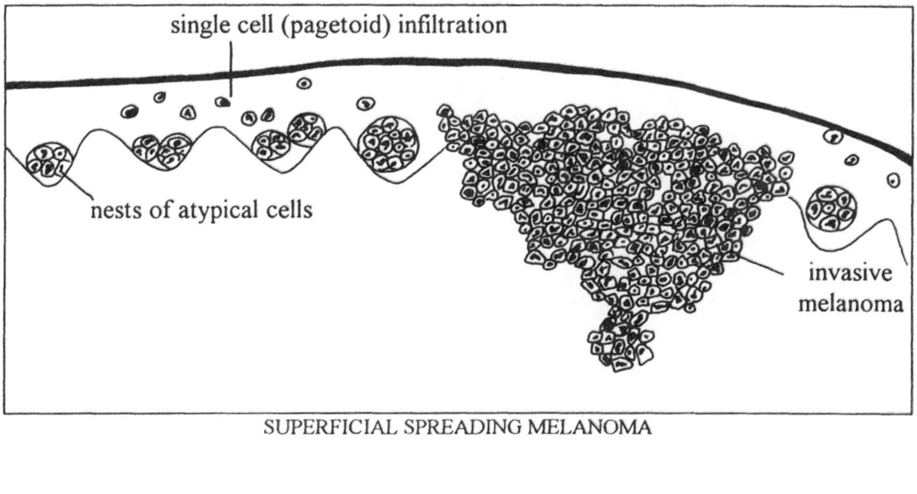

SUPERFICIAL SPREADING MELANOMA

- **Nodular melanoma** does not seem to have an appreciable radial growth phase and invasion takes place early. This type of lesion carries a worse prognosis because of its invasiveness.

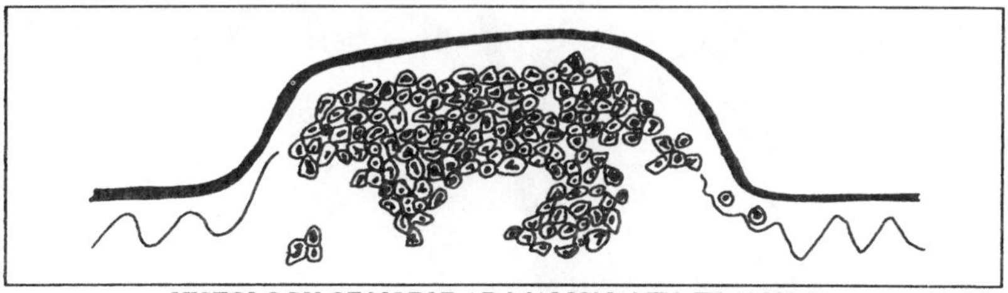

HISTOLOGY OF NODULAR MALIGNANT MELANOMA

- **Acral lentiginous melanoma** is a variant arising on the palms and soles which may show the pattern similar to that of lentigo maligna in the radial growth phase. Unfortunately invasion takes place at an earlier stage and the prognosis is not as good as for lentigo maligna. (It should be noted that superficial spreading melanomas and nodular melanomas can also occur on the extremities).

- **The depth of invasion**: All types except nodular melanoma have an in-situ stage, when excision will result in cure. The prognosis of invasive melanoma can be related to the depth of the lesion (measured from the granular layer of the epidermis vertically to the deepest part of invasive tumour). Thin lesions less than 0,76mm in thickness have an almost 100 per cent cure rate following surgery. The prognosis worsens with increasing thickness, so that five year survival in lesions 4mm or more in thickness is down to about forty per cent.
- **Mitotic rate** : The higher the mitotic rate, the worse the outlook

Thick lesions spread via the lymphatics and the blood stream, which is responsible for their poor prognosis. The survival rate can be much improved if there is a high index of clinical suspicion, with consequent early diagnosis and treatment.

Finally, as was mentioned previously, there may sometimes be apparent cure with recurrence of the tumour manifested by the appearance of metastases occurring many years later (particularly where the primary site for the tumour was the eye).

GENERAL PATHOLOGY - CHAPTER TWENTY NINE

SOME TUMOURS OF CONNECTIVE TISSUE

Tumours of the various types of connective tissue, or soft tissue tumours as they are sometimes called, can be found in most parts of the body. It is therefore useful to know something about them as they should be thought about in the differential diagnosis of almost any tumour mass.

Generally speaking, benign soft tisue tumours are common and relatively unimportant in terms of their clinical effects.

Malignant soft tissue tumours can occur at any age but are fortunately rare, accounting for only a few per cent of all cancers (one or two per cent). Thus carcinomas of the lung, for example, are much more important in terms of human disease and suffering than all the sarcomas put together. However high grade sarcomas tend to grow rapidly and spread via the blood stream at an early stage. They also tend to be more resistant to treatment with X-rays or chemotherapy. Histologically the structure of soft tissue tumours is less distinct than that of epithelial tumours. There is no marked division between tumour tissue and background stroma as such. Tumour cells may be separated from each other by extracellular matrix, and it is sometimes the amount and nature of this which determines the degree of differentiation of the tumour.

Sarcomas may be graded according to their histological characteristics:

- Overall cellularity
- Pleomorphism of cells
- Number of mitoses (per ten HPF)
- Extent of necrosis

The latter two points presumably indicate the rate of growth of the tumour.

Sarcomas may be staged according to the TNM classification according to the size of the tumour (larger than 5cm in diameter = T_2), the absence or presence of lymph node involvement (N_0or N_1) and the absence or presence of distant metasases (M_0 or M_1).

Low grade sarcomas respond well to adequate excision with an over 95 per cent ten year survival whereas high grade sarcomas, characterised by large size and much necrosis together with cellular anaplasia, have a much worse outlook. Male sex is also a factor which worsens the prognosis.

Soft tissue tumours can be classified according to their histogenesis, that is, according to their tissue of origin, although this may not always be immediately apparent, especially in the case of malignant tumours.

TUMOURS OF FIBROUS TISSUE

True benign tumours of fibrous tissue, that is fibromas, are rare although they do occur, for example, in the ovary. More common are unencapsulated overgrowths of fibrous tissue which are termed fibromatoses. Examples include:

- **Palmar and plantar fibromatosis:** There is overgrowth of cellular fibrous tissue around the tendons on the palms and soles. In the hand a characteristic flexion deformity of the fourth and fifth digits is known as Dupuytren's contracture

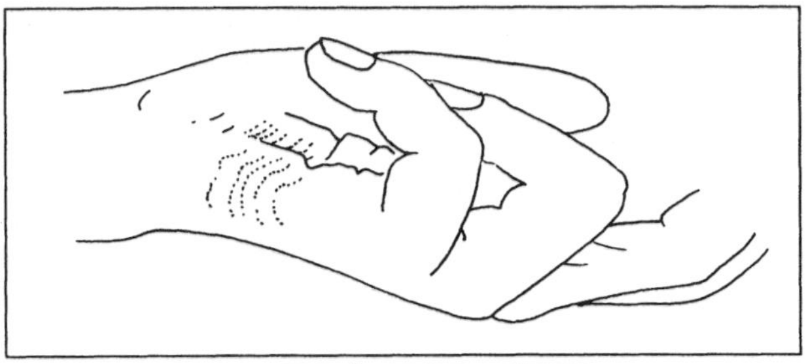

DUPUYTREN'S CONTRACTURE

- **Nodular fasciitis:** This is probably a reactive condition rather than a true neoplasm, characterised by marked fibroblastic proliferation which gives a pseudosarcomatous appearance. Generally lesions are small and situated in the subcutaneous tissues close to the fascia.

There are also lesions which are locally aggressive tumours which show local invasion and may recur after excision, but which do not metastasise. **Desmoid tumours,** for example, are fibrous proliferations occurring in skeletal muscle, often in the muscles of the anterior abdominal wall.

True pure **fibrosarcomas** are rare and many lesions formerly diagnosed as such are now categorised differently. As for many sarcomas the soft tissues of the lower limb or retroperitoneum are common sites. The true fibrosarcoma is a cellular tumour composed of spindle cells which are often arranged in a 'herring-bone' pattern. Haemorrhage and necrosis are common, and blood borne metastases may occur at an early stage. Five year survivals are in the region of 60 per cent.

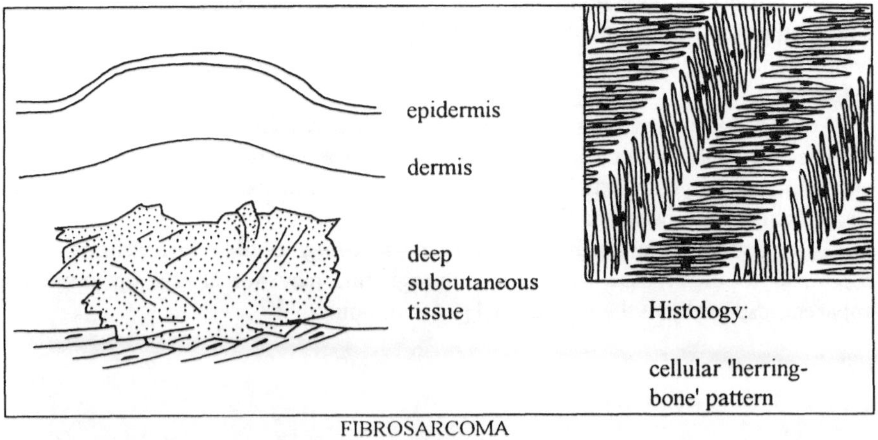

FIBROSARCOMA

FIBROUS HISTIOCYTOMAS

These are now thought to be more common than pure fibrous timours. They are composed of a mixture of cells including fibroblasts and myofibroblasts, histiocytes and intermediate cells.

Benign fibrous histiocytomas: These tumours occur in the dermis and may represent reaction to trauma of some sort rather than being true neoplasms. However the histological pattern of spindle cell proliferation in a 'storiform' or radiating cartwheel pattern is characteristic of all the tumours of this group. Recognisable macrophages or histiocytes may also be seen. The lesions are ill-defined. Sometimes there is proliferation or excessive pigmentation of the overlying dermis.

Dermatofibrosarcoma protruberans: also occurs in the dermis of the skin but is of intermediate malignancy, tending to show local aggression and recurrence after excision, although it does not metastasise. The tumour may extend or invade into the subcutaneous fat. The overlying epidermis is often stretched over the tumour and ulceration may occur.

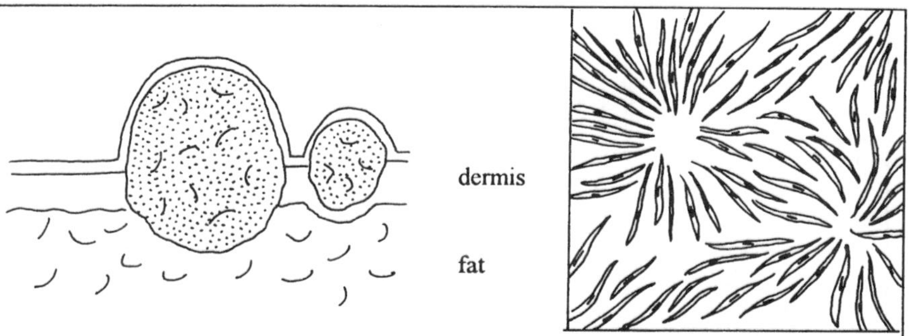

Histology:
'storiform'
pattern of fibroblasts

DERMATOFIBROSARCOMA PROTRUBERANS

Malignant fibrous histiocytoma is thought to arise from a primitive mesenchymal cell which can differentiate along a number of lines. Different histological patterns are described but the most common variant, the storiform-pleomorphic, has spindle cells arranged in a storiform pattern, some histiocytoid cells and also numbers of multinucleate giant cells. There are also myxoid, giant cell, inflammatory and angiomatoid types. These tumours occur in the deep subcutaneous tissues of the limbs or abdomen (retroperitoneally) and are now regarded as one of the most common types of sarcoma accounting for one quarter to one third of all sarcomas. Although the myxoid variant may have a slightly better prognosis these are highly malignant tumours which metastasise; the five year survival is in the region of fifty per cent.

TUMOURS OF NERVE SHEATH ORIGIN

These are derived from Schwann cells and have a resemblance to fibroblastic tumours. They may be differentiated by immunohistochemical staining as they are S100 positive. There are two main types of benign tumour of nerve sheath origin:

The schwannoma (or neurilemmoma) is generally a solitary encapsulated tumour which arises to one side of a nerve trunk. No nerve fibres course through the tumour, thus allowing removal without interference with the nerve function. Whilst these tumours can arise anywhere, one site of particular significance is in relation to the sheath of the auditory nerve intra-cranially. This benign tumour may cause problems because of its situation, but can be removed by neurosurgery with preservation of the faculty of hearing.

The histological pattern is fairly characteristic. In the more cellular part (Antoni A area) there may be palisading of the spindle shaped nuclei around hyaline material - the so-called Verocay bodies. The less cellular areas (Antoni B) appear more oedematous. Blood vessels within the tumour may show hyaline thickening.

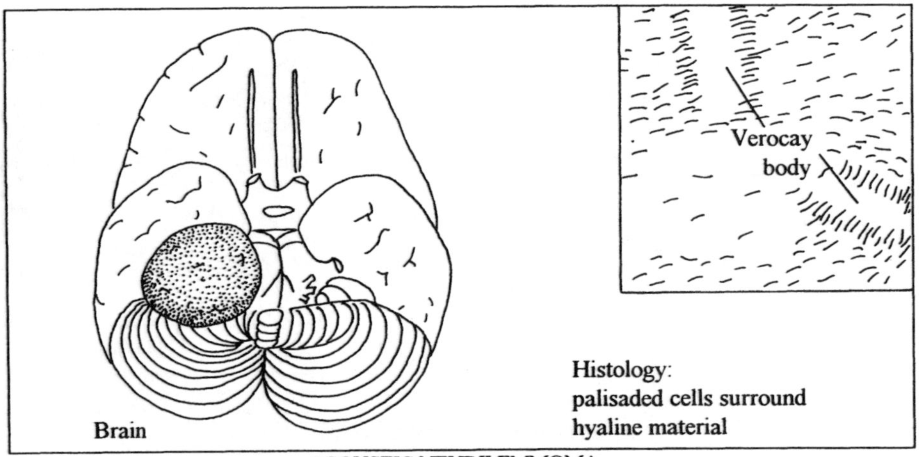

ACOUSTIC NEURILEMMOMA

Neurofibromas may be multiple, are not encapsulated and involve the nerve fibres in the proliferation of spindle-shaped cells with wavy nuclei set in a loose connective tissue background. Thus this tumour cannot be removed without interrupting nerve function. In the condition of neurofibromatosis multiple neurofibromas occur. This is an inherited condition (autosomal dominant) in half the cases in which multiple neurofibromas are associated with patches of pigmentation of the skin ('cafe-au-lait spots'). There are also pigmented hamartomas of the iris (Lisch nodules) The tumours may be very numerous and grow very large, and may even undergo malignant transformation. The gene has been mapped to chromosome 17 and there seems to be loss of a tumour suppressor gene (NF-1).

Rarely the liability to develop bilateral acoustic neurilemmomas may be inherited. This condition is known as type 2 neurofibromatosis, the gene for which has been mapped to chromosome 22.

Neurofibrosarcoma is a poorly differentiated malignant tumour which may arise *de novo* or from a neurofibroma in cases of neurofibromatosis.

MYXOID OR MYXOMATOUS TUMOURS

A number of tumours, including neurofibromas and one form of malignant fibrous histiocytoma, may show what is called myxomatous change, that is, areas in which the cells have a stellate appearance and are set in a loose matrix which is rich in glycosaminoglycans (like embryonic connective tissue). Some tumours appear to consist almost entirely of this sort of tissue and are referred to as myxomas, but it is doubtful whether these are separate entities in their own right. It is more likely that they are variants of one of the other sorts of tumour. Very often liposarcomas may have a myxoid differentiated appearance.

TUMOURS OF FAT

A lipoma is a benign tumour which consists of a localised collection of normal mature adult fat cells. Lipomas may be encapsulated. they are commonly found in the subcutaneous fat, particularly on the back or the back of the neck. grossly and microscopically they consist of normal appearing fat. Some lipomas may have a prominent fibrous or vascular component.

Liposarcomas arise *de novo* rather than from malignant transformation of a lipoma. They tend to occur in older age groups (40-60 years) in the deeper tissues of the lower limbs or retroperitoneally. Macroscopically they are yellowish white or grey timours, which may appear gelatinous. Cyst formation, haemorrhage and necrosis are common. Histologically there are a number of types:

- Well-differentiated liposarcoma consists of fat in which there may be a few areas of myxomatous change, some nuclear atypia, a few mitoses and some lipoblasts containing some fat vacuoles in abundant cytoplasm. Fibrous areas may resemble malignant fibrous histiocytoma.
- Myxoid differentiated liposarcoma contains primitive cells of mesenchymal type set in a myxoid stroma. Mature fat cells and lipoblasts may be seen together with occasional giant cells and mitoses. Small vessels are prominent, branching and giving rise to a 'chicken-wire' pattern.
- Round cell liposarcoma may be confused with lyphoma or an embryonal tumour, but is distinguished by the presence of lipoblasts.
- Pleomorphic liposarcoma is a highly anaplastic lesion but is also characterised by the presence of lipoblasts.

The first two forms, the well-differentiated and the myxoid are locally aggressive but rarely metastasise. The latter two are highly malignant and often have metastasised by the time the diagnosis is made.

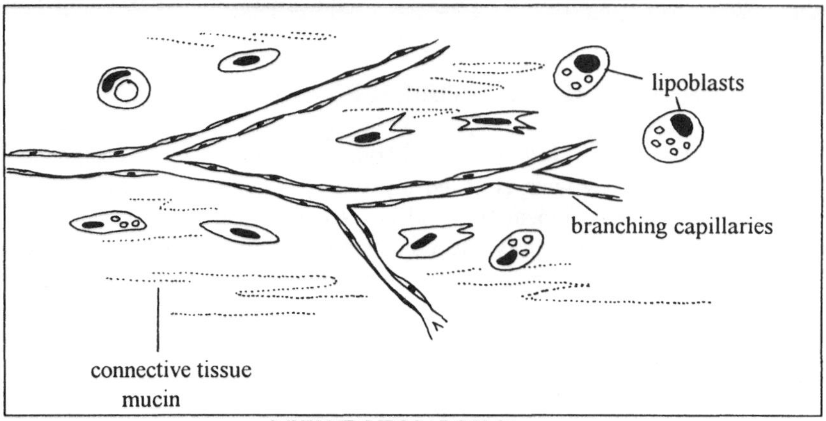

MYXOID LIPOSARCOMA

TUMOURS OF SMOOTH MUSCLE

Leiomyomas are perhaps the most common of all soft tissue tumours. About one in five women over the age of 30 may have a leiomyoma of the uterus (uterine fibroid). These tumours are not so frequently found at other sites, but can occur in the smooth muscle of the stomach and small intestine and may also be found in the skin and subcutaneous tissues (derived from smooth muscles of the hair follicles or blood vessel walls).

The tumour is well encapsulated and consists of interlacing bundles of smooth smooth muscle fibres. There may be interstitial fibrosis at a later stage, hence the name fibroid. Later still, hyaline change and calcification may occur, especially within uterine leiomyomas.

Leiomyomas probably very rarely undergo malignant change, but difficulties may arise in differentiating between a leiomyoma which is somewhat atypical and a well differentiated leiomyosarcoma. Where the mitotic rate is greater than ten mitoses per ten high power fields (HPF) of the microscope the tumour is regarded as a leiomyosarcoma, however well-differentiated the cells may appear. If there is cellular atypia it is regarded as a leiomyosarcoma if there are five mitoses per ten HPF.

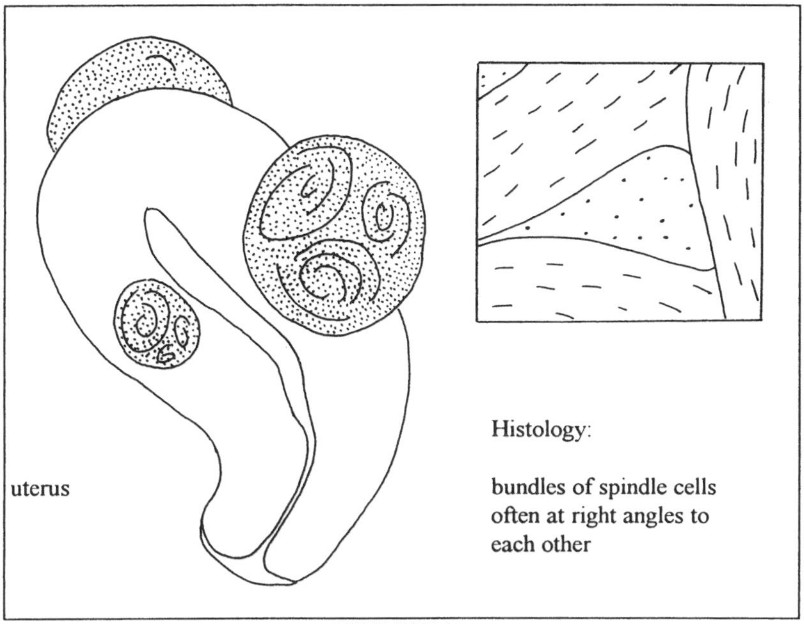

LEIOMYOMA

Leiomyosarcomas are much more rare. They are more likely to occur in extra-uterine sites, such as the wall of the stomach or intestine. There is a considerable range in the degree of differentiation. Poorly differentiated tumours may be characterised by very bizarre forms of cell and multinucleate tumour giant cells. Very anaplastic leiomyosarcomas may be difficult to diagnose, but special stains to demonstrate myofibrils or immunohistochemistry to show the presence of desmin and smooth muscle actin within the cells may be helpful.

TUMOURS OF STRIATED MUSCLE

Rhabdomyoma or benign tumour of skeletal muscle is extraordinarily rare. The so-called rhabdomyoma of the heart occurring in children is probably hamartomatous in nature.

Rhabdomyosarcomas are somewhat more common. Three major types have been described:

- **The embryonal form** occurs in the head and neck and genitourinary tract of infants and children. One particular form presents as a multilobulated mass resembling a bunch of grapes (sarcoma botryoides). Histologically there may be small primitive tumour cells set against a somewhat myxoid background. The diagnosis is made by finding larger, strap-like cells with atypical nuclei and cross-striations present in the cytoplasm. Immunohistochemistry may demonstrate desmin and myoglobin within these cells.

  Although a poorly differentiated sarcoma, it is responsive to combined therapy and up to 75 per cent ten year survival has been reported.

- **Alveolar rhabdomyosarcoma** occurs in adolescents and young adults, more commonly in males, and usually in the muscles of the extremities. Primitive malignant cells are arranged in groups separated by fibrous septa. Some rhabdomyoblasts can be seen.
- **Pleomorphic rhabdomyosarcoma** is the least common form, occurring in older adults in the skeletal muscles. It is extremely pleomorphic, with many bizarre cells of poor differentiation and the diagnosis may prove to be very difficult to establish.

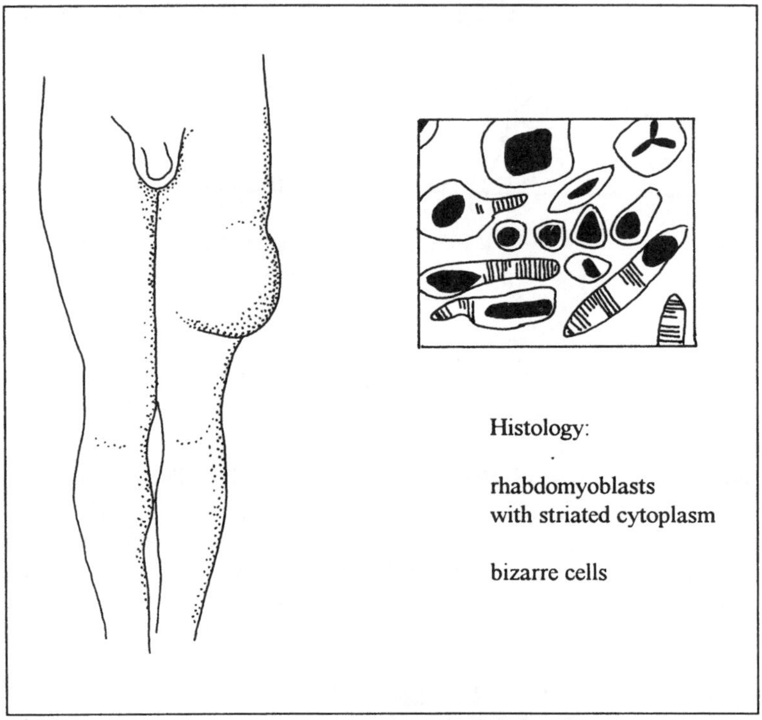

RHABDOMYOSARCOMA (Adult form)

TUMOURS OF CARTILAGE AND BONE

Occasionally tumours of cartilage (chondromas and chondrosarcomas) and tumours of bone (osteomas and osteogenic sarcomas) may be found in the soft tissues in extra-osseous locations.

TUMOURS OF SYNOVIUM

The benign synovioma is probably reactive rather than truly neoplastic. It arises from tendon sheaths and contains fibrocytic and fibrohistiocytic cells as well as histiocytes, multinucleate giant cells and haemosiderin pigment. It is probably related to the histiocytoma rather than to the synovium proper. This tumour is also known as pigmented villonodular synovitis when occurring within a joint, and giant cell tumour of tendon sheath when it occurs outside joints, commonly on the digits.

Synovial sarcoma, despite its name, does not occur in joints and is probably derived from primitive mesenchymal cells rather than synovial cells. The classical histological appearance is of a biphasic tumour composed of a mixture of cells of spindle cell type and epithelial like cells which form clefts or microcystic spaces. Some of these tumours are predominantly monophasic which may make diagnosis difficult. These are aggressive tumours with a poor prognosis (less than 50 per cent five year survival). A chromosomal translocation may be found in tumour cells (X;18) (p11.2;q11.2). Somewhat similar in appearance are the malignant tumours of mesothelium or mesotheliomas.

TUMOURS OF MESOTHELIUM

Malignant mesothelioma may resemble synovial sarcoma in its histological appearance, or it too may be monophasic of predominantly either fibrosacromatous or adenocarcinoamtous pattern. This tumour can be related to exposure to asbestos fibres. The pleura is chiefly affected by tumour, which appears to ensheath the lungs. Less frequently there may be peritoneal mesothelioma.

Biphasic forms of both synovial sarcoma and mesothelioma may show immunocytochemical reactions for keratins, epithelial membrane antigen *and* vimentin. Malignant mesothelioma is carcinoembryonic antigen (CEA) negative.

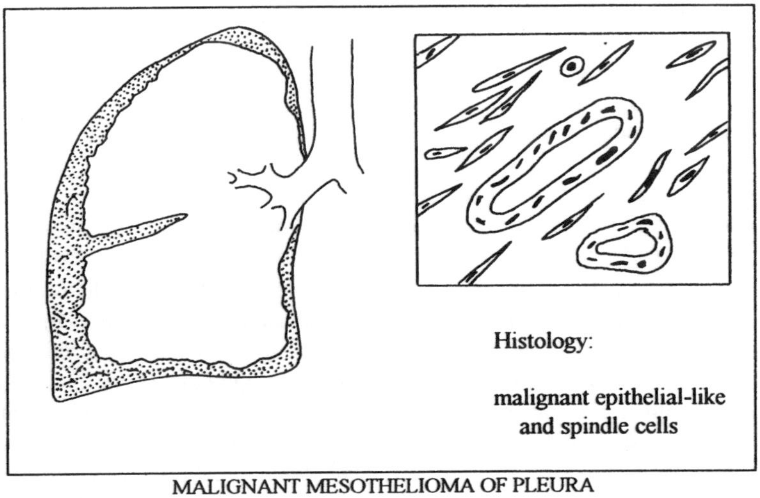

MALIGNANT MESOTHELIOMA OF PLEURA

TUMOURS OF VESSELS

There are a number of tumours of vessels which range from the reactive to the highly malignant. There is little doubt that vascular proliferations occurring in response to infection or irritation such as granuloma pyogenicum (haemangioma of granulation tissue type) are reactive. A similar lesion may occur on the gums during pregnancy. Also dilatations of vessels, known as telangiectases may occur in response to hormonal stimulation or irradiation.

There is some doubt as to whether benign lesions of vessels, which may be present at or shortly after birth and which appear to grow with the host and regress in later life, are genuine neoplasms or developmental abnormalities, that is, hamartomas. They will be described as neoplasms for the purposes of this chapter.

Benign haemangiomas are of a number of types:

- **Cavernous angiomas** consist of dilated vascular spaces lined by a single layer of epithelium and separated by fibrous septa. Common sites are on the skin or in the liver.
- **Capillary haemangioma** is a mass of tightly packed small capillaries with little intervening connective tissue. The skin is a common site. The strawberry naevus, a red prominent localised capillary haemangioma found in infants usually fades after a few years.
- **Glomus tumour** is a tumour of glomus bodies, vascular receptors most frequently found in toes and fingers. Branching vascular channels are present surrounded by aggregates of regular even rounded glomus cells. These tumours are often exquisitely tender.

There is also a tumour of intermediate malignancy.

- **Haemangioendothelioma** may be described as a haemangioma in which there is excessive proliferation of endothelial cells which may be many layers thick.

There are also three major types of malignant tumours of blood vessels:

- **Haemangiosarcoma** or angiosarcoma may occur anywhere in the body. In the liver these tumours may follow exposure to polyvinyl chloride (PVC). Any vascular tumour in the breast should be suspected of malignancy as angiosarcomas in this site may appear deceptively well differentiated. Histologically, there is marked proliferation of endothelial cells which are large and atypical, and new vessels appear with the formation of a multitude of anastomotic channels. These tumours are highly malignant.
- **Haemangiopericytoma** appears to be a tumour of pericytes which are situated externally to the endothelial cells. They tend to occur on the lower extremities or in the retroperitoneal region. Microscopically, numerous vascular channels may be seen within a spindle cell tumour. Reticulin stains will show that the pericytes are situated outside the capillary basement membrane.
- **Kaposi's sarcoma** is a rare tumour now becoming more common as it occurs in patients suffering from AIDS. There are four types described:

  - The European type in which reddish nodules appear on the skin of the feet and legs. These are locally aggressive and slowly spread proximally; visceral involvement occurs in a minority of patients.

- The African type occurs in children in disseminated lymph node form and in young adults as larger skin tumours which may invade locally, for example into the underlying bone, and spread to involve nodes and viscera.
- Kaposi's sarcoma in association with renal transplantation is associated with immunosuppressive therapy
- Kaposi's sarcoma in AIDS is characterised by the occurrence of widespread red or purple plaques or nodules in the skin, which rapidly spread to involve mucous membranes, the gut and lymph nodes.

Histologically, in the early stages there is proliferation of small vessels with some inflammatory cell infiltration and some extravasation of red cells which closely resembles granulation tissue. The nodules that form are composed of spindle shaped cells with prominent cytoplasm. Some mitoses may be seen. The spindle cells contain clefts that are filled with red cells. There is also new vessel formation, which may be inconspicuous as vessels are compressed by tumour cells. Haemosiderin deposition may be prominent in the tumour and at its edges.

Tissue	**Degree of malignancy**			
	Benign	**Intermediate**	**Malignant**	**Non-neoplastic**
Fibrous	Fibroma	Desmoid	Fibrosarcoma	Fibromatosis
Fibro-histiocytic		Dermatofibro-sarcoma protruberans	Malignant fibrous histiocytoma	Histiocytoma
Nerve sheath	Schwannoma Neurofibroma		Neurofibro-sarcoma	Amputation neuroma
Fat	Lipoma		Liposarcoma	
Smooth muscle	Leiomyoma		Leiomyo-sarcoma	
Striated muscle			Rhabdomyo-sarcoma	Rhabdo-myoma
Synovium			?Synovial sarcoma	?Benign synovioma
Mesothelium			Malignant mesothelioma	
Vessels	Cavernous angioma Capillary angioma Glomus tumour	Haemangio-endothelioma	Angiosarcoma Haemangio-pericytoma Kaposi's sarcoma	Granuloma pyogenicum

SOME TUMOURS OF SOFT TISSUES

GENERAL PATHOLOGY - CHAPTER THIRTY

SOME TUMOURS OF LYMPHOID TISSUE AND BONE MARROW

Having considered some of the tumours of epithelial tissues and of connective tissues it is appropriate to think about tumours of lymphoid tissue and bone marrow, which fall into three major groups and one minor group:

- **Malignant lymphomas,** including Hodgkin's disease and the non-Hodgkin's lymphomas, which are cancers of lymphoid tissue usually arising in lymph nodes.
- **Leukaemias and the myeloproliferative disorders** which arise from haemopoietic stem cells in the bone marrow.
- **Plasma cell dyscrasias** in which there is malignant proliferation of immunoglobulin antibody-producing cells. Multiple myeloma is the best known example of these tumours which usually arise in the bone marrow.
- **The histiocytoses** include rare malignant proliferations of histiocytes which present as lymphomas and also localised and generalised proliferations of Langerhans cells.

These tumours will be considered in greater detail in the course on systematic pathology, but as these tumours may be disseminated throughout the body at an early stage and as they may have generalised immunological and biochemical effects an overview of them at this stage is timely.

THE MALIGNANT LYMPHOMAS

It should be emphasised at the outset that the most common malignant tumours found in lymph nodes are secondary carcinomas. Primary malignant tumours of lymph nodes are the malignant lymphomas.

HODGKIN'S DISEASE

This disease arises in a node or group of nodes and tends to spread to adjacent nodes, and then to more distant ones before spreading to extranodal sites such as liver and bone marrow. It is characterised by the presence of a histologically diagnostic and distinctive cell, the Reed-Sternberg cell. This is a large cell which has either a lobed nucleus or multiple nuclei. Classically there are two mirror-image nuclei each containing a prominent eosinophilic nucleolus surrounded by a clear zone or halo, giving rise to the so-called 'owl-eye' appearance.

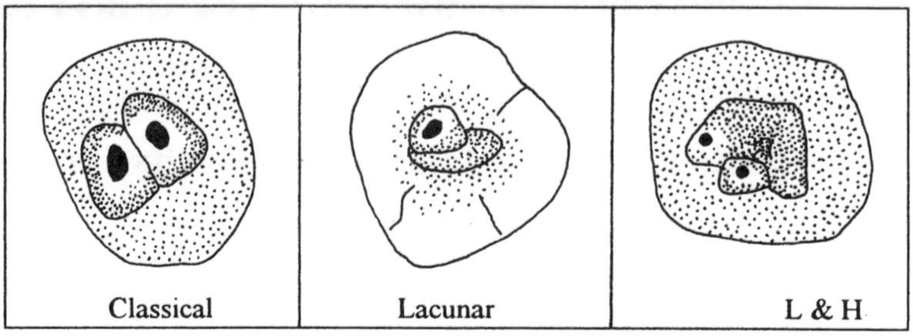

CLASSICAL REED-STERNBERG CELL AND VARIANTS

Variants of this cell may be found in different subtypes of Hodgkin's disease. The Reed-Sternberg cells appear to be the neoplastic cells, but the normal architecture of nodes is destroyed and there is a mixed infiltrate of lymphocytes, plasma cells, macrophages and eosinophils throughout the node. There may also be fibrosis.

There are four stages of disease spread, and four different histological subtypes:

	Staging of Hodgkin's disease	**Histological classification**
I	involving single node or group or single extra-nodal site	Lymphocyte predominance (L & H cells)
II	involving two or more groups of lymph nodes on the same side of the diaphragm	Nodular sclerosis (lacunar cells) - type I - type II (lymphocyte-depleted nodules)
III	involving lymph nodes on both sides of the diaphragm +/- spleen	Mixed cellularity
IV	diffuse extranodal involvement, for example,bone marrow or liver	Lymphocyte depletion

*** A** No systemic symptoms

B Symptoms: fever, night sweats, unexplained loss of weight of more than ten per cent

The prognosis of Hodgkin's disease can be related to the spread of the disease, as estimated according to the clinical staging, and the histological type.

Lymphocyte predominance. This type forms about ten per cent of cases of Hodgkin's disease and occurs mainly in young adult males. Affected lymph nodes have lost their normal architecture and appear to consist almost entirely of mature lymphocytes, with some benign histiocytes and with few Reed-Sternberg cells. A multilobed variant the L & H (lymphocytic and histiocytic) or 'popcorn' cell seems to be the main neoplastic cell. The prognosis of this variant is very good.

Nodular sclerosis. This forms about 60-70 per cent cases and affects young adults. Unlike the other types it is more common in females. Affected lymph nodes are divided into nodules by fibrous bands.

There are some Reed-Sternberg cells, but the characteristic cell of this type is the lacunar cell, an atypical cell with peripheral cytoplasmic vacuolation, so that the cell appears to be sitting in a clear space. The prognosis is generally good (especially of type I) if the disease has not progressed to the later clinical stages (III and IV).

Mixed cellularity. This is the second most common type (about 20 per cent of cases) and tends to be even more common in third world countries. Histologically, Reed-Sternberg cells and mononuclear variants are prominent and set against a background of mixed cells (lymphocytes, histiocytes and epithelioid cells, plasma cells and eosinophils). This type is often more advanced at the time of presentation and the outlook is less good.

Lymphocyte depletion. This type forms only five per cent of cases. Histologically there are few lymphocytes and atypical cells resembling Reed-Sternberg cells are prominent. Sometimes there is diffuse fibrosis with a paucity of cells of all types. Patients presenting with this form tend to be older, to have more advanced and aggressive disease and to have a generally worse outlook.

Overall the presence of lymphocytes and of nodular sclerosis are good prognostic features. However with current therapy any type may respond and the clinical stage is the more important prognostic indicator.

NON-HODGKIN'S LYMPHOMAS

These are a varied group of tumours mostly arising in lymph nodes, although up to one third may occur in extra-nodal lymphoid tissue such as in bone marrow, gut, respiratory tract and skin. There is likely to be diffuse nodal involvement at a relatively early stage because of the circulation of lymphocytes.

The tumours are monoclonal proliferations of B cells (two thirds of cases) or T cells, and can be recognised as such by special techniques, including immunohistochemical staining. Lymphomas from immunoglobulin producing B cells will express only one type of immunoglobulin light chain (all κ or all λ). All cells will show similar rearrangement of the immunoglobulin genes. T cell tumours will show clonal rearrangement of the T cell antigen receptor genes.

The histological classification of the non-Hodgkin's lymphomas is based upon:

- **The architecture of the tumour**: Nodular or follicular tumours (all of B cell origin) have a better prognosis than those with a diffuse pattern of growth.
- **The size and shape of the cells**: Small cells and cleaved cells are generally associated with a better outlook and less aggressive behaviour.

Microscopically, the normal architecture of the node is destroyed in lymphoma and there may be invasion of neoplastic cells through the capsule of the node. Follicular tumours are composed of poorly defined enlarged abnormal follicles, with atypical cells and mitotic figures inside and outside the follicle. In diffuse tumours there is usually a uniform diffuse monotonous infiltrate of neoplastic cells replacing the node.

These observations have been used in what is a generally accepted classification of the non-Hodgkin's lymphomas for clinical purposes.

These observations have been used in what is a generally accepted classification of the non-Hodgkin's lymphomas for clinical purposes.
More complex histological classifications exist (such as the Kiel classification) and these make use of morphological features to assess the degree of differentiation or maturation of cells (for example, centroblasts and centrocytes). These lymphomas may also be staged in a fashion similar to Hodgkin's disease.

International Working Formulation for the classification of non-Hodgkin's Lymphomas (NHLs)

LOW GRADE survive 7-10 yrs with or without treatment	3-4%	95%B-cells	Small lymphocytic; chronic lymphocytic leukaemia; plasmacytoid
		all B-cells	Follicular: predominantly small cleaved cell
	40% of NHLs follicular	all B-cells	Follicular: mixed small cleaved cell and large cell
INTERMEDIATE GRADE survive 2-4 yrs without treatment		all B-cells	Follicular: predominantly large cell
		B-cells	Diffuse: small cleaved cell
		70% B-cells	Diffuse: small and large cell
	40-50% NHLs large cell	70% B-cells	Diffuse: large cell
HIGH GRADE die in 1 yr without treatment		70% B-cells	Large cell: immunoblastic, plasmacytoid, clear cell, polymorphous
	4% of NHLs	95% T-cells	Lymphoblastic: convoluted or non-convoluted
	<1% of NHLs	all B-cells	Small non-cleaved cell: Burkitt's lymphoma

To this classification a miscellaneous group may be added:

- Mycosis fungoides, a T cell lymphoma of the skin
- Adult T cell leukaemia/lymphoma following infection with human T cell leukaemia virus type-1 (HTLV-1).

In B cell follicular lymphomas there is almost always a chromosomal abnormality with a translocation t(14;18); the break point on chromosome 18 is near the site of the protooncogene *bcl-2* which is thought to be moved to an area of greater transcriptional activity. This gene operates by preventing apoptosis or programmed cell death and extends cell survival.

THE LEUKAEMIAS AND MYELOPROLIFERATIVE DISORDERS

Very little will be said here about these haematological disorders which are neoplastic monoclonal proliferations of the cells of the bone marrow. The tumour cells escape from the bone marrow into the blood, and there is some overlap with the non-Hodgkin's lymphomas where spill over into the blood can occur late in the stage of the disease if there is marked marrow involvement.

Acute leukaemias can be divided into

- Acute lymphoblastic leukaemia (ALL)
- Acute non-lymphoblastic leukaemia (ANLL) which may be of myeloid, monocytic, erythroid or megakaryocytic differentiation.

The acute leukaemias are proliferations of haemopoietic stem cells which are unable to mature and remain in undifferentiated form as blast cells.
Chronic leukaemias are of two main types:

- Chronic lymphocytic leukaemia (CLL)
- Chronic myeloid (granulocytic) leukaemia (CML)

Chronic leukaemias are proliferations of stem cells which then proceed to undergo some degree of maturation or differentiation.

Chronic lymphocytic leukaemia has some features in common with diffuse small lymphocytic lymphoma; for example, the appearances in the lymph nodes are identical in these two conditions.

Chronic myeloid leukaemia is one of the **myeloproliferative disorders**, together with polycythaemia vera (proliferation of red blood cells); myeloid metaplasia with myelofibrosis in which the marrow becomes fibrosed whilst haemopoiesis takes place in other sites such as spleen and liver; and thrombocythaemia or proliferation of platelets.

The leukaemias are more common in older patients, with the exception of acute lymphoblastic leukaemia, which is the most frequent leukaemia in children.

The leukaemias infiltrate and replace the bone marrow and there is a resultant shortage of red cells, white cells and platelets. Death may ensue from infection due to deficiency in normal white cells or from bleeding due to lack of platelets.

The spleen is diffusely infiltrated and may attain a very large size, especially in chronic myeloid leukaemia. The lymph nodes may be enlarged, especially in chronic lymphocytic leukaemia. The liver is infiltrated (periportally in the lymphocytic leukaemias; more diffusely in myeloid leukaemias).

Chromosomal abnormalities may be found in a number of the lymphomas (for example, t(14;18) in the follicular B cell lymphomas). In 90 per cent of patients with chronic myeloid leukaemia there is a chromosomal abnormality known as the Ph^1 (Philadelphia) chromosome in which there is a reciprocal translocation from 22q to 9q which may be recognised from karyotyping.

PLASMA CELL DYSCRASIAS - MULTIPLE MYELOMA

These are B cell neoplasms in which there is malignant proliferation of a clone of immunoglobulin-producing cells, resulting in the excessive production of a specific abnormal protein or paraprotein, the M-protein (Myeloma protein). This protein may be identified by electrophoresis as it forms a narrow, very dense band usually in the γ-globulin region. The protein is usually IgG, less often IgA and rarely another immunoglobulin.

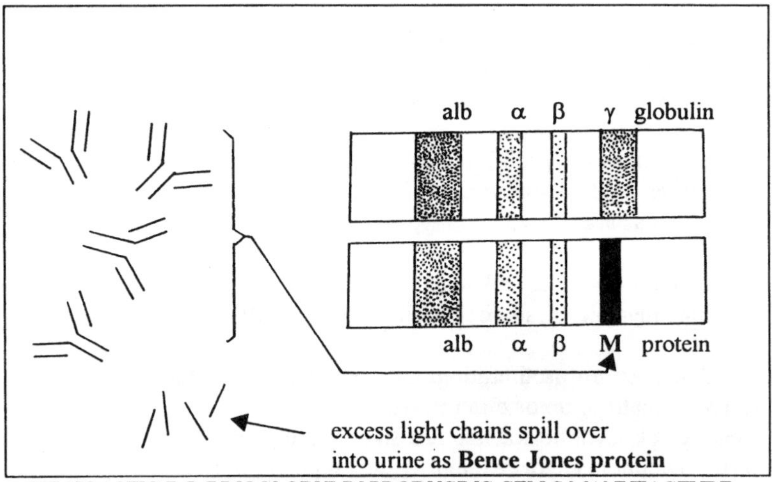

NEOPLASTIC IMMUNOGLOBULIN PRODUCING CELLS MANUFACTURE M-PROTEIN AND EXCESS LIGHT CHAINS

The malignant cells may form immunoglobulin imperfectly, producing an excess of light chains (of one type only, either all κ or all λ). Free light chains may spill over into the urine where they can be recognised as Bence Jones protein. This can be recognised simply because it precipitates when urine is heated to 55°C or 60°C and redissolves at a higher temperature just before the urine boils. This is unlike other types of protein in the urine which are usually precipitated on boiling and cannot be redissolved. AL amyloid is derived from light chains or Bence Jones protein.

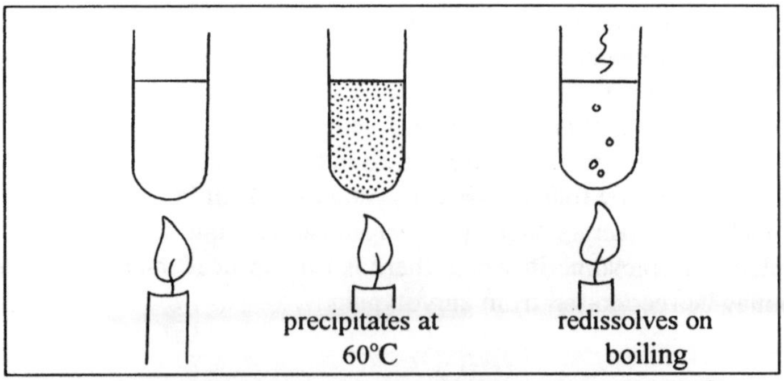

BENCE JONES PROTEIN IN URINE (EXCESS LIGHT CHAINS)

MULTIPLE MYELOMA.

This is a malignant growth or cancer of more or less mature plasma cells which grow multifocally in the bone marrow.
The clinical effects are due to two main causes:

- Proliferation of neoplastic plasma cells within bone and elsewhere. The cancer cells destroy and replace bone, causing pain, swelling, deformity and even pathological fracture. The skull, vertebrae, ribs, pelvis and femur are the bones most often affected. X-rays show radiolucent punched-out holes in the bones where the plasma cells are proliferating. As the bone marrow is replaced by the growth, normal haemopoietic marrow is destroyed and deficiency of the formed elements of the blood occurs.

  Thus there is: deficiency of red cells - **anaemia**
  deficiency of white cells - **leucopenia**
  deficiency of platelets - **thrombocytopenia**

  as there may be in the leukaemias or in any condition causing widespread marrow infiltration.

  There may also be hypercalcaemia due to destruction and lysis of bone.

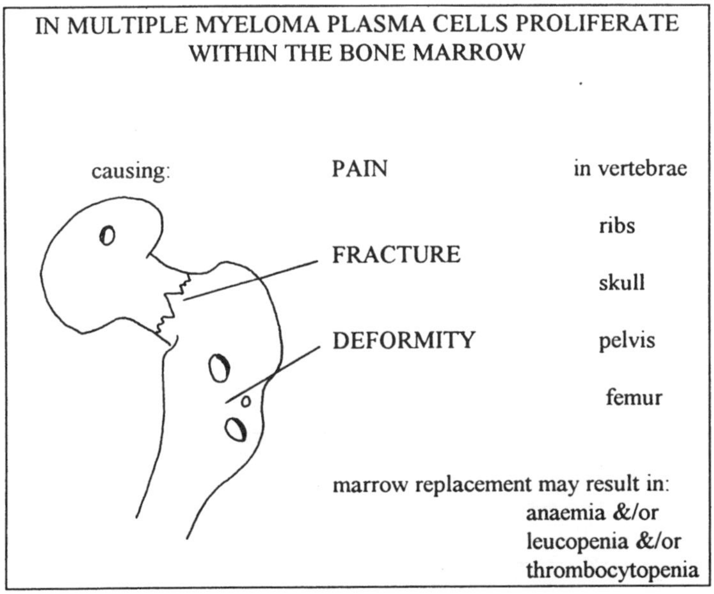

Patients are prone to infection because of the reduction in white cells. There may be dangerous bleeding or haemorrhage due to platelet deficiency.

- Abnormal protein production results in the presence of M protein in the circulating blood and/or Bence Jones protein in the urine in virtually all cases. In up to 20 per cent of cases Bence Jones proteinuria may occur alone. Electrophoresis of serum and urine will usually establish the diagnosis which can be confirmed by X-ray and marrow biopsy.

The rapidly growing proliferation of abnormal plasma cells is also related to deficient production of normal immunoglobulins, which are reduced and there may also be decreased resistance to infection on this account.

Bence Jones protein may precipitate in the kidney, blocking the tubules and interfering with renal function.

Amyloid deposition of AL or light change type may be noted in about ten per cent of cases

Increased viscosity of the blood due to hyperglobulinaemia may cause ischaemia or worsen already existing ischaemic conditions.

The kidney may show changes which exemplify many of the effects of this disease.

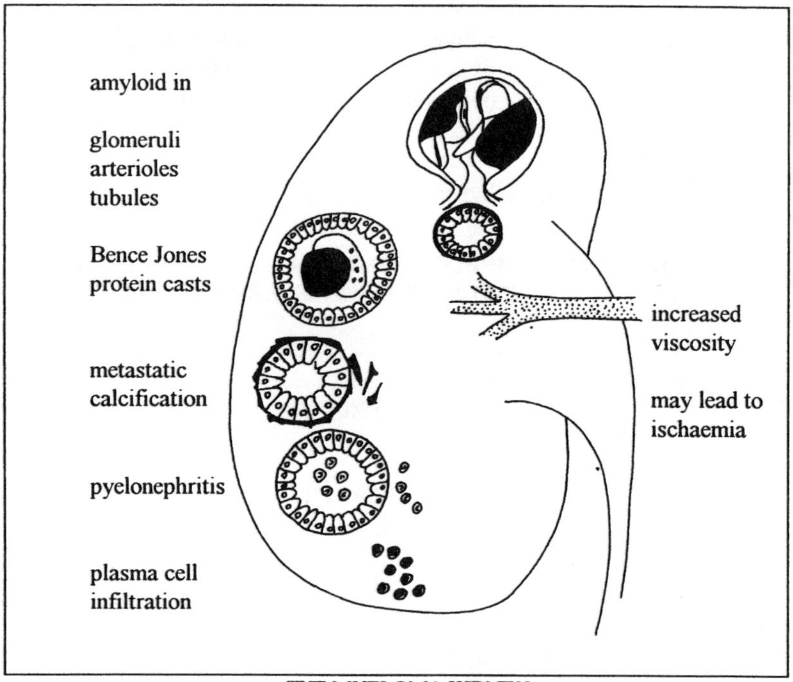

THE MYELOMA KIDNEY

Multiple myeloma is a serious condition with a poor survival, even with treatment.

OTHER PLASMA CELL DYSCRASIAS include:

Solitary plasmacytoma in which a single lesion may occur in the skeleton or in the soft tissues. Bony lesions tend to progress into multiple myeloma with time, but soft tissue lesions are less likely to disseminate and may be curable by surgery.

Waldenstrom's macroglobulinaemia can be looked at as a sort of cross between small lymphocytic lymphoma and multiple myeloma.

There is a diffuse infiltrate of marrow, lymph nodes, liver and spleen by lymphocytes and plasma cells and intermediate plasmacytoid lymphocytes. The high molecular weight proteins are monoclonal IgM which can be demonstrated on serum electrophoresis.

Heavy chain disease is very rare. There may be production of *only* heavy chains:

- γ-chain disease resembles lymphoma
- α-chain disease is associated with plasma cell proliferation in the intestine, which may develop into full blown lymphoma, usually of immunoblastic type
- μ-chain disease is very rare but may sometimes occur in chronic lymphatic leukaemia.

Monoclonal gammopathy or the appearance of M protein may sometimes occur in the apparent absence of other disease. A proportion of these cases will go on to develop multiple myeloma or other plasma cell dyscrasia.

THE HISTIOCYTOSES

Very occasionally a true lymphoma may develop from histiocytes.

Accumulations of histiocytes occur in lysosomal storage disease, where glycogen or fatty materials such as sphingomyelin may accumulate within lysosomes in greatly enlarged cells.

One group of diseases in which there is proliferation of a histiocytic cells goes by the designation of **histiocytosis X.** The cells involved are Langerhans cells, which are probably antigen-presenting cells. These cells are normally found in the epidermis. These cells have a lobulated nucleus, and on electron microscopy can be seen to contain inclusions, shaped like tennis rackets, which are called Birbeck granules. Langerhans cells contain S-100 protein which can be demonstrated by immunohistochemistry. In histiocytosis X many eosinophils are often present in the lesions.

There are a number of forms of the disease

Letterer-Siwe disease is the name given to the acute disseminated form which behaves like a malignant tumour. This disease occurs in infants and young children. There is proliferation of Langerhans cells in the skin and reticulo- endothelial system. Patients present with multiple skin nodules, generalised lymphadenopathy and later bony lesions, which may lead to problems with anaemia, leucopenia and thrombocytopenia. The liver and spleen are usually enlarged. The clinical picture is not unlike that of acute leukaemia. Patients usually die within a year of diagnosis without treatment, although there may be a good response to chemotherapy.

Hand-Schüller-Christian disease is the name given to multifocal lesions of proliferating histiocytes (often with foamy cytoplasm) mixed with eosinophils, lymphocytes, plasma cells and even a few neutrophils usually found in the bones, especially of the skull. the classical presentation is of :

- Radiolucencies of the skull due to the bony lesions.
- Diabetes insipidus due to involvement of the posterior pituitary or hypothalamus
- Due to orbital deposits.

This form tends to occur in older patients, and even if the lesions are disseminated with involvement of, for example, skin, liver and spleen, the prognosis is better than that of Letterer-Siwe disease. Half the patients will recover and others respond to chemotherapy.

Eosinophilic granuloma Solitary lesions of histiocytes, eosinophils and mixed inflammatory cells may occur in bone, or less commonly lymph node. The nodular collections of histiocytes resemble granulomatous inflammation, hence the name. This form occurs in adolescents or young adults, and may regress spontaneously or respond to local surgical treatment.

Although the cell of origin is now known, the nature of the disease and its pathogenesis are still obscure.

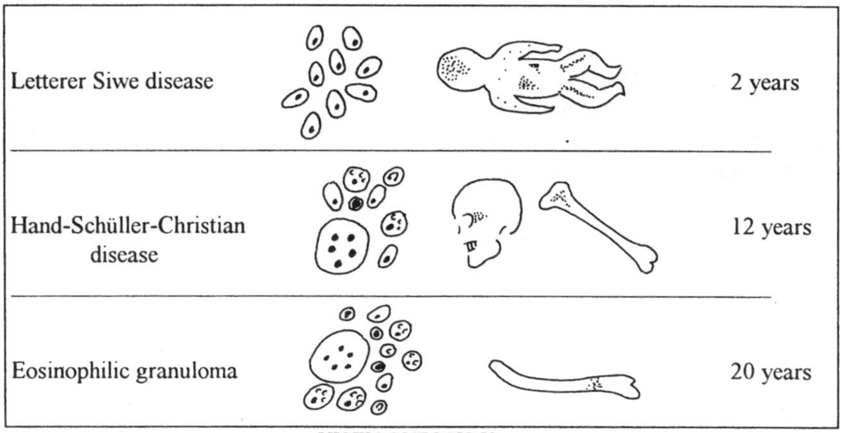

HISTIOCYTOSIS X

GENERAL PATHOLOGY - CHAPTER THIRTY ONE

SOME TUMOURS OF CHILDHOOD

A number of tumours, benign and malignant, occur in childhood. Some of these are tumorous developmental malformations and some are true neoplasms. Sometimes the child may be born with the lesion.

Childhood tumours are not as unimportant as they may seem, because although malignant tumours of childhood are distinctly uncommon, they nevertheless now rank as one of the major causes of death in infancy and childhood, second only to accidental deaths in most Western countries.

DEVELOPMENTAL TUMOURS

When one considers the miracle of embryonic development it always seems surprising that mistakes occur so infrequently.

Any anatomical error of development may give rise to a congenital malformation. These are usually obvious at birth, but not always so. It is not possible here to consider all the various types of malformation.
have already been mentioned. These do not give rise to tumour formation.

Some, such as **agenesis** (the complete failure of development of a part) and **hypoplasia** (the incomplete development of a part) have already been mentioned. These do not give rise to tumour formation.

Choristoma is the name given to a tumour-like mass of normal, non-neoplastic ectopic tissue in an abnormal location, for example, adrenal cells in kidney and pancreatic tissue in the wall of the stomach or intestine. This is growth of normal tissue in an abnormal (ectopic) site.

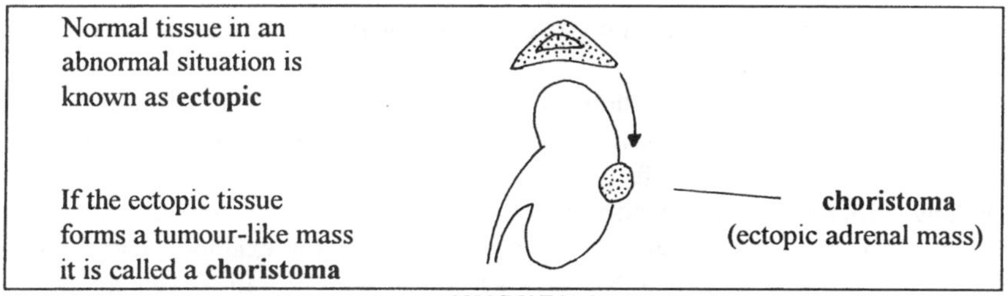

CHORISTOMA

A **hamartoma** is a tumour-like malformation in which the tissues of a particular part of the body are arranged haphazardly, usually with an excess of one or more of its components.

There is abnormal growth and arrangement of cells or tissues normally found in that particular part of the body. This results in a tumour-like mass or hamartoma, for example, chondroid hamartoma of the lung in which the chief component is cartilage (of bronchial cartilage origin).

Hamartomas may be difficult to distinguish from true neoplasms, especially if the latter are benign:

- Hamartomas consist of tissues specific to the part in which they arise.
- They have no tendency to excessive growth. Their growth is co-ordinated with that of the organism, therefore they do not produce pressure effects or induce capsule formation, and growth usually stops at adolescence when normal growth stops.

Many benign tumours which are apparently neoplastic appear to be of congenital origin. They may be present at birth or occur during childhood. Sometimes they remain small and grow in later life. There is argument as to whether these are hamartomas or true neoplasms. Examples are naevi and angiomas.

- **Naevi.** Naevus (the plural of naevus is naevi) is a word which can be used to describe any developmental blemish of the skin. It is used most often to describe pigmented cellular lesions (naevi) or moles. Vascular lesions and lesions of sebaceous glands can also sometimes be called naevi.

  Moles or pigmented naevi are probably present at birth, although they may not grow appreciably until puberty or later life.
- **Angiomas** are tumours of vessels, either haemangiomas of blood vessels or lymphangiomas of lymphatics. They consist of masses of vascular channels arranged haphazardly and again, benign angiomas are thought by many to be hamartomatous.

There can be little doubt that the malignant forms of these are true neoplasms.

- **Malignant melanoma** is a malignant tumour of melanocytes
- **Haemangiosarcoma** and **lymphangiosarcoma** are malignant tumours of blood vessels and lymphatic vessels respectively.

Malignancies of these types are, however, extremely rare in children and progression from benign neoplasm or hamartoma to a malignant tumour takes place only very infrequently, and then in later life.

Finally, other congenital malformations may result in the development of true tumours which may then show malignant proliferation, for example, teratomas and embryonic tumours. Unfortunately, there are many gaps in our knowledge and we are far from being able to categorise tumours exactly as to their origin.

Teratomas are tumours consisting of multiple tissues foreign to the part from which they arise, and usually containing representatives of the three primitive germ layers.

They may be considered to be congenital tumours or true neoplasms if it is believed that they are derived from persistence of toti-potential embryonic cells which have not developed with the rest of the body, but have become activated at a later stage. This theory may certainly account for the origin of midline teratomas, for example, mediastinal or sacrococcygeal. These may be malignant in about ten per cent of cases.

Gonadal teratomas are, however, more likely to result from an unsuccessful attempt at development by an unfertilised germ cell in late childhood or early adulthood.

THE EMBRYONIC TUMOURS OF INFANCY

The embryonic tumours are those arising in still immature tissues, in which a part of a developing area undergoes neoplastic transformation, which may result in a highly malignant, rapidly growing tumour. These tumours may be present at birth, but usually develop in children less than three or four years of age, although they may occur later. Whilst not as common as leukaemia (usually acute lymphoblastic leukaemia), which accounts for the majority of deaths in childhood from malignancy, these tumours as a group are none the less important. They are rapidly growing, high grade tumours but are susceptible to treatment with combined therapy of surgery, chemotherapy and radiotherapy. Surprisingly good results and greatly improved cure rates have been achieved in recent years.

Genetic studies on these tumours have increased our knowledge of the inheritance of tumours and of the genetic influences at work in their causation.

Organs or tissues which may be affected include :

• Adrenal and sympathetic nervous tissue -	**neuroblastoma**
• Kidney -	**nephroblastoma** (Wilm's tumour)
• Retina of eye -	**retinoblastoma**
• Brain-cerebellum -	**medulloblastoma**
• Liver -	**hepatoblastoma**
• Muscle -	**embryonal rhabdomyosarcoma**

Neuroblastoma is a tumour of the adrenal medulla or of sympathetic nervous tissue (sympathetic ganglia); neuroblastomas rarely may arise in the brain.

It may present as an abdominal swelling or with liver or bony secondary deposits.

Microscopically, it is composed of small, tightly packed, darkly staining primitive cells resembling lymphocytes.

Sometimes these cells are arranged around a central space - forming 'rosettes'. Other evidence of differentiation includes fibril formation (like primitive nerve fibres) and differentiation of the cells into ganglion cells.

Occasionally the whole tumour may differentiate into fibres and nerve cells. It is then called a ganglioneuroma. This differentiated form is no longer malignant.

Only a few neuroblastomas differentiate into the benign form. Some appear to regress spontaneously, but the majority are highly malignant tumours which spread locally and metastasise, especially to the liver and the bones. Later in the course of spread any site may develop metastases.

Differential diagnosis may be difficult as the tumours consist of sheets of small poorly differentiated cells. Distinction must be made from lymphoma, Ewing's sarcoma and peripheral neuroectodermal tumours, as well as from small cell carcinoma of any sort.

Tumour cells may secrete catecholamines. They contain membrane-bound granules on electron microscopy. Immunocytochemistry for neuro-specific enolase is positive.

Cytogenetic studies of advanced tumours have shown chromosomal abnormalities affecting the short arm of chromosome 1 (1p32). Homogeneous staining regions and double minute chromosomes are formed from multiple copies of the protooncogene N-*myc*.

Treatment is with surgery and radiotherapy. Generally speaking, the earlier the diagnosis is made, the better is the prognosis.

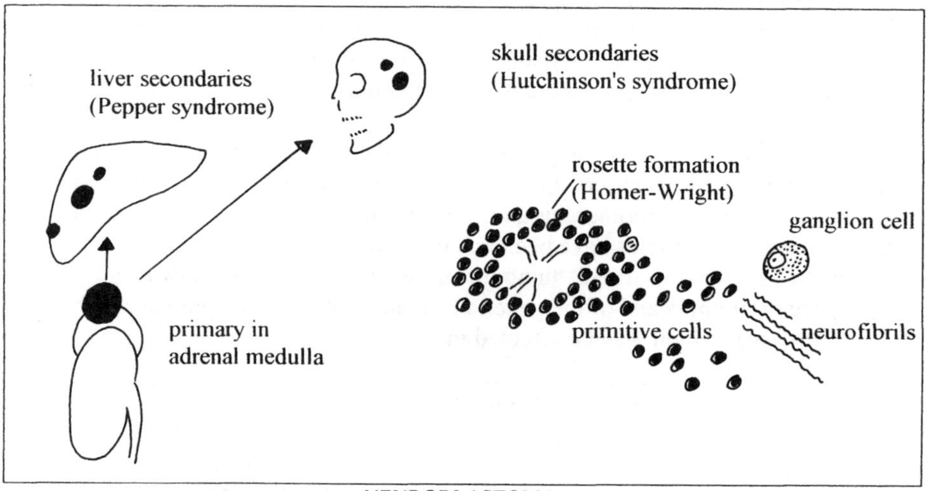

NEUROBLASTOMA

Nephroblastoma (Wilm's tumour) appears to be a malignant tumour arising from the primitive blastema of the embryonic kidney.

It presents as an abdominal mass in infancy or early childhood.

Macroscopically, the tumour is large, rounded or bosselated and may appear to compress the remaining normal kidney. Most of the kidney may be replaced by the tumour.

Microscopically, blastema cells are small dark primitive cells. They may show differentiation into tubules, or less frequently, into attempts at primitive glomerulus formation.

The blastema cells are set in a primitive mesenchymal stroma, which may form the bulk of the tissue. This generally has a myxomatous look, but differentiation into mesodermal structures such as smooth and striated muscle, cartilage and bone can occur.

The tumour is highly malignant and spreads locally via lymph nodes, and may metastasise to lungs and liver and brain. In less than ten per cent of cases both kidneys may be involved.

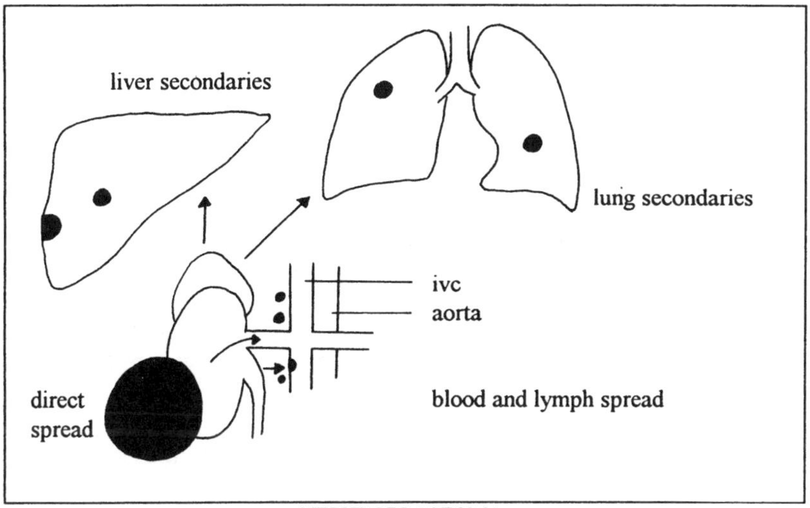

NEPHROBLASTOMA

However, this primitive tumour is sensitive to modern methods of therapy and a combination of surgery, radiotherapy and chemotherapy can result in a remarkable number of cures (there may be up to 90 per cent survival for two years which usually means complete cure).

Nephroblastoma may be familial in some instances, like retinoblastoma.

An abnormality or deletion in the short arm of chromosome 11 (11p13) together with loss of the cancer suppressor gene WT-1 has been described, but it is likely that other loci could also be involved.

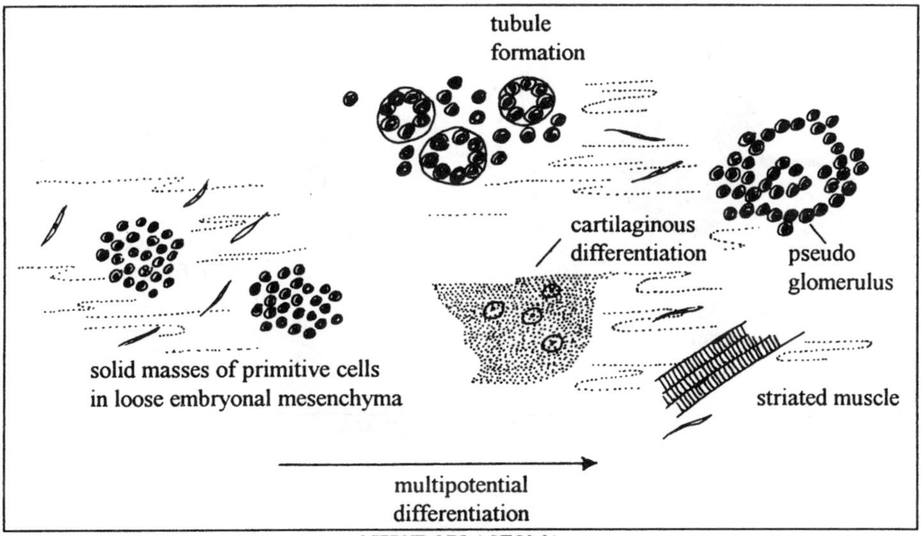

NEPHROBLASTOMA

Retinoblastoma is a tumour of primitive retinal cells which appears to be inherited as an autosomal dominant in about 40 per cent of cases. It appears that a tumour suppressor gene (the RB gene) is located on the long arm of chromosome 13 (13q14). In inherited transmission one allele of this gene is defective. This defect is then transmitted as an autosomal dominant. The recipient of the defective gene only then needs a somatic mutation of some sort to inactivate the other gene within the retinoblasts for the tumour to develop. In sporadic cases it is necessary for local mutations within retinoblasts to affect both alleles before retinoblastoma will occur. Familial cases may be bilateral and multifocal, whereas sporadic cases are unilateral and unifocal.

It should be noted that loss of the RB tumour suppressor genes is not confined to retinoblastoma, and has been noted in other tumours including osteosarcoma and carcinoma of breast and lung. Patients with retinoblastoma are also at risk of developing other cancers.

The tumour causes blindness and presents as a mass behind the lens of the eye. In up to one third of cases the tumour may be bilateral.

Microscopically, small round darkly staining cells form the tumour. Rosettes may be formed and these may have an internal limiting membrane.

Spread may be intra-ocular, or extra-ocular into the orbit via direct extension or into the brain via the optic nerve. Orbital lymph nodes may be involved, but systemic spread and metastases elsewhere are unusual.

Treatment is by enucleation of the eye and radiotherapy by which cure rates of up to 90 per cent may be achieved.

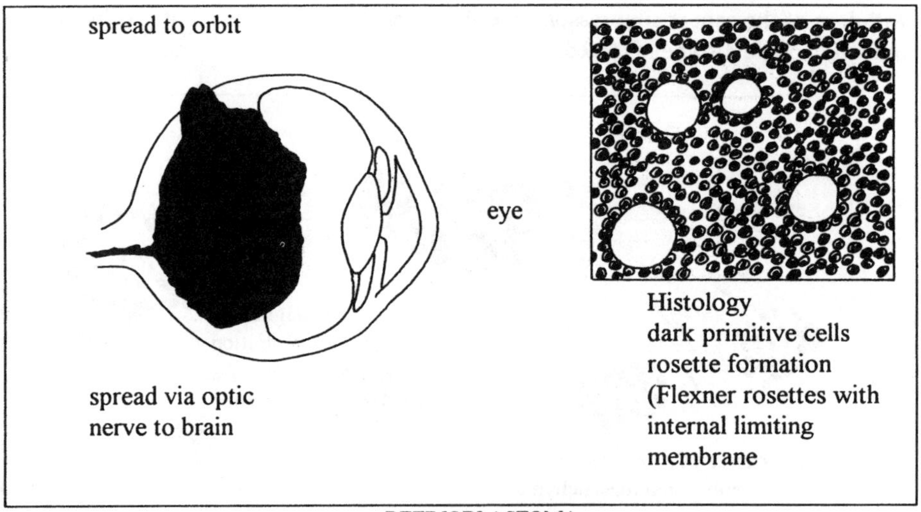

RETINOBLASTOMA

Medulloblastoma is a tumour thought to arise from the foetal cells of the external granular layer of the cerebellum. It may occur in older children or adolescents.

It occurs in the midline of the cerebellum as a fleshy gray mass which usually projects into the fourth ventricle, blocking the flow of cerebro-spinal fluid, and causing raised intracranial pressure.

Microscopically, it consists of small darkly staining, tightly packed cells with oval or carrot shaped nuclei. These may show rosette formation occasionally (evidence of primitive neuronal differentiation). There may also be evidence of glial differentiation with spindle cells containing Glial Fibrillary Acidic Protein (GFAP).

Spread is via the cerebrospinal fluid, with seeding deposits formed anywhere in the central nervous system. Metastasis does not occur outside the nervous system.

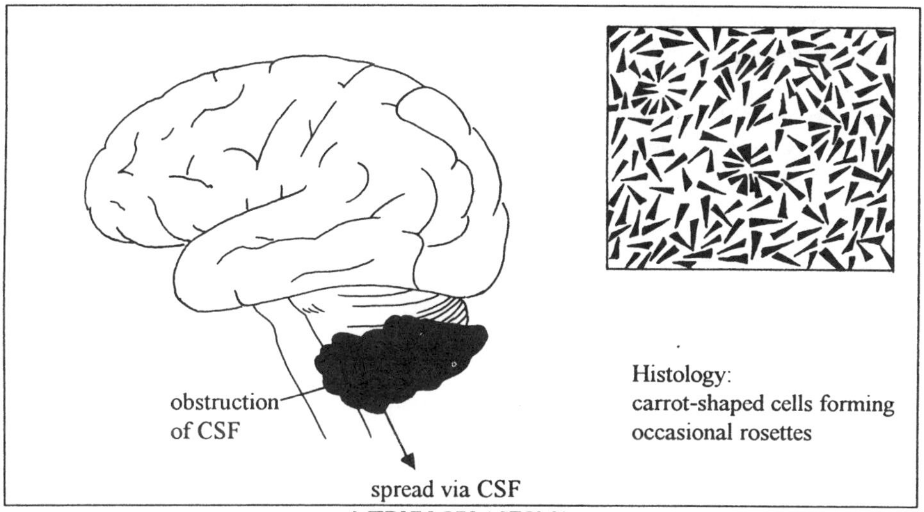

MEDULLOBLASTOMA

Hepatoblastoma is a very rare developmental tumour of the liver apparently from hepatic blastema. It may or may not have a mesenchymal component similar to that described for nephroblastoma.

The embryonic sarcomas can be mentioned here. The embryonal form of rhabdomyosarcoma presents in young children. This is probably a tumour of embryonic mesenchymal tissue.

The sites involved include the genital region (for example, vagina and bladder) and the nasopharynx.

The tumour may appear as a polypoid mass with grape-like projections (sarcoma botryoides).

Microscopically, sarcoma cells may be seen in a myxoid stroma. Some cells may have ample eosinophilic cytoplasm which can show striations similar to those of striated muscle, especially if special stains are used. It may be possible to demonstrate desmin and myoglobin in tumour cells by immunohistochemical methods.

Although the tumour is an aggressive one, it responds to combined therapy and cure rates of up to 75 per cent may be achieved.

OTHER MALIGNANT TUMOURS OF CHILDHOOD

Tumours similar to those occurring in adults may occur in infancy and childhood. Childhood tumours of importance include :

Leukaemia/Lymphoma is probably the most important group accounting for 30-50 per cent of deaths in childhood from malignant disease. Leukaemias are usually of the acute type and have a poor outlook.

Burkitt's lymphoma is one particular form of lymphoma which occurs predominantly in young people and children, especially in Central Africa.

Brain tumours. In addition to the embryonic medulloblastoma, other brain tumours such as astrocytomas may also occur in children.

Bone tumours of all sorts are more likely to occur in growing bone. Osteogenic sarcoma is a very malignant tumour which occurs predominantly in children.

GENERAL PATHOLOGY - CHAPTER THIRTY TWO

THE EFFECTS OF TUMOURS

It is difficult to generalise about the effects of tumours, as so much depends upon the nature, site and size of each individual tumour. However there are some general features concerning the effects and complications of tumours which are worth considering together. They may be localised or generalised, only some of these effects can easily be explained in terms of tumour growth and spread. Others consist of groups of symptoms (a collection of symptoms occurring together is known as a syndrome) known as paraneoplastic syndromes, which are less easy to explain.

LOCAL EFFECTS OF TUMOURS

BENIGN TUMOURS may rarely cause significant clinical disease locally:

- **Pressure and obstruction**: Even a small benign tumour, such as the rare papilloma of the bile duct, can cause obstruction leading to obstructive jaundice which could be fatal. Similarly, a benign tumour of the pituitary can grow within the sella turcica causing destruction of normal pituitary tissue and severe endocrine effects. Acoustic neurilemmomas within the cranium can cause deafness and more severe effects due to raised intracranial pressure. Pressure effects of benign neoplasms are uncommon, but can be significant because of the site of the tumour.
- **Functional activity**: Excessive hormone production can occur within a benign tumour of endocrine glands, causing severe or even fatal disease. Examples include adenoma of the parathyroid causing bone disease, or adenoma of the β cells of the islets of the pancreas which can bring about severe hypoglycaemia due to excessive insulin production.
- **Local anatomical complications**: Haemorrhage and ulceration with secondary infection are much more likely to occur in malignant than benign tumours. A benign tumour, however, may become pedunculated and then twisted around the stalk, cutting off the blood supply and leading to ischaemic necrosis, for example, cystic teratoma of the ovary and juvenile polyp of the rectum. A benign tumour of the gut (such as adenoma) may lead to the condition of intussusception. This occurs when the proximal bowel slides inside the more distal part bringing about intestinal obstruction.
- **Malignant transformation** of benign tumours is rare but can happen. Malignant change, for example, may occur in a benign pigmented cellular naevus with the development of malignant melanoma.

MALIGNANT TUMOURS may also show similar effects but usually to an even greater degree.

- **Pressure and obstruction:** The obstruction produced by a malignant tumour is more pronounced and complete than that produced by a benign tumour, for example carcinoma of the colon causing intestinal obstruction. Not only the primary tumour but also the metastases can cause trouble, for example, secondary deposits of tumour in the lymph nodes situated at the porta hepatis of the liver can obstruct the bile ducts giving rise to obstructive jaundice.
- **Destruction of tissue**: This may be due to infiltration by primary or secondary malignant tumour. Disseminated secondary carcinoma in bones can cause pathological fractures, or may destroy haemopoietic marrow leading to anaemia and pancytopenia.
- **Ulceration, haemorrhage and secondary infection**: These are common in destructive malignant tumours and may be fatal in themselves. Bleeding may be the first sign of a malignant tumour of the gut or uterus. Chronic bleeding can give rise to anaemia. Infection is often a consequence of ulceration of a surface. Obstruction can also predispose to infection, for example, bronchopneumonia and even lung abscess can arise in lung tissue distal to the obstruction of the bronchus by carcinoma.
- **Pain**: This is more likely to be a feature of malignant than of benign tumours. In terminal cancer it may be very severe, especially if there are secondary deposits of tumour in bone.

GENERALISED EFFECTS OF MALIGNANT TUMOURS

- **Starvation** can occur following obstruction of the oesophagus or stomach by cancer. Carcinoma of the large bowel may be associated with diarrhoea or with fistula formation (a channel formed between loops of bowel causing a by-pass), in which case there may be malabsorption.
- **Cachexia** is the name given to the extreme wasting which may be seen in terminal cases of cancer. Whilst there may be a number of contributory factors this does not seem to be directly due to starvation. Cytokines which are released from macrophages such as Tumour Necrosis Factor-α (TNF-α) and Interleukin-1 (IL-1) may play a part. TNF-α suppresses appetite and affects fat metabolism by mobilising adipose tissue.
- **Fever.** This may also be due to the action of cytokines IL-1 and TNF-α. IL-1 from leucocytes can bring about the release of prostaglandins which will act on the thermosensory centre in the brain to raise body temperature in cancer, even in the absence of secondary infection.
- **Haematological changes.** Anaemia may result from haemorrhage, malabsorption or bone marrow replacement by tumour.

  If there is widespread bone marrow infiltration, an anaemia together with loss of white cells and platelets will follow and hence there may be an increased liability to infection or bleeding.
- **Immunological effects.** A number of tumours, particularly lymphomas, may be associated with depression of the immune response. Treatment with radiotherapy or chemotherapy may also have an immunosuppressive effect.

PARANEOPLASTIC SYNDROMES

This name is given to a collection of symptoms occurring in patients with cancer which cannot be explained in terms of the direct growth of the tumour, locally or by metastatic spread, or due to hormones which are made by the tissues from which the tumour appears to be derived.

These do not seem to be the result of direct physical growth of the tumour, although there may be inappropriate secretion, for example, of hormones, by tumour cells. These effects seem to be chiefly endocrine, neurological, cutaneous or haematological. The most common syndromes are hypercalcaemia, Cushing's syndrome and nonbacterial thrombotic endocarditis. The most common cancers are carcinomas of the lung (most often small cell carcinoma) and the breast. It is worth noting that one cancer may produce more than one abnormal hormone, and hence cause more than one syndrome. The mechanism would appear to be deregulation of the genetic programme within the cell for hormone production.

The paraneoplastic syndromes may be the first sign of a malignant tumour and may cause severe clinical effects or cause death in their own right. They may cause difficulties in treatment.

ENDOCRINE:

- **Cushing's syndrome** may be produced by, for example, small cell carcinoma of the lung, islet cell carcinoma of the pancreas and carcinoma of the breast. It is thought to be due to the secretion of ACTH or an ACTH-like substance.
- **Carcinoid syndrome** is usually produced by carcinoid tumours which secrete 5-hydroxytryptamine; there is periodic flushing, diarrhoea, bronchoconstriction and eventually fibrosis of the right side of the heart. Non-carcinoid tumours of lung, pancreas and stomach can cause this syndrome.
- **Hypoglycaemia** due to the secretion of insulin or an insulin-like substance may be found in some liver cancers and in fibrosarcomas.
- **Hyponatraemia** due to water retention may be found if there is secretion of antidiuretic hormone (ADH) by small cell cancer of the lung
- **Hypercalcaemia** may be due to the secretion of a parathyroid hormone-like substance and can occur in the absence of secondary deposits in the bones in some cancers of the lung, breast and kidney.
- **Polycythaemia** or increased numbers of red cells can occur in renal carcinoma, in cerebellar haemangioblastoma or in liver cancer and is thought to be due to increased secretion of erythropoietin.

NEUROMYOPATHIC:

- **Neuropathy:** Either neuronal degeneration, as in the cerebellar cortex, or demyelination of nerves giving rise to a peripheral neuropathy may occur.
- **Myopathy:** Muscular weakness occurring in either polymyositis-like form or presenting as myasthenia gravis may accompany certain tumours.

The reasons for these neuromyopathic effects are unknown

CUTANEOUS:

Sometimes an internal cancer maybe suspected because of the presence of a widespread skin eruption.

- **Acanthosis nigricans** is a condition in which thick dark brown papillomatous lesions of the skin develop in flexures such as the axillae and groins. This may be an indication of underlying deep visceral cancer.
- **Dermatomyositis** is an inflammatory disease in which there is inflammation and vasculitis affecting skin and muscle. It is a connective tissue disease in which a certain proportion of cases is associated with cancer.

HAEMATOLOGICAL:

- **Coagulation disorders:** Migratory thrombophlebitis has been described in deep-seated carcinomas, especially of the pancreas, but also of other sites such as lung and stomach. Disseminated intravascular coagulation and fibrinolysis giving rise to a haemorrhagic condition can occur.

  Nonbacterial thrombotic endocarditis is a condition in which fibrin thrombi are deposited on the heart valves along the line of closure, possibly as a result of a hypercoagulable state.

MISCELLANEOUS COMPLICATIONS

- **Clubbing of the fingers** with swelling of the terminal phalanges is due to soft tissue swelling and sometimes also accompanying bony overgrowth. This may occur in some cases of carcinoma of the lung.
- **Amyloid deposition** may be generalised in multiple myeloma (AL) or in some cases of Hodgkin's disease or renal carcinoma (AA) or localised to the site of the tumour as in medullary carcinoma of the thyroid.

PARANEOPLASTIC SYNDROMES		
Syndrome	**Hormone or mechanism**	**Type of tumour**
Endocrine		
Cushing's syndrome	ACTH or ACTH-like	small cell Ca lung, islet cell Ca pancreas, Ca breast
Carcinoid syndrome	5-hydroxytryptamine	Ca lung, pancreas and stomach
Hypoglycaemia	insulin or insulin-like	fibrosarcoma, hepatocellular Ca
Hyponatraemia	ADH	small cell Ca lung
Hypercalcaemia	parathyroid hormone-like substance	Ca lung, breast and kidney
Polycythaemia	erythropoioetin	Ca liver or kidney cerebellar haemangioblastoma
Neuromyopathic		
Neuropathy	neuronal degeneration }	
	demyelination of nerves }	e.g. Ca lung, breast
Myopathy	polymyositis }	
	myasthenia gravis }	
Cutaneous		
Acanthosis nigricans		Ca lung, stomach & uterus
Dermatomyositis		Ca lung and breast
Haematological		
Migratory thrombophlebitis	hypercoagulability	Ca pancreas, lung etc.
Nonbacterial endocarditis	hypercoagulability	late stage of many cancers
Miscellaneous		
Clubbing of fingers	?	Ca lung
Amyloid deposition	AL AA	multiple myeloma local in thyroid Ca Hodgkin's disease & Ca kidney

GENERAL PATHOLOGY - CHAPTER THIRTY THREE

TISSUE RESPONSES TO IONISING RADIATION

Ionising radiation occurs naturally in low dosage and is referred to as background radiation. Radiation is also produced by nuclear reactors used to produce energy. It may also be used in low dosage for diagnostic X-rays; high dosage megavoltage X-rays are used to treat cancer. But what can be a valuable tool when used properly, can prove very dangerous in certain circumstances. Like fire, radiation can burn and kill, but it also has a hidden delayed action with tissue damage which may persist over many years. In addition, as well as curing cancer, the mutagenic effect of radiation is such that it may actually cause cancer.

Naturally occurring elements like radium may emit radiations such as:

- α-particles: These are helium nuclei with a mass of four and two positive charges. They have little penetrating power.
- β-particles: These are electrons with negligible mass and one negative charge.

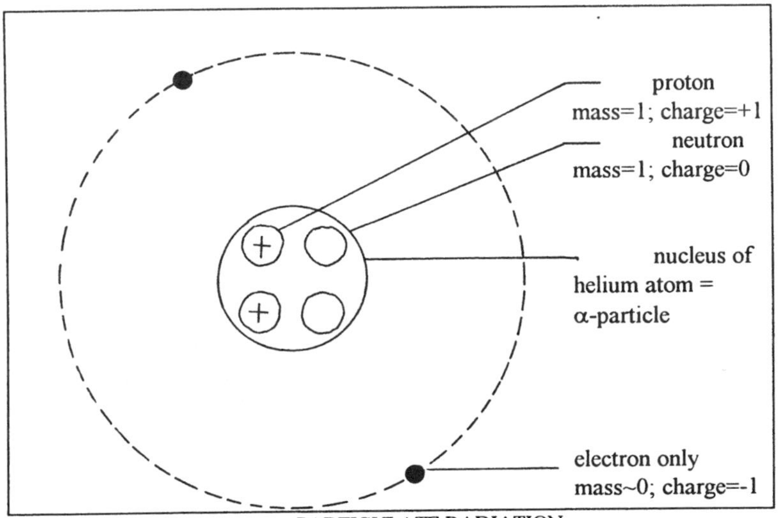

PARTICULATE RADIATION

- γ-rays: These are electromagnetic waves of short wave-length. They are similar to X-rays but produced naturally, for example, by radium decay.
- X-rays: These are produced artificially. They are emitted by a target which has been bombarded by electrons in a vacuum. The higher the voltage the greater the energy of the electrons and hence the shorter the wave-length of the X-rays, the greater their penetrating power.

- Protons: are intermediate between γ-rays and β-particles in penetrating power.
- Neutrons: These can penetrate with ease, only being stopped when they collide with atomic nuclei. Fast neutrons will knock a proton out of the nucleus; slow neutrons will be trapped to form an isotope, which may be radio-active.

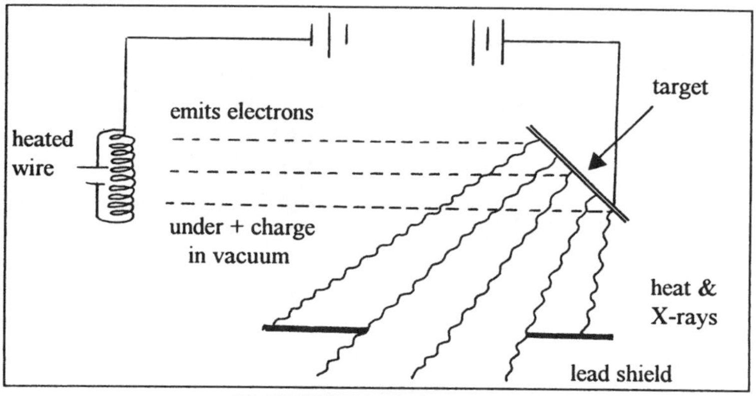

PRODUCTION OF X-RAYS

THE EFFECTS OF IONISING RADIATION

- **Physical:** The effect of radiation is to ionise or excite molecules.
- **Chemical:** Radiation may have a **direct** effect on nuclear DNA. Ionised water may give off free radicals such as OH· (hydroxyl radicals) and these may damage cells by an **indirect** effect. Particulate radiation has a more marked direct effect whereas γ-rays and X-rays cause damage more by indirect effect.
- **Biological:** The nuclear DNA of cells may be damaged in some way, either by breakage of strands or the formation of cross links causing chromosomal abnormalities and interfering with cell division.
- The **effect on the cell:** Heavy doses of radiation in a sensitive cell may cause immediate death. DNA synthesis may be inhibited or interfered with, and mitosis may be delayed or abnormal, for example, with chromosomal breaks, deletions and translocations. Heritable mutations (which may be carcinogenic) can occur.

The overall effects depend on

- The nature of the radiation and its degree of penetration
- The total dose and the time over which it is administered, for example, large doses given over a short period produce maximal effect.
- The irradiated tissue, including the amount of tissue irradiated, its sensitivity to radiation and its ability to carry out repair.

These effects may be obvious in the short term or the long term.

MEASUREMENT OF RADIATION

One roentgen (R) of X-rays or γ-rays causes a certain amount of ionisation in air (induces in 1cc air an emission equivalent to one electrostatic unit of charge).

The absorption of energy by the irradiated material is measured by the rad (r). One rad is a dose which is equivalent to an absorption of 100ergs per gram of material irradiated. (One roentgen delivered in air will result in a dose of about 0.87 rads to most tissues.) One Gray (Gy) is equal to 100 rads.

Radioisotope emissions are measured in curies (ci) and one curie is the quantity of any radioisotope (radionuclide) which undergoes 3.7×10^{10} disintegrations per second.

To assess the effects on tissues and cells the Relative Biological Effectiveness (RBE) may be used to equate the different biological effects of different forms of radiation to those of γ-rays or X-rays.

Linear Energy Transfer (LET) is a measure of energy loss per unit distance travelled. Radiation with a high linear energy transfer, for example, heavily charged, slow moving particles such as α-particles produces ionisations grouped closely together. Such radiation is more likely to cause cellular damage than radiation of low linear energy transfer, for example γ-rays and X-rays which are penetrating and lose their energy over a longer distance, causing less severe cellular damage in the process.

SENSITIVITY OF CELLS TO RADIATION

The sensitivity of cells to irradiation is directly proportional to their rate of division. Thus tissues composed of labile cells, with a high cell turnover such as the bone marrow or the intestinal mucosa, can be damaged by low doses and show signs of this damage at an early stage. Tissues of intermediate sensitivity are those in which the cells are stable, with many in the resting phase, and it takes time before the effects of injury to dividing cells is apparent. Tissues composed of permanent cells in the terminal differentiated phase are the most radio-resistant, for example, muscle and nerve cells.

RADIOSENSITIVITY OF VARIOUS TISSUES	
Radiosensitivity	**Tissues**
High	Haemopoietic tissue, lymphoid tissue, gonads, intestinal mucosa
Medium	Liver, pancreas, endocrine glands, Connective tissues
Low	Heart muscle and skeletal muscle, nerve cells and brain, mature bone and cartilage

The effects will depend upon the **dose** of radiation and the **time** over which it is given.

In radiotherapy it is usual to give divided doses. This allows normal tissues time to repair damage to DNA between doses, whereas the tumour being treated may not be able to repair its DNA so effectively.

It also affords the opportunity for more tumour cells to enter the cell cycle into the dividing stages and so to become susceptible to destruction by the X-rays.

Radiation of tissues which are hypoxic causes much less effect and conversely in the presence of high oxygen presures radiation is even more damaging to tissues.

THE EFFECTS OF IRRADIATION ON TISSUES

As has been stated, the effect of radiation on the cell may be to bring about immediate death, to prevent further cell division or to cause mutations as a result of non-fatal chromosomal alterations. Mutations which are passed on to daughter cells may cause abnormalities in offspring if germ cells are affected, or be carcinogenic in somatic cells.

Morphologically, at cellular level there may be changes in both cytoplasm and nucleus, with swelling, vacuolation and alterations in cell or nuclear membranes.

Cell enlargement with giant cells and bizarre nuclear changes are characteristic. This atypia may occur in epithelial or connective tissue cells when there has been radiation damage short of that required to cause necrosis

Changes in the vessels include:

- Early dilatation
- Endothelial damage and swelling
- Focal necrosis (fibrinoid) of small vessel walls, and at a later stage,
- Telangiectasia and
- Endothelial proliferation with hyaline thickening of vessel walls leading to narrowing or obliteration.

It is probably the vascular changes and resultant ischaemia which are responsible for the late and long term effects of radiation on the tissues.

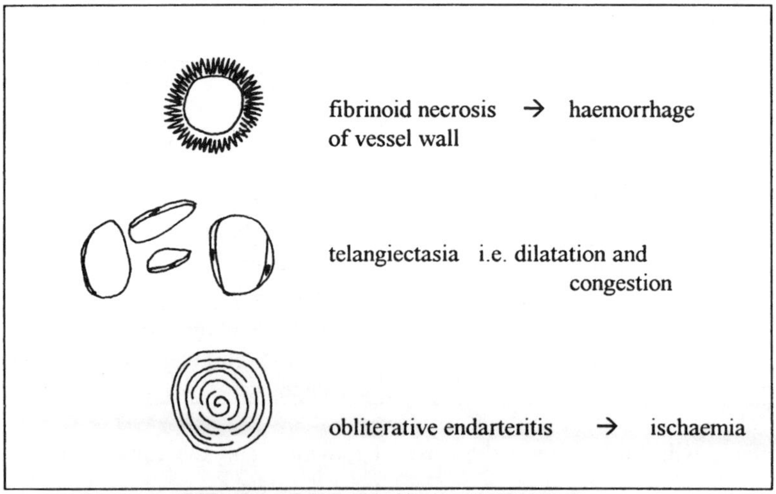

EFFECTS OF IRRADIATION ON VESSELS

The effects on the tissues can be summarised as **necrosis**, **atrophy** and **fibrosis** in that order of occurrence. The effects of irradiation may take at least ten days to two weeks to become apparent, but may take much longer. Necrosis of cells and breakdown of extracellular matrix can occur, and there may be an inflammatory reaction.

The skin is invariably affected if there is external radiation, and shows oedema and haemorrhage in the first instance due to initial vascular changes. Blistering may occur later when epidermal damage occurs. Months to years later there is thinning or atrophy of the epidermis, telangiectasia of superficial small vessels and loss of elasticity of the dermis. Ulceration, presumed to be ischaemic, may take place in the later stages. Skin cancer may develop as a late complication.

The haemopoietic and lymphoid systems are very susceptible to radiation damage with immediate lymphopenia and later (seven to ten days) granulocytopenia. There may also be thrombocytopenia. Red cell depression is a late effect because red cells are long lived and radio resistant in their mature forms. The overall effects depend on the extent of the bone marrow subjected to irradiation. Whole body irradiation may be fatal because of marrow suppression. Otherwise marrow regeneration from surviving unaffected marrow cells takes place after a few months.

The gonads are susceptible to the effects of radiation, and testicular atrophy of spermatogenic cells may occur. Germ cells in the ovaries may be destroyed leading to sterility. The destruction of ovarian follicular cells can bring about an artificial menopause. Minor heritable changes to germ cells may result in foetal abnormalities.

The lungs are vulnerable because of their rich blood supply. Early vascular changes may give rise to oedema, with protein rich fluid, fibrin deposition and hyaline membrane formation producing the picture of the adult respiratory distress syndrome. Later, fibrous thickening of the vessels and alveolar walls can cause alveolar-capillary block leading to respiratory failure.

The gastro-intestinal tract mucosa is vulnerable because it is rapidly dividing. Nausea, vomiting and diarrhoea are often unwelcome accompaniments of radiotherapy. There may be ulceration with haemorrhage. Later atrophy and fibrosis may lead to stricture formation and obstruction.

The kidney may show glomerular fibrosis.

Mature bone may undergo radionecrosis.

Immunological responses are depressed by radiation, presumably due to the destruction of lymphoid tissue.

WHOLE BODY IRRADIATION

Heavy doses of irradiation in the order of 4,000 rads maybe delivered to one part of the body in the radiotherapy of tumours, but very much lower doses than this can cause serious symptoms if the whole of the body is exposed to radiation

Non-fatal whole body irradiation has occurred in those exposed to radioactive fall-out. Survivors of the atom-bomb blasts in Japan have been observed to have a twenty-fold increase in leukaemia. Children who were exposed have shown an increased liability to many different types of cancer, especially cancer of the thyroid gland.

Dosage	Effect
300-500 rads	**Haemopoietic syndrome** with leucocyte depletion and thrombocytopenia will bring about death from infection or bleeding in a few weeks. There is virtual disappearance of haemopoietic marrow.
800-1,500 rads	**Gastro-intestinal syndrome**. There is intense vomiting and diarrhoea, resulting in dehydration and marked electrolyte disturbances. Ulceration, infection and haemorrhage of the gastro-intestinal tract occur. Death is brought about within 1 week.
more than 5,000 rads	**Cerebral syndrome**. There is rapid onset of drowsiness followed by convulsions, coma and death within a few hours.

Theoretically, irradiation to germ cells should increase the risk of genetic damage which would be passed on to the offspring, but there is little evidence of specific genetic abnormalities in the children of those exposed to radiation. This may be because recessive abnormalities may not be obvious for generations.

There is also little evidence of danger from long-term low-level exposure to radiation, although studies are continuing. Risks are probably small. However this does not mean that every care should not be taken to prevent exposure to unnecessary radiation in any circumstances, especially as the effects of radiation can be cumulative.

RADIOTHERAPY

Radiotherapy is frequently used in the treatment of a wide range of malignant tumours. Tumours differ markedly in their response, having differing degrees of sensitivity, which is apparently inherent. Lymphomas and germ cell tumours such as seminoma of the testis are very sensitive, as are some embryonal tumours. Most visceral carcinomas are of medium sensitivity. Some tumours such as tumours of the brain, melanomas, renal carcinomas and osteosarcomas are relatively insensitive as are many of the soft tissue sarcomas.

Localised tumours (such as carcinoma of the cervix) can be cured, and in more disseminated disease symptomatic relief from pain and pressure effects may be obtained. Radiotherapy may be used in combination with surgery and chemotherapy with good effect, for example in the embryonic tumours of childhood, where quite high cure rates can be achieved.

Radio-active isotopes, for example, radio-active iodine or radio-active phosphorus which localise in thyroid and bone marrow respectively may be used to treat functioning tumours in these areas.

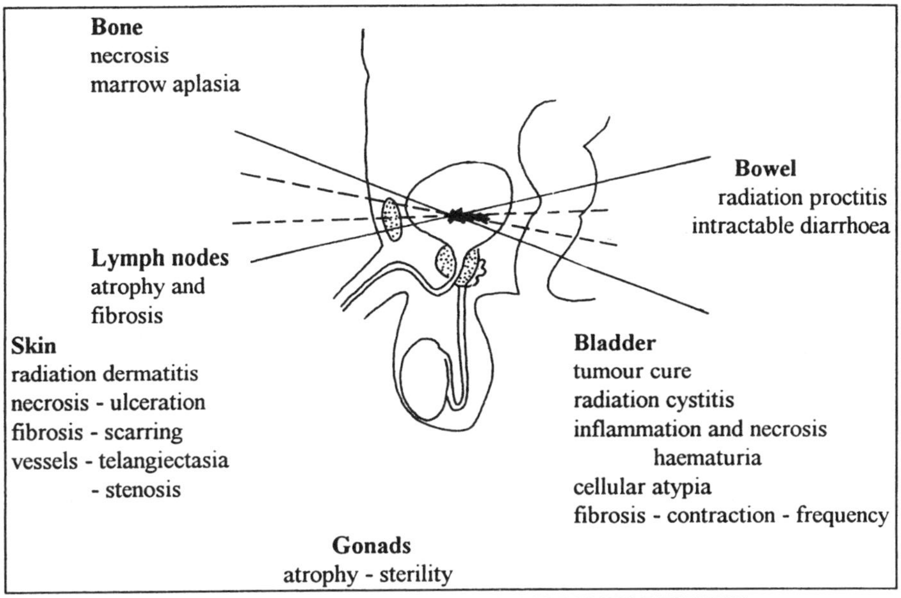

IRRADIATION OF CA BLADDER - SOME POSSIBLE EFFECTS

GENERAL PATHOLOGY - CHAPTER THIRTY FOUR

DISORDERS OF CARBOHYDRATE METABOLISM: DIABETES MELLITUS

Diabetes mellitus is a chronic metabolic disorder in which there is a relative or absolute lack of insulin, or failure of its action. There is impaired carbohydrate metabolism and a raised blood glucose level (hyperglycaemia) is a diagnostic feature. The clinical diagnosis is made in terms of plasma glucose levels when there is:

- A fasting plasma glucose level of more than 7.8mmol/l
- A plasma glucose level of more than 11.1mmol/l more than two hours after food was last taken. (75mg of oral glucose)

THE ACTION OF INSULIN

Insulin is manufactured by the beta cells of the islets of Langerhans in the pancreas. It is secreted and stored in membrane bound vesicles in the cells and released on demand, usually in response to an increase in plasma glucose levels. Glucose also stimulates insulin synthesis by the cells.

The most important action of insulin is to facilitate the entry of glucose into cells, especially voluntary muscle cells, fibroblasts and fat cells. Insulin also:

- increases glycogen formation.
- promotes the synthesis of proteins from amino acids (rather than gluconeogenesis).
- and in fat cells insulin is necessary for the conversion of glucose to triglycerides. The breakdown of neutral fat is inhibited.

Insulin and insulin-like growth factors can bring about DNA synthesis in cells, and stimulate growth and differentiation.

Some intestinal hormones, for example, glucagon, gastrin and drugs such as sulphonylureas will stimulate the release of insulin (but not its synthesis)

Glucose is maintained at a plasma glucose level of 3.0-5.0mmol/l by the action of insulin which lowers the level and the activity of the liver. The liver can store glucose as glycogen or produce glucose from glycogen or non-carbohydrate sources such as amino acids from protein and fat.

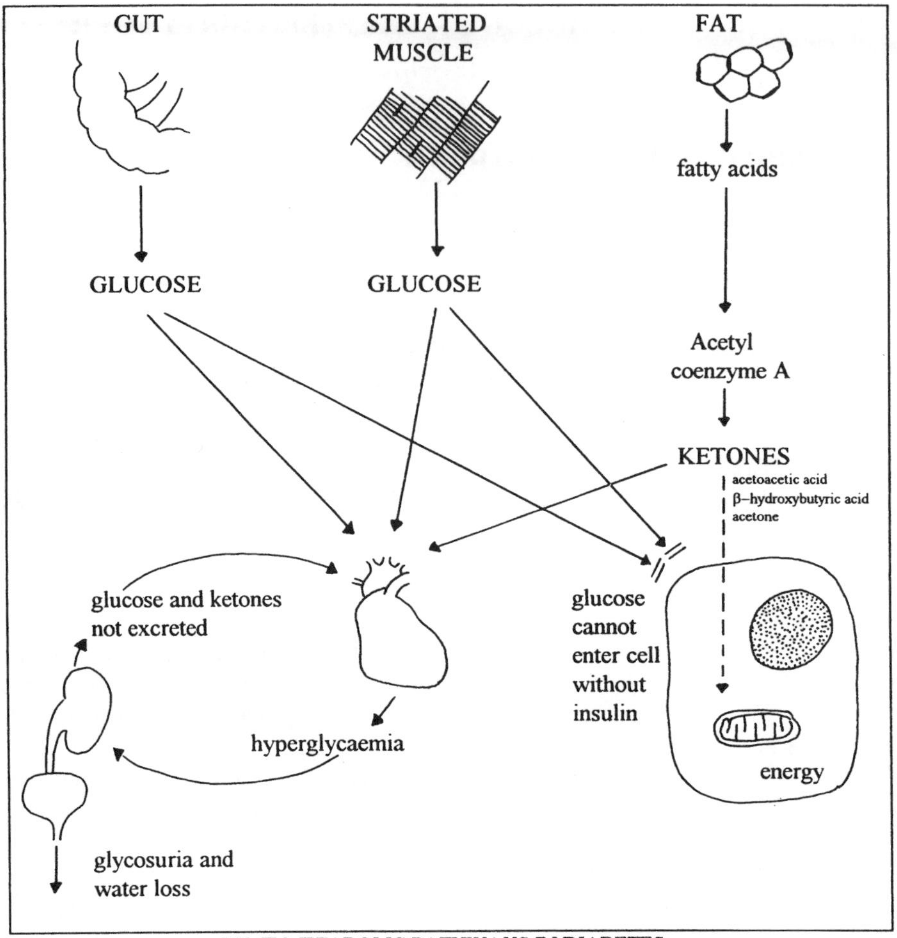

SOME METABOLIC PATHWAYS IN DIABETES

TYPES OF DIABETES

- Diabetes in its most common form is apparently primary or spontaneous. This type accounts for about 99 per cent of cases.
- Glucagon, adrenaline, growth hormone and adrenal glucocorticoids oppose some of the actions of insulin, and will bring about an increase in plasma glucose. Some forms of diabetes may be associated with sustained overproduction of these hormones, for example,
  - Islet cell tumour of α-cells of the pancreas producing glucagon
  - Phaeochromocytoma of the adrenal medulla, producing adrenaline
  - Acromegaly with increased production of growth hormone
  - Cushing's syndrome with over secretion of adrenal glucocorticoids.
- One form of diabetes may also be associated with the destruction of islets as may occur in chronic pancreatitis, following surgery for carcinoma of the pancreas and in iron overload (for example, haemochromatosis).

PRIMARY DIABETES MELLITUS

There are two major types of primary diabetes mellitus :

Insulin-dependent diabetes mellitus (IDDM) or **Type I** used to be known as juvenile diabetes. It occurs in younger patients, who will die in ketoacidotic coma unless they receive insulin treatment. Insulin levels are low, due to the destruction of B cells by what is thought to be an autoimmune process. Lymphocytic infiltration of the islets may be seen in the early stages, followed later by atrophy and fibrosis. The majority of patients (about 80 per cent) have circulating islet cell antibodies. It is thought that there is a genetic predisposition (linked to genes associated with the HLA-D class II major histocompatibility complex) and that in these individuals the autoimmune reaction is triggered by some outside event such as a virus infection (Coxsackie B virus and rubella are amongst those which have been suggested).

There is marked hyperglycaemia as absorbed glucose is not entering the cells. The glucose levels exceed the renal threshold and there is loss of glucose in the urine (glycosuria). This brings about an osmotic diuresis and there is polyuria (excessive production of urine), leading to dehydration, extracellular at first, and later intracellular. Thirst leads to increased fluid intake (polydipsia). There is also an increase in appetite (for unknown reasons) and loss of weight.

The presenting symptoms of type I diabetes may be polyuria, polydipsia and increased appetite with loss of weight.

As glucose cannot be used for energy production there is breakdown of fat with release of free fatty acids, which are used to form acetylcoenzyme A in the liver. In the absence of insulin the acetylcoenzyme A cannot enter the Krebs cycle, and is converted into ketone bodies (β-hydroxybutyrate, acetoacetic acid and acetone). Again with low insulin, these cannot all be used by muscles and other tissues for energy production, and they accumulate to give ketosis and metabolic acidosis, which are responsible for diabetic coma and death. Although some fats and proteins may be converted into glucose, this cannot be used and there is thus loss of weight and wasting.

Non-insulin-dependent diabetes (NIDDM) or **Type II** occurs in older persons and is often associated with obesity. It is the most common form of diabetes with 80-90 per cent of cases being of this type. There is a familial incidence. B cells are preserved in the islets, no islet cell antibodies are present and there may be amyloid (of endocrine type) deposited in the islets. Insulin is present but may be secreted in lesser quantity than is required; the tissues also develop resistance to the action of insulin. with decreased numbers of insulin receptors and glucose transport units. Obesity seems to be associated with the development of insulin resistance by cells. Oral hypoglycaemic agents stimulate the release of insulin.

Some of the presenting symptoms are the same as type I diabetes, with polyuria and polydipsia, but the patients are not wasted and the disease is milder. Ketone bodies do not build up but the osmotic diuresis may lead to dehydration. If water cannot be replaced a hyperosmolar non-ketotic coma may develop, as water is drawn from inside the cells, especially those of the brain. There may be hyperlipidaemia and hypercholesterolaemia, but there seems to be enough insulin to prevent ketone bodies from building up in the type II diabetes.

	IDDM - type I	NIDDM - type II
Prevalence	10-20% of diabetics ~0.25% of population	80-90% of diabetics ~2% of population
Onset	Usually under 30 years old	Over 40 years of age
Build	Thin	Usually obese
Genetics	HLA-D associated	Familial
Islet cell antibodies	Present	Absent
Pathology of pancreas	Islets inflamed;' B cells destroyed with atrophy and fibrosis	No inflammation B cells present may be amyloid deposits
Metabolic complications	ketoacidosis coma	non-ketotic hyperosmolar coma

In fact primary diabetes is probably more than just two diseases, as there seem to be variants of the type II form.

Diabetes is a common disease affecting about two per cent of the population in Western world. It is an important cause of death. The mortality and the morbidity of diabetes are now due chiefly to the complications which seem to be common to the two major types, and which affect blood vessels, kidney, eye and nerves.

The development of complications seems to be related to hyperglycaemia, for example, kidneys transplanted into diabetics will develop diabetic nephropathy in a few years, while the reverse is also true, a diabetic kidney transplanted into a normal individual will show a diminution or disappearance of the lesions.

Two mechanisms have been postulated as possibly being responsible for the complications:

- **Nonenzymatic glycosylation** by which glucose attaches to the amino group of proteins. This is increased in diabetes and it is possible that glycosylation products on collagen in vessel walls can accumulate, encouraging lipid deposits in larger vessels and trapping protein in smaller vessels, such as those of the kidney, which could account for basement membrane thickening and increased permeability in glomerular vessels. This could explain the enhanced tendency to develop atheroma and the renal lesions.
- **Sorbitol accumulation** can occur in cells or tissues which do not need insulin for glucose to enter them (for example, nerves, blood vessels, the kidney and the lens of the eye), as intracellular glucose excess is metabolised to sorbitol. This accumulation may interfere with the metabolism of the cell, leading to damage to Schwann cells, retinal vessels and the lens of the eye.

MORPHOLOGICAL CHANGES IN DIABETES

Pancreas: There may be lymphocytic infiltration of the islets (insulitis) in the early stages of type I diabetes, with later reduction in size of islets, loss of B cells, atrophy and fibrosis.

In type II diabetes the islets may often appear normal (30-50 per cent of cases) or there may be hyalinisation with amyloid (of endocrine type) deposition in some instances.

Vascular complications: These are of two forms

- **Diabetic microangiopathy**. This is characterised by diffuse thickening of the basement membranes of small vessels or capillaries. These can be demonstrated by the PAS stain. Vascular permeability to proteins is increased. These changes are important in contributing to the renal changes and those in the retina.

  There is also marked hyalinisation of arterioles. This lesion can occur in hypertension in the absence of diabetes, but is more pronounced in diabetics. Hyalinisation of both afferent *and* efferent glomerular arterioles is virtually pathognomonic of diabetes.
- **Atheroma**. Diabetics are more prone to develop atheroma of the larger arteries earlier in life and in a more severe form. Most diabetic deaths result from the complications of atheroma, such as myocardial infarction (five times more common in diabetics), cerebrovascular accidents, aortic aneurysm and gangrene of the lower extremities (100-fold more common in diabetics). Some of this increased risk is due to diabetes *per se*, but diabetics have hyperlipidaemia, and many are obese and hypertensive, all of which are risk factors for the development of atheroma. Nevertheless, it is a fact that all diabetics who have had the disease for ten years or more will have a significant degree of atheroma.

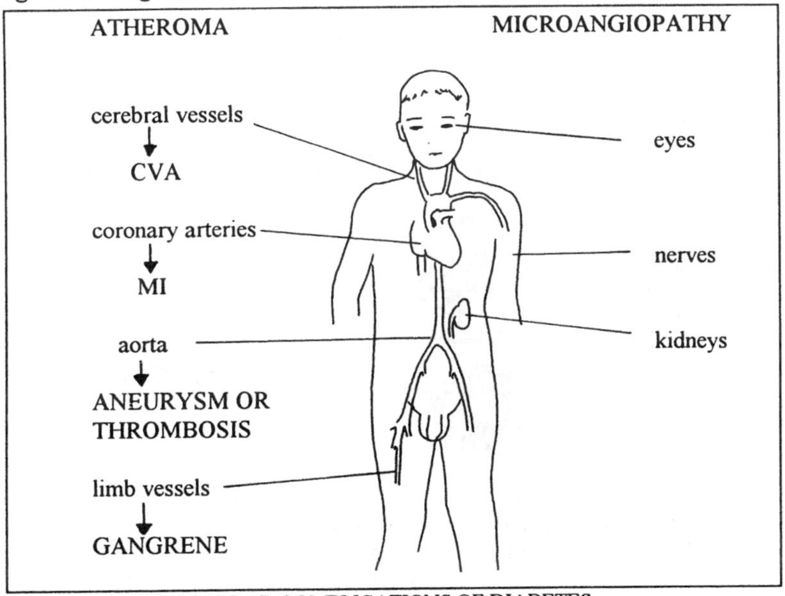

VASCULAR COMPLICATIONS OF DIABETES

Renal complications: Many different lesions may be found within the kidney:

- Glomerular lesions:
  - Diffuse glomerulosclerosis involves thickening of the basement membrane with the deposition of PAS positive basement membrane-like material in the mesangium.
  - Nodular glomerulosclerosis is virtually pathognomonic of diabetes (Kimmelstiel-Wilson disease). There are nodular rounded focal depositions of basement-membrane like material in the peripheral parts of the glomeruli.

  These two glomerular lesions may be associated with the development of a clinical nephrotic syndrome (heavy proteinuria, hypoalbuminaemia and generalised oedema).
  - Exudative lesions of fibrinous material in Bowman's space form capsular drops and fibrin caps.
- Renal vascular lesions: Atheroma of the main renal arteries may cause a degree of renal ischaemia. More marked is the hyaline arteriolosclerosis affecting both afferent and efferent glomerular arterioles.
- Pyelonephritis: Diabetics are predisposed to infection and acute and chronic pyelonephritis may be found more commonly in diabetics than in the general population. Diabetics are also particularly prone to develop necrotising papillitis, in which the renal papillae undergo infarct-like necrosis and slough off.
- Renal glycogen infiltration may be found in the tubular cells in uncontrolled diabetes and fatty change is sometimes present in the proximal convoluted tubules.

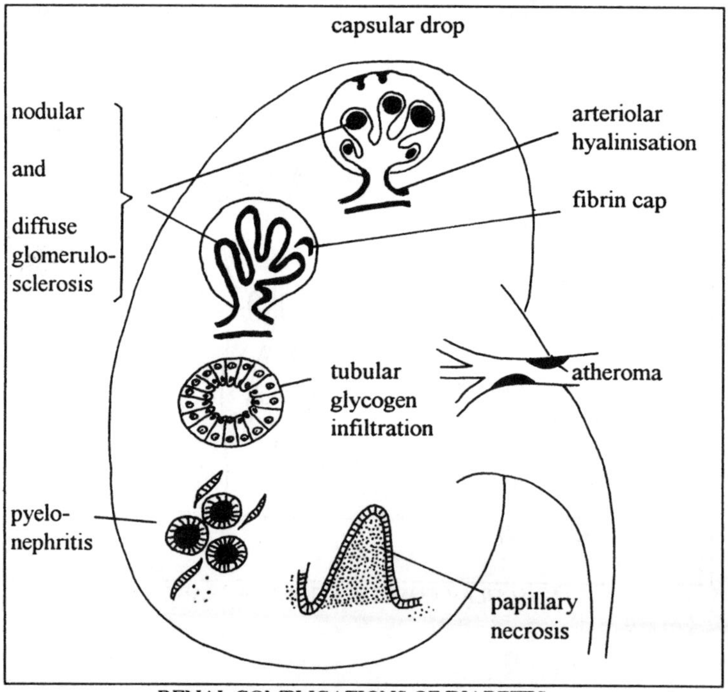

RENAL COMPLICATIONS OF DIABETES

Ocular complications: Small vessels have thickened basement membranes and are narrowed. There may be exudates and haemorrhages (due to increased permeability). Focal dilatation of small vessels gives rise to micro-aneurysms, which are another diagnostic feature of diabetes mellitus, and are almost always present in patients who have nodular glomerulosclerosis of the kidneys. These features are known as diabetic retinopathy

New vessel formation with fibrosis may occur, and this is known as proliferative retinopathy.

In addition cataract formation is an important complication of diabetes.

Diabetes is a major cause of blindness in the world today.

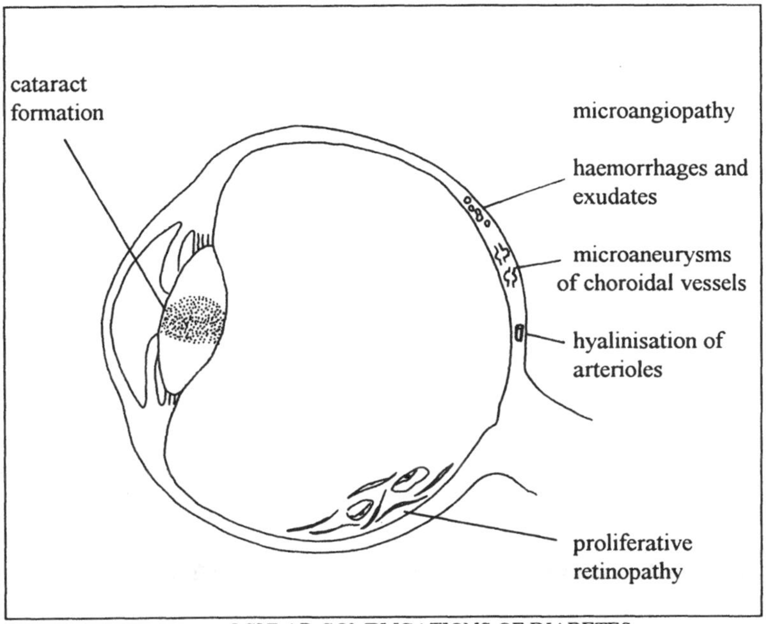

OCULAR COMPLICATIONS OF DIABETES

Nervous system: The neurologic disorders are not specific. There may be a symmetrical peripheral neuropathy of the lower limbs, affecting sensory function particularly. There may also be disturbance of autonomic nerve function affecting the bladder and bowel (incontinence and diarrhoea); isolated nerve paralyses (diabetic mononeuropathy) also may occur.

Infarcts and haemorrhages into the brain are related to the vascular complications of the disease.

Skin complications: Infections may be found. Xanthomas are yellow tumour-like nodules composed of fat-laden macrophages. These are related to the hyperlipidaemia which may be found in diabetics.

Liver complications: There may be fatty change. Glycogen infiltration of liver cells with nuclear vacuolation may be seen in a proportion of cases.

CLINICAL COURSE

Type I diabetics suffer mainly from the metabolic disturbances due to diabetes. They may present with polydipsia and polyuria and weight loss in the presence of a normal appetite or they may present with ketoacidosis and coma, often precipitated by an infection of some sort. Once insulin treatment is started they may fluctuate between hypoglycaemia and hyperglycaemia with ketoacidosis.

Type II diabetics may have some polydipsia and polyuria, but their diabetes is often discovered incidentally on medical examination. Plasma glucose levels can usually be controlled by dieting and oral hypoglycaemic agents. Rarely, they may become dehydrated and go into non-ketotic hyperosmolar coma.

In both types of diabetes the complications of atheroma with myocardial infarction, cerebrovascular accident, aortic aneurysm or gangrene of the lower extremities are the most likely to cause severe illness or death. The nephrotic syndrome with heavy protein loss may result from the changes in the kidneys and may lead to renal failure.

CAUSES OF DEATH IN DIABETES

Cause of death	Percentage of deaths
Vascular disease (myocardial infarction, cerebrovascular accident, gangrene)	76%
Renal failure	7%
Infection (especially bronchopneumonia)	5%
Coma (usually ketoacidotic)	1%
Other causes	11%
	100%

OTHER DISORDERS OF CARBOHYDRATE METABOLISM

Other disorders of carbohydrate metabolism, which need only be mentioned briefly, fall into the category of inherited errors of metabolism. They include:

Galactosaemia: This is an autosomal recessive disorder of galactose metabolism. Lactose in milk is split into glucose and galactose and the galactose is then converted into glucose. In the absence of the necessary enzymes for this to happen, particularly galactose-1-phosphate uridyl transferase, galactose-1-phosphate can accumulate in various parts of the body, particularly the liver, the eyes and the brain.

In the liver there is early fatty change followed by widespread scarring amounting to cirrhosis. Clinically there is early jaundice. Lens opacities (cataracts) develop. Non-specific changes in the brain include loss of nerve cells and gliosis resulting in mental retardation.

The diagnosis can be made by the detection of galactose in the urine. There may also be aminoaciduria due to kidney damage.

Once the diagnosis has been made, a lactose-free diet can prevent further damage but cannot reverse the mental retardation.

Glycogen storage diseases. Most of these diseases are inherited as autosomal recessive. Deficiency of one of the many enzymes involved in the synthesis or breakdown of glycogen can cause excessive accumulation of glycogen in some form.

There are three groups of glycogen storage disease:

- **The hepatic form.** Enzymes in the liver are concerned with glycogen metabolism, and if they are deficient there may be liver enlargement (hepatomegaly) due to excess accumulation of glycogen in the cytoplasm and nuclei of hepatocytes, together with hypoglycaemia due to failure of glucose mobilisation. Glycogen also accumulates in renal tubular cells. Von Gierke's disease (type I) is the best known example and this is due to a deficiency of glucose-6-phosphatase.

  Affected children fail to thrive and do not grow properly. There is enlargement of liver and kidneys. Hyperlipidaemia and hyperuricaemia may be present to make up for the lack of glucose in the blood as an energy source. There is a high mortality.
- **The myopathic form**. Glycogen is important in muscle as a source of energy. If it cannot be broken down it will accumulate within muscle cells and there will be muscular weakness. McArdle's disease (type V) is due to lack of muscle phosphorylase.

  Glycogen accumulates in skeletal muscle. There may be pain on exercise: lactic acid does not accumulate after exercise as there is a block in glycolysis. There may be myoglobinuria. McArdle's disease often only appears in adult life and is not a fatal condition.
- **Generalised glycogen accumulation** can occur in Pompe's disease (type II), where there is a deficiency of a lysosomal enzyme, glucosidase (acid maltase). Glycogen is deposited in virtually every organ, but especially in heart muscle and skeletal muscle as well as in the liver. Glycogen accumulates within lysosomes. Death usually occurs in infancy from cardiorespiratory failure.

GENERAL PATHOLOGY - CHAPTER THIRTY FIVE

SOME INHERITED AND METABOLIC DISORDERS

Whilst genetics is the subject of a separate course in medical education, it is still worthwhile to look at some of the genetically transmitted diseases with regard to the observable pathological changes that they cause in the body. The relationship between an inborn error of metabolism, in which perhaps only one amino acid in a protein chain is abnormal, and a disease state in which there are widespread visible changes causing severe malfunction is remarkable. Some of these disorders are considered elsewhere, for example, glycogen storage diseases under disorders of carbohydrate metabolism. In other conditions such as insulin-dependent diabetes mellitus, a genetic predisposition is but one factor in the complicated aetiology of the disease.

In this chapter it is intended to tie up a few loose ends and to discuss some diseases of general interest which are important on account of the pathological changes associated with them.

HYPERLIPIDAEMIA

This common disorder is of interest because of the relationship between increased levels of cholesterol and triglycerides in the blood and the development of atheroma in arteries.

Lipids are transported in the blood plasma attached to apoproteins which make them soluble. They are in microparticulate form and can be classified as to the size and density of the lipoprotein and the nature of the carrier apoprotein. As size decreases density increases.

Apoprotein abnormalities, lipoprotein enzyme deficiencies and LDL receptor defects are all mechanisms which can be associated with hyperlipidaemia of one sort or another

Familial hypercholesterolaemia is a condition which arises from defects in the cell receptors for LDL. Heterozygotes with one abnormal gene are fairly common and have plasma cholesterol levels two or three time normal. Homozygotes are even more severely affected with very high levels of cholesterol and can die from ischaemic heart disease and myocardial infarction at a very early age. Most cells have LDL receptors, but they are more frequent in hepatocytes. The number of receptors depends on the amount of cholesterol in the cell; increasing the cholesterol content switches off the synthesis of LDL receptors, thus increasing the plasma cholesterol level. LDL receptor synthesis is increased when there is low cholesterol absorption or when resorption of bile acids is prevented, reducing plasma cholesterol levels.

Although lipoproteins can be detected electrophoretically, a simple measure of blood lipid status and liability to atheromatous vascular and heart disease can be obtained from the estimation of plasma cholesterol levels, which reflect LDL levels. The higher the level, the greater the risk. It has been shown that cholesterol levels above 6.8mmol/l (265mg/dl) are associated with a five times higher risk of coronary artery disease than levels of 5.6mmol/l(220mg/dl). The median value of cholesterol in Australia is 5.5mmol/l and the National Heart Foundation recommends that efforts be made to keep serum or plasma cholesterol levels below this by means of diet and exercise. No level can be said to be absolutely safe, but risks are much less in individuals (or communities) with levels below 5mmol/l (about 200mg/dl).

Hypertriglyceridaemia with high VLDL levels is also associated with increased risk of vascular disease, although perhaps to a lesser extent. The plasma levels of HDL are inversely related to risk, in other words high HDL levels appear to exert a protective effect.

Plasma lipoproteins

Particle	Diameter (nm)	Triglyceride / cholesterol content	Apolipoprotein
Chylomicron	80-1200	80-95% TG	A-1, A-2, A-4, B-48 and C
Very low density lipo-protein (VLDL)	30-80	45-65% TG 25% Chol	B-48, C and E
Intermediate density lipo-protein (IDL)	25-35	45% Chol	B-100, E
Low density lipoprotein (LDL)	15-25	70% Chol	B-100
High density lipoprotein (HDL)	5-15	<20% Chol	A-1, A-2

These lipoproteins can be related to Friedrickson's classification of the primary hyperlipidaemias

Type	Lipoprotein	Lipid	Defect
I	Chylomicrons	TG	lipoprotein lipase or apoprotein CII deficiency
IIa IIb	LDL LDL and VLDL	Chol Chol and TG	LDL receptor defect LDL receptor defect
III	IDL	TG and Chol	abnormal apoprotein E
IV	VLDL	TG	unknown; ? hepatic lipase deficiency
V	VLDL and chylomicrons	TG and Chol	lipoprotein lipase or apoprotein CII deficiency

This classification refers to inherited hyperlipidaemias but there can be acquired hypercholesterolaemia and hypertriglyceridaemia and these too predispose towards the development of atheromatous arterial degeneration.

CAUSES OF HYPERCHOLESTEROLAEMIA AND HYPERTRIGLYCERIDAEMIA

	Cholesterol	Triglycerides
Condition		
Familial	+ + +	+ + +
Diabetes mellitus	+ + +	+ + +
Hypothyroidism	+ + +	+ + +
Nephrotic syndrome	+ + +	+ + +
Pregnancy	+ + +	+ + +
Biliary obstruction	+ + +	+/-
Dietary intake	+ + + or +/-	+ + + or +/-
Pancreatitis	+/-	+ + +
Excessive alcohol	+/-	+ + +
Glycogen storage disease	+/-	+ + +

Xanthomas are benign yellow tumours that are composed of aggregates of foamy macrophages filled with lipid. Touton giant cells may be formed in which a ring of nuclei in the middle of the cell is surrounded by foamy lipid vacuoles. Xanthomas are associated with severe hypercholesterolaemia (and to a lesser extent with hypertriglyceridaemia) and may be found in the skin around the knees and elbows, and in tendons. Xanthomatous deposits may also be seen in the form of yellow plaques in the upper eyelids, although these may sometimes also be formed in the presence of normal cholesterol levels.

MARFAN'S SYNDROME

This is an autosomally dominant inherited condition which is due to a defect in the connective tissues of the body.

There are abnormalities in the **skeleton.** Individuals are tall with long limbs and the arm span may exceed the overall height. Long fingers are referred to as arachnodactyly. There may be a high arched palate and spinal and chest deformities. Joints are hyperextensible, particularly in the hands and feet.

In the **eyes** there may be dislocation of the lens, due to lax suspensory ligaments.

It is in the **heart** that the most serious complications arise. There may be mitral valve prolapse with a 'floppy mitral valve' More importantly, there is cystic medionecrosis of the aorta which leads to aortic valve dilatation and aortic incompetence and also predisposes to the formation of a dissecting aneurysm, which may be fatal.

No defect in collagen or elastin has been demonstrated but the microfibrillar fibres of connective tissue which serve as a scaffolding for the laying down of elastin appear to have an abnormal component of the glycoprotein fibrillin. The defect for this condition has been located on chromosome 15q21.

CYSTIC FIBROSIS

This is an autosomally recessive transmitted condition which occurs in one in 2000 live births in Whites. The prevalence of heterozygous carriers is approximately one in 25.

The disease is characterised by the production of thick viscid secretions by exocrine glands in the pancreas and intestine, and by mucinous glands of the respiratory tract and gastrointestinal tract. There may also be secretory abnormalities of, for example, sweat glands, liver and salivary glands.

An affected child may present at birth with intestinal obstruction due to thick intestinal mucin - a condition known as **meconium ileus** (meconium is the name given to the contents of the bowel in the foetus before birth).

In infancy there may be an inability to absorb fat because of a lack of pancreatic enzymes, resulting in fatty, foul smelling faeces which float (**steatorrhoea**) and malabsorption with failure to gain weight, together with deficiencies of fat soluble vitamins A, D and K.

The main complication is the susceptibility to severe and repeated **pulmonary infections** following blockage of bronchi and small air passages by thick viscid mucin. Chronic bronchitis, dilatation and infection of bronchi, that is, bronchiectasis, bronchopneumonia and lung abscesses, are possible complications and most patients die as a result of pulmonary infection.

Pathologically:

- **The pancreas** shows dilated plugged ducts with atrophy of the exocrine glands. These may be replaced by fibrous tissue, which together with the cystically dilated ducts gives the picture of 'fibrocystic disease of the pancreas'. The islets of Langerhans are spared and persist.
- **The liver** may show areas in which small bile ducts are dilated and filled with inspissated secretion and eventually biliary cirrhosis may develop.
- The **lungs** show the plugged dilated bronchi and the changes of infection as mentioned above.

The diagnosis can be confirmed by a sweat test as the sweat contains increased concentrations of chloride. The basic defect is in the transport of chloride ions across the cell membrane of epithelial cells. This leads to increased sodium and chloride excretion in the sweat but in the airways, for example, chloride and hence sodium and hence water are removed from the secretions rendering them more viscid.

The gene for cystic fibrosis has been mapped to chromosome 7 (7q31). This gene has been found to produce a membrane-associated protein (the cystic fibrosis transmembrane conductance regulator - CFTR) that either acts as a chloride channel or regulates other chloride channels. The most common mutation of this gene is an F508 deletion which encodes a phenylalanine which is absent from the regulator protein.

THE LYSOSOMAL STORAGE DISEASES

There are a number of diseases in which an inherited enzyme deficiency leads to the accumulation of substrate which cannot be metabolised. These include the glycogen storage diseases which were discussed under disease of carbohydrate metabolism. In one of these (Pompe's disease - type II) the deficient enzyme is lysosomal glucosidase. Other diseases or groups of diseases are also due to inherited deficiencies of lysosomal enzymes of various sorts. These diseases are transmitted as autosomal recessive characteristics and are fortunately rare.

These diseases can be categorised according to the nature of the substance which accumulates, for example, glycogenoses, sphingolipidoses and mucopolysaccharidoses.

THE SPHINGOLIPIDOSES

The sphingolipidoses are diseases due to deficiency of lysosomal gangliosidases. The brain is rich in gangliosides and unmetabolised gangliosides accumulate within neurones. The diseases are mostly characterised by neurologic problems. These diseases are particularly prevalent among Jews of Ashkenazi origin. Three are worth special mention. These are

Tay-Sachs disease. This is the most common form and is characterised by a deficiency of an enzyme, hexosaminidase A, which is necessary for the breakdown of G_{M2}-ganglioside, which therefore accumulates in the tissues (including the brain). Affected neurones are enlarged and contain swollen lysosomes, bulging with ganglioside. These give the cells a foamy appearance on light microscopy. The material stains positively for fat (Oil Red O). Electron microscopy shows a whorled appearance within the enlarged lysosomes. Neurones are destroyed and the lipid substances accumulate in macrophages throughout the brain.

Ganglion cells in the retina are swollen at the margins of the macula, giving the retina a paler appearance than normal and accentuating the red colouration of the macular choroid, which is described as a cherry red spot.

Affected infants undergo progressive motor and mental deterioration and become blind and die in early childhood.

Antenatal diagnosis is now possible, as absence of the enzyme can be detected in cultures of foetal fibroblasts obtained from amniotic fluid. Hexaminidase A can also be estimated in the serum and carriers of the disease, who have low levels of the enzyme can be identified.

Niemann-Pick disease: Biochemically this designation describes a group of diseases in which sphingomyelin and cholesterol accumulate in lysosomes in cells throughout the body. In the sphingomyelinase deficient type A variant, sphingomyelin accumulates in neurones and in phagocytic cells of the mononuclear phagocyte system (MPS) in, for example, the spleen, liver, bone marrow and lymph nodes. The cells are enlarged with foamy vacuolated cytoplasm, which stains positively for fat. On electron microscopy lysosomes contain lamellar structures (zebra bodies).

Patients present in infancy with livers enlarged to three or four times normal size, massive splenomegaly (up to ten times normal size) and generalised lymphadenopathy. There is mental retardation due to involvement of brain cells. Death occurs in the first few years of life. Diagnosis can be made by an enzyme assay of involved tissue from liver or bone marrow. Antenatal diagnosis may also be made from fibroblast cultures.

Gaucher's disease. In this disease there is deficiency of glucocerebrosidase. Glucocerebroside accumulates in the cells of the MPS throughout the body. There are a number of types:

- In type I or the adult type glucocerebroside accumulates in the cells of the MPS without nervous system involvement. Signs of the disease, such as splenomegaly or bony swelling and pain, appear in adulthood.
- In type II or the infantile type cerebrosidase deficiency is almost total and there is nervous system involvement leading to early death.
- There is also an intermediate variant known as type III.

Pathologically, Gaucher cells may be seen on bone marrow aspirate as very large macrophages (up to 100μm) with eccentric nuclei. The cytoplasm has a fibrillar or wrinkled appearance and stains positively with the PAS stain, but does not stain for fat. On electron microscopy lysosomes can be seen to be greatly elongated and enlarged. As well as enlargement of the spleen, there is also hepatomegaly and lymphadenopathy. Involved bone may be deformed or may suffer from pathological fracture. Nevertheless in the adult form there may be normal longevity.

In this disease the mutant gene has been analysed and has been shown to vary from the normal by a single base substitution (proline for leucine).

As in Tay-Sachs disease and Niemann-Pick disease, the enzyme can be assayed in tissue cultures and antenatal diagnosis is possible.

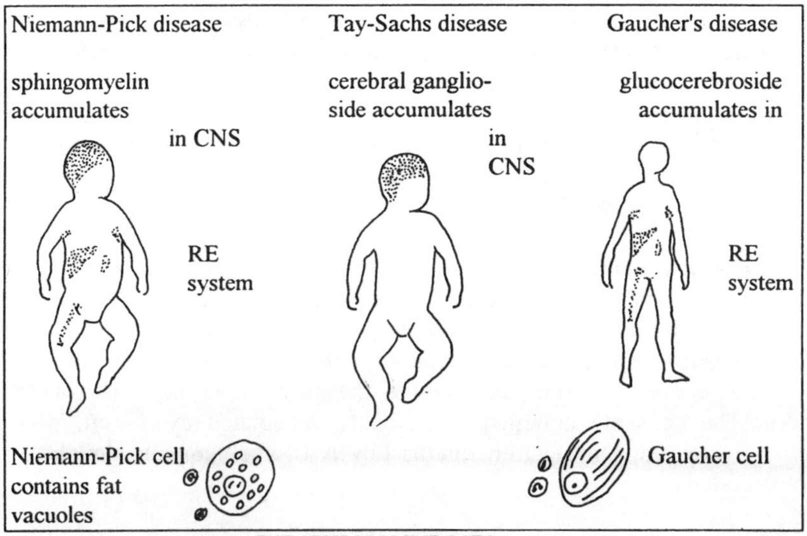

THE SPHINGOLIPIDOSES

THE MUCOPOLYSACCHARIDOSES

These are a group of disorders in which there is faulty breakdown of mucopolysaccharides which accumulate within lysosomes. Various deficiencies of lysosomal enzymes are responsible and the disease can take a number of forms. Inheritance is usually as an autosomal recessive (although one type is X-linked).

Mucopolysaccharides are part of the ground substance of connective tissues which are in part degraded by lysosomal enzymes. Mucopolysaccharides accumulate in cells of the MPS, vascular endothelium, smooth muscle cells, fibroblasts, parenchymal cells and neurones. This leads to hepatosplenomegaly. Bony deformities with dwarfism can occur. There is often mental deficiency (for example, in type I, Hurler's syndrome). There may be valvular lesions of the heart or myocardial ischaemia resulting from subendothelial deposits in the cardiovascular system.

Skeletal deformities and coarsening of the features in type I led to the term gargoylism being applied to this disease. (Gargoyles are horrific figures which used to be carved on the exterior of old churches - perhaps to frighten evil spirits away).

Affected cells are swollen or ballooned and contain great numbers of very small vacuoles which stain positively with PAS, giving a finely granular appearance. On electron microscopy lamellated zebra bodies (as in Niemann-Pick disease) may be seen in lysosomes.

Excess mucopolysaccharides may be excreted in the urine.

GOUT

Gout is a disturbance of purine metabolism. Purines are constituents of nucleic acids which make up RNA and DNA, and hence of nucleoproteins. Uric acid and urates are the end results of purine metabolism.

Gout may occur when there is a raised serum uric acid level (above 0.45mmol/l or 7mg/dl). In fact, hyperuricaemia of this level is fairly common in developed countries, but only a small proportion of patients with a raised uric acid level actually develop gout.

Most cases of gout are primary and appear to be due to one of a number of inherited defects. Males are mostly affected and it is believed that inheritance is multifactorial or polygenic.

Secondary gout occurs when a reason for raised uric acid levels exists, such as:

- Increased production of uric acid as in nucleic acid breakdown, for example, in malignant tumours such as leukaemias, especially after treatment.
- Reduced urate excretion, as in renal failure.

Clinically there may be:

- Acute arthritis due to the precipitation of sodium urate in joint fluid. This usually affects the smaller, more peripheral and colder joints (classically the big toe) causing an acute inflammatory reaction. The crystals that are formed are chemotactic for neutrophils, and when phagocytosed bring about the release of many different mediators of the acute inflammatory reaction which persists until the crystals redissolve.

Clinically the affected joint is swollen and red and exquisitely tender and painful. Microscopically there is acute inflammation of the synovial membranes of the joint.

- Chronic arthritis may follow on repeated attacks of acute arthritis, with permanent deposition of sodium urate in and around the joint causing degeneration and deformity.

  Gouty **tophi** are masses of urate crystals surrounded by a foreign body giant cell reaction, occurring in and around joints and also in other sites such as tendons and the cartilages of the ear.
- Renal lesions may be due to:
  - Deposition of sodium urate crystals in the medulla with the eventual development of microtophi
  - Intratubular precipitation of uric acid crystals when sudden high levels of uric acid occur (as in treated leukaemia) can cause obstruction and renal failure
  - Uric acid stones may be formed in the kidney.

Patients with gout may be hypertensive and there is also an association with obesity.

Renal failure may be a late complication in a number of cases. However efficient treatment exists to control the hyperuricaemia by increasing the rate of renal excretion and reducing the formation of urate; colchicine has been known for many years to be a drug which is very effective in treating the acute attacks of arthritis.

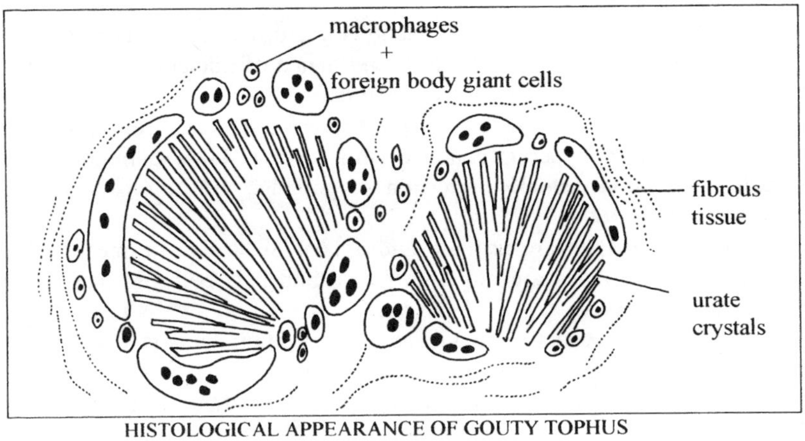

HISTOLOGICAL APPEARANCE OF GOUTY TOPHUS

GENERAL PATHOLOGY - CHAPTER THIRTY SIX

NUTRITIONAL DISORDERS

When talking about malnutrition it is usual to consider the deficiency diseases which result from the lack of one or more vital ingredients in the diet. Indeed mineral or vitamin deficiencies can be related to specific disorders, but it is relatively uncommon for a diet to fail to contain just one vitamin or mineral. In this day and age in developed countries it is almost more common for there to be an excess of one or more vitamins due to overenthusiastic consumption by those seeking improved health. But the most important aspects of malnutrition stem from the quantity of calories in the diet, relative to bodily needs, and diets may be deficient in calories or there may be an excessive intake of calories.

OBESITY

Obesity results from an excessive intake of calories relative to the needs of the body. Excess calories or kilojoules are stored as fat. It is not easy to measure the amount of body fat *per se* and obesity is usually defined in terms of overall body weight. Thus:

- Normal weight plus ten per cent equals **overweight**
- Normal weight plus twenty per cent equals **obesity**

20 per cent of middle aged males and 40 per cent of middle aged females in U.S.A are considered obese where normal body weight is taken from tables of normal weights for height and age and sex.

A more useful measurement is the body mass index (BMI): $\frac{\text{Weight in kg}}{(\text{Height in m})^2}$

The ideal BMI is between 20 and 25kg/m² and obesity can be said to occur when the BMI is greater than 30kg/m². Those with a BMI higher than this can be shown to have an increased mortality rate.

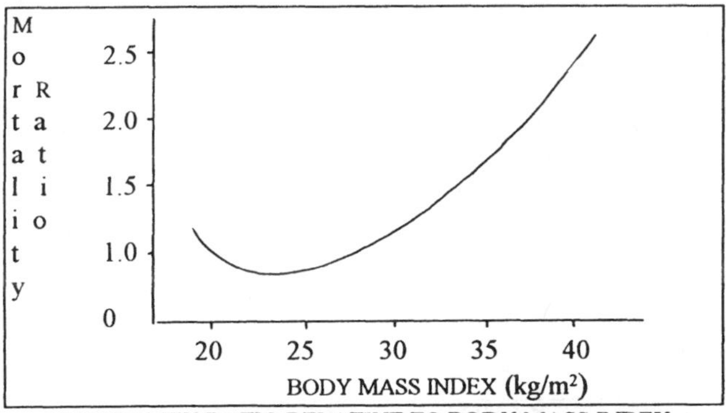

MORTALITY RATIO RELATIVE TO BODY MASS INDEX

The distribution of body fat is also important. Central obesity, with fat concentrated in the abdomen and on the trunk seems to be associated with higher risks of disease than when fat is more evenly distributed.

The cause of obesity is excessive ingestion of calories. Put more simply, it is overeating. There may be genetic factors relating to appetite and to the efficiency of metabolism, psychiatric reasons for compulsive overeating, and endocrine disorders, such as hypothyroidism, which play a part, but in the long run it comes down to:

- **Calories ingested** should equal **calorie output**, or
- **Food intake** should equal **energy expenditure**, that is, physical activity.

for body weight to be maintained at a constant level. Overemphasis of the left hand side of this equation will result in obesity.

The direct pathological effects of obesity are slight. There may be stromal fatty infiltration of some organs. There may be mild fatty change in the liver. The importance of obesity is that it is associated with:

- **Diabetes mellitus**: There is increased resistance of tissues to insulin in obesity and 80 per centof patients with NIDDM are obese.
- **Hypertension**: The obese are some eight times as likely to be hypertensive as the non-obese.
- **Atherosclerosis** resulting in part from hyperlipoproteinaemia is more common in the obese, and myocardial infarction is also more common.
- **Gall stones** and gall bladder disease.
- **Osteoarthritis.**
- **Complications following surgery**
- **Respiratory complications**: such as hypoventilation syndrome may occur in the very obese.

Obesity has even been implicated in the causation of cancer, such as cancer of the endometrium and breast, but the nature of the association is not yet established.

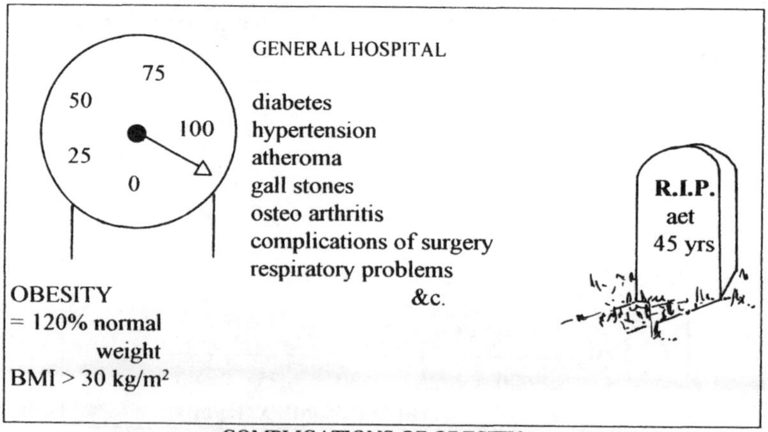

COMPLICATIONS OF OBESITY

PROTEIN-ENERGY MALNUTRITION

Undernutrition is far more important than obesity if the world is considered as a whole, and starvation may be a fact of life (or of death) in many parts of Africa and Asia. Unfortunately it is the children who are hardest hit. In developed countries there may also be malnutrition of this type in the poor and homeless, in the chronically ill, in alcoholics and in those with specific eating disorders.

There is a range of clinical presentations resulting from dietary deficiency of protein (the most expensive constituent of the diet) to total calorie deficiency and these are grouped under the heading of:

The state of nutrition can be measured by:

- Body weight (compared with standard tables)
- Subcutaneous fat as measured by skinfold thickness
- Muscle mass as measured by midarm circumference
- Serum proteins.

The latter two measurements give an indication of the body's protein content. Overall:

- Less than eighty per cent of normal body weight for age is **malnutrition**
- Lack of protein with more or less adequate calorie intake is **kwashiorkor**
- Less than sixty per cent of normal body weight for age is **marasmus.**

Thus kwashiorkor is mainly caused by protein deficiency, and marasmus is caused by a deficiency in total calorie intake, which is basically a state of starvation. Co-existent vitamin deficiencies of various sorts are likely to occur in these conditions.

KWASHIORKOR

This condition often develops after a baby is weaned from its mother's milk and has to exist on a mainly carbohydrate diet.

With protein deficiency there is muscle wasting and failure to grow, but subcutaneous fat may be preserved if the calorie intake is sufficient. There is hypoproteinaemia, mainly hypoalbuminaemia and severe oedema, often with ascites. As a result of persistence of body fat and oedema affected children may actually appear quite plump.

Enzymes are proteins and most enzyme levels are low, accounting for widespread metabolic upsets. The problem may be exacerbated by inefficent intestinal absorption.

There may be anaemia due to inadequate synthesis of haemoglobin, because of insufficient protein

Depigmentation of the skin and hair occur and there is a 'flaky paint' dermatosis of the face and extremities.

The immune response is reduced in both kwashiorkor and marasmus, with thymic atrophy. This may explain the susceptibility to diseases such as tuberculosis and viral diseases, such as measles.

Pathological examination shows that there is persistence of body fat. The main features are:

- Atrophy of protein containing tissue such as muscle
- Atrophy of organs secreting protein enzymes, and
- Fatty change in the liver. As a result:

The skeletal muscles appear wasted and pale.
The heart may be small. Heart failure may occur.
The pancreas is markedly atrophic with loss of cytoplasm in the cells of the acini.
Other glands such as salivary glands and enzyme secreting glands of the gut appear atrophic as may the endocrine glands. In the intestine there may be villous atrophy.

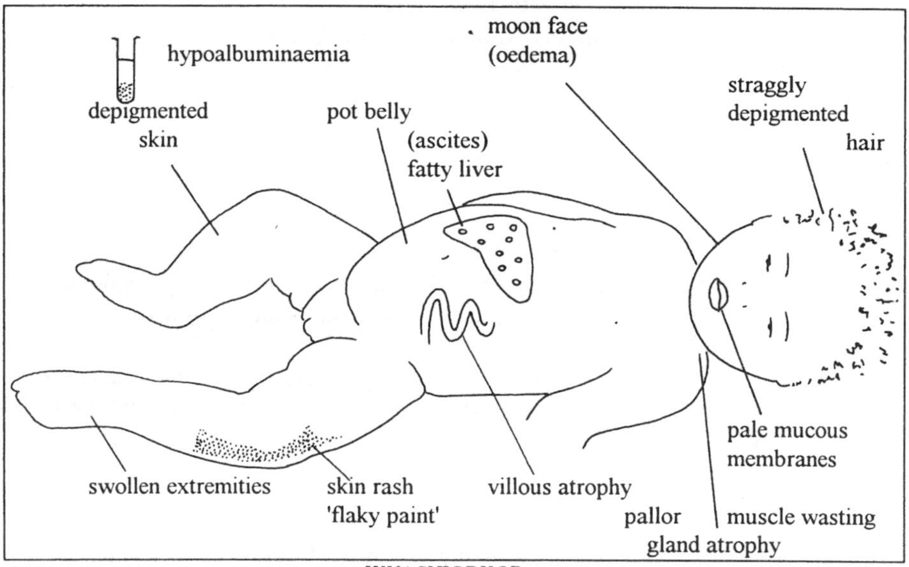

KWASHIORKOR

The liver may show very marked fatty change with some increase in portal fibrosis. Fat may be brought to the liver from fat depots or synthesised from glucose, but because of insufficient protein to make apoprotein carriers the fat cannot be transported from the liver cells.
Infections of all sorts are common in both kwashiorkor and marasmus. There may be gastrointestinal or pulmonary infections and infestation with parasites and worms.

Kwashiorkor can be a fatal disease but is easily cured by adding protein to the diet, usually as milk or skim milk powder. Recovery after treatment appears to be complete, there is not permanent stunting of growth. The fatty liver reverts to normal without lasting damage.

MARASMUS

Marasmus in its extreme form presents a very different picture from kwashiorkor. It must be remembered that these two diseases are at opposite ends of a spectrum and many intermediate forms may be found, including a so-called marasmic kwashiorkor.

This is a wasting disease with loss of muscle bulk and of fat stores. The skin is wrinkled and the appearance is of a shrunken prematurely old person. There is no oedema or hepatomegaly. Growth failure is marked.

In contrast to kwashiorkor in which there is mental apathy, marasmic children may be quite alert.

The diagram sums up the changes characteristic of marasmus.

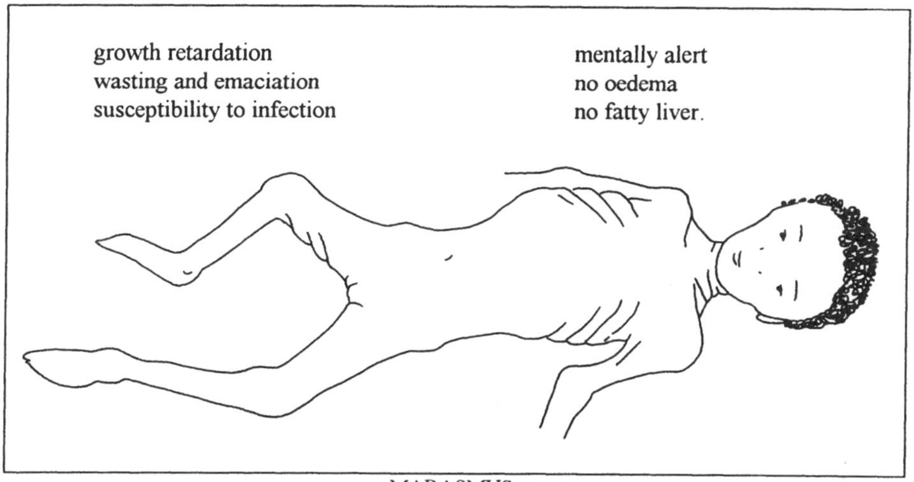

MARASMUS

In South Africa marasmus would appear more likely to be fatal than kwashiorkor.

These diseases chiefly affect children, and undernutrition of such severity would be rare in adults except in times of severe famine or in prison camps. If such appearances are seen, serious underlying disease such as TB, AIDS or malignancy would have to be excluded.

DEFICIENCY DISEASES

Deficiencies of specific substances may occur, that is deficiencies of accessory food substances which are necessary for metabolism and growth and which cannot be synthesised by the body.

These substances include:

- **Minerals.** These may be necessary in considerable quantities such as calcium, phosphorus and iron or as trace elements such as zinc, copper, selenium and iodine.
- **Essential amino acids** include tryptophan, lysine, phenylalanine, leucine, isoleucine, threonine, methionine and valine (and possibly histidine and arginine).
- **Essential fatty acids** include linoleic, linolenic and arachidonic.
- **Vitamins.** Vitamins A, D, E and K are fat soluble and the B complex (thiamine, riboflavine, niacin, B_6, folic acid and B_{12}) and vitamin C are water soluble.

Deficiency states may arise from:

- Reduced dietary intake
- Availability in diet (for example, may be destroyed during cooking)
- Impaired absorption from gut (as in the malabsorption syndrome)
- Failure of production by organisms in gut (such as vitamin K in newborn)
- Increased demands at times of stress (such as puberty or pregnancy)
- Following surgery (for example, intestinal bypass)
- Following drugs (such as cholestyramine, which is a vitamin D antagonist)

It should be noted that it is only in experimental animals that deficiencies of a single substance can easily be identified. Most of the accessory food substances are present in the diet in adequate quantities and deficiencies when they occur are often found to be multiple and mixed.

DEFICIENCIES CAUSING ANAEMIA:

Iron deficiency results in a hypochromic microcytic anaemia. This is a common type of anaemia. Only small quantities of iron are absorbed and most iron in the body is contained in the blood. Thus chronic blood loss can drain iron stores and give rise to anaemia.

Vitamin B_{12} is a water soluble vitamin present in liver, milk and eggs, which is absorbed in the presence of intrinsic factor secreted by the stomach. In certain persons there may be gastric atrophy with failure to produce gastric acid or intrinsic factor, and the vitamin is not absorbed.

After many months the body's stores of Vitamin B_{12} will be used up. This vitamin is necessary for nucleic acid and DNA formation. Without it there is abnormal maturation of red blood cells, with few large abnormal nucleated forms called megaloblasts, being made in the marrow. This results in megaloblastic anaemia.

Folic acid deficiency also results in a type of megaloblastic anaemia.

Vitamin B_6 or pyridoxine may give rise to a type of anaemia in which iron cannot be used properly in the formation of haemoglobin (sideroblastic anaemia).

OTHER VITAMIN DEFICIENCIES. These will be mentioned very briefly:

Vitamin A deficiency is characterised by ocular complications which may cause permanent blindness:

- Night blindness due to decreased rhodopsin regeneration.
- Squamous metaplasia of conjunctiva leads to xerophthalmia (dryness) and
- In the cornea, keratomalacia (softening) may occur.

Other complications due to squamous metaplasia of epithelium in other parts of the body include:

- In the bronchus, bronchopneumonia
- In the urinary tract, kidney stones
- In the skin, follicular hyperkeratosis

Vitamin D deficiency results in

- **Rickets** in children. There is excessive osteoid production and deficient mineralisation causing softening of bones, with deformity and enlargement.
- **Osteomalacia** in adults, with softening and deformity of bone due to demineralisation.

In both rickets and osteomalacia there is a low serum calcium and a high alkaline phosphatase.

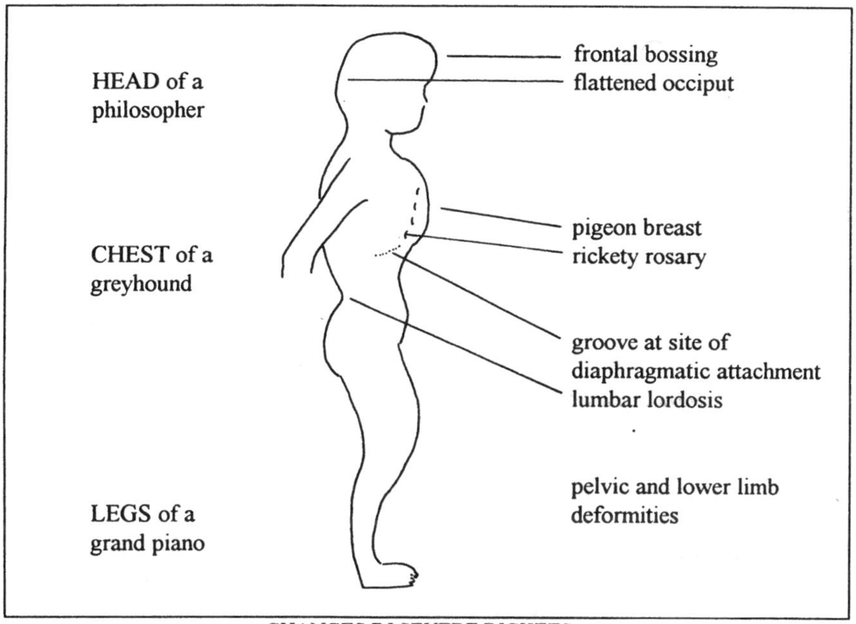

CHANGES IN SEVERE RICKETS

Vitamin K is also fat soluble. It is synthesised by gut bacteria and absorbed in the presence of bile. It is necessary for the formation of prothrombin and factors V and VII by the liver. Deficiency of Vitamin K in the newborn or in cases of obstructive jaundice can lead to a haemorrhagic state.

Vitamin B Complex is water soluble.

- **Thiamine or Vitamin B_1** is a vitamin generally widely available in the diet. However it is not present in polished rice which forms a major part of the diet in some Asian countries, and deficiency can also occur in chronic alcoholics. There are three main methods of presentation:
  - as a polyneuropathy with myelin degeneration and motor and sensory loss in a symmetrical glove and stocking distribution (dry beri-beri).
  - as high output cardiac failure with oedema (wet beri-beri).
  - Wernicke-Korsakoff syndrome. Wernicke's encephalopathy involves paralysis of ocular muscles, nystagmus, abnormal gait and mental confusion. The associated Korsakoff psychosis is a serious memory defect, with learning disability.

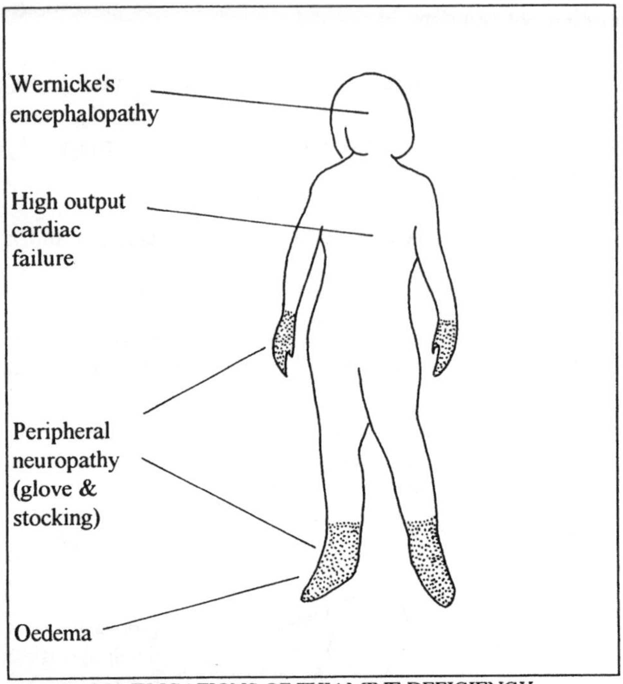

COMPLICATIONS OF THIAMINE DEFICIENCY

- **Riboflavine deficiency** causes a sore tongue and fissuring at the corners of the mouth. There may also be a dermatitis and inflammation of the cornea with vascularisation (interstitial keratitis).
- **Niacin deficiency** causes pellagra. The clinical features include:
  - **D**ermatitis of the exposed parts
  - **D**iarrhoea due to inflammation of the mucous membranes of the gut
  - **D**ementia resulting from degeneration of ganglion cells of the brain and spinal cord tracts, which can result in
  - **D**eath if untreated.
- **Vitamin C deficiency** is known as scurvy. This was one of the first deficiency diseases to be discovered and afflicted sailors who were at sea for long periods without fresh vegetables. Lime juice was found to cure the condition.

  Vitamin C is essential for the formation of stable and strong collagen via the hydroxylation of proline. Most of the complications stem from weakness of collagen in blood vessel walls with:
  - Bleeding into the skin and from the gums
  - Subperiosteal haematomas and bleeding into joints which are painful
  - Defective osteoid formation and bony deformities in children
  - Impaired wound healing
  - Anaemia may follow bleeding.

GENERAL PATHOLOGY - CHAPTER THIRTY SEVEN

TRANSPLANT REJECTION AND LOCALISED AUTO IMMUNEDISEASE

As has already been seen in a number of instances there is a close relationship between anatomical pathology and other branches of pathology such as microbiology. In this chapter and the next the morphological changes accompanying some diseases brought about by disturbances of the immune system will be reviewed. It is of some interest that the diseases were known about long before there was so much knowledge about their aetiology. However once the nature of the disease and its actual causation becomes clearer, the way is open to institute more specific therapeutic measures. For the moment it is the pathological changes and the clinical picture that they produce that will be the focus of attention, and these can then be related to the changes in the immune system which may have brought them about.

TRANSPLANT REJECTION

This is a matter of concern at a time when transplantation of many organs takes place on a regular basis and it may be vital for the transplanted organ to be accepted by the recipient. There may be both a humoral antibody-mediated response and a cell-mediated response to HLA antigens in the graft. The kidney is the organ in which most studies have taken place, and the descriptions will be of those occurring in the kidney, although similar changes take place in other vascularised organs. Skin grafting has been practised for many years, but allografts have a high rejection rate, although serving a useful function to cover defects in the skin whilst healing takes place. At the other end of the scale blood transfusion may be regarded as a sort of transplant which is almost universally accepted, if ABO compatibility is ensured.

T cell-mediated reaction is a type IV reaction where both delayed type hypersensitivity and T cell-mediated cytotoxicity is involved.

Antibody-mediated rejection also plays a part in the rejection reaction and although it is probably of less overall importance than T cell-mediated immunity, is not responsive to immunosuppressive therapy. Thus antibody formation can continue and slowly cause damage by antibody-dependent cell-mediated cytotoxicity (ADCC) (type II hypersensitivity) and also by the fixation of complement with the formation of immune complexes (type III hypersensitivity)

Three main types of rejection reaction have been described:

Hyperacute rejection occurs rapidly within minutes or hours of the graft operation. It occurs when there are **preformed** antibodies to the graft HLA antigens already present in the circulation of the recipient. This may occur in second transplants, in mothers who have developed antibodies to paternal HLA antigens in the foetus, or following blood transfusions in which HLA incompatible white cells or platelets may have stimulated antibody formation.

A widespread Arthus type reaction due to antigen/antibody immune complex formation within the vessel walls takes place. There is fixation of complement, activation of the complement cascade and an acute inflammatory reaction with vasculitis. Histologically, the earliest changes are of adherence of neutrophils to the endothelium of arterioles and glomerular capillaries. Small fibrin and platelet thrombi form. These changes intensify with fibrinoid necrosis of the wall of arteries and arterioles and the formation of occlusive thrombi within them. Ischaemic necrosis of the tissues follows, and the whole renal cortex undergoes infarction. At macroscopic level the kidney looks cyanotic and mottled (instead of a healthy uniform pink colour) and fails to excrete urine.

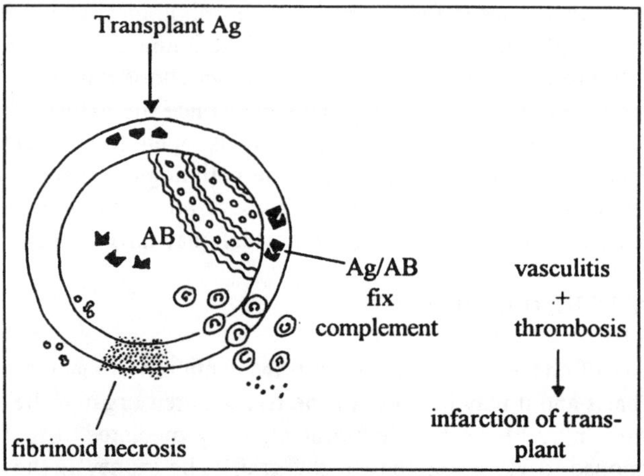

HYPERACUTE REJECTION in the presence of preformed antibodies

Acute rejection is a combined response in which there may be a cellular and a humoral component. It may develop only days after transplantation or may appear years later.

- **Acute cellular rejection** may present with the sudden onset of renal failure. Histologically, there may be interstitial mononuclear infiltration with small and large lymphocytes (both CD4+ and CD8+). These are present in glomerular capillaries and around tubules. There may also be some oedema and haemorrhage.

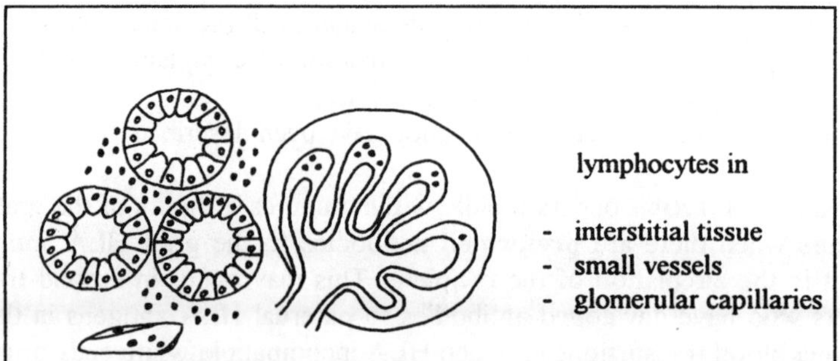

ACUTE CELLULAR REJECTION

In the absence of vascular changes this type of rejection should respond well to immunosuppressive therapy.

- **Acute humoral rejection** is characterised by vasculitis with necrosis of the vessel walls and endothelium, neutrophil infiltration of the wall and the deposition of immunoglobulin, fibrin and complement. Thrombosis will result in necrosis and infarction of cortical renal tissue. Patients present with poor renal function, not responding to immunosuppressive treatment. There is usually also some component of cellular rejection.

  The process may be less acute if there are repeated incidents of alteration of renal function clinically. In this case vessels show intimal proliferation of myofibroblasts with narrowing of the vessel lumen. Scattered inflammatory cells and some foamy macrophages may be seen within the intima. Deposits of immunoglobulin and complement may be seen within the vessel wall on immunofluorescent staining. These deposits tend to confirm the nature of the process.

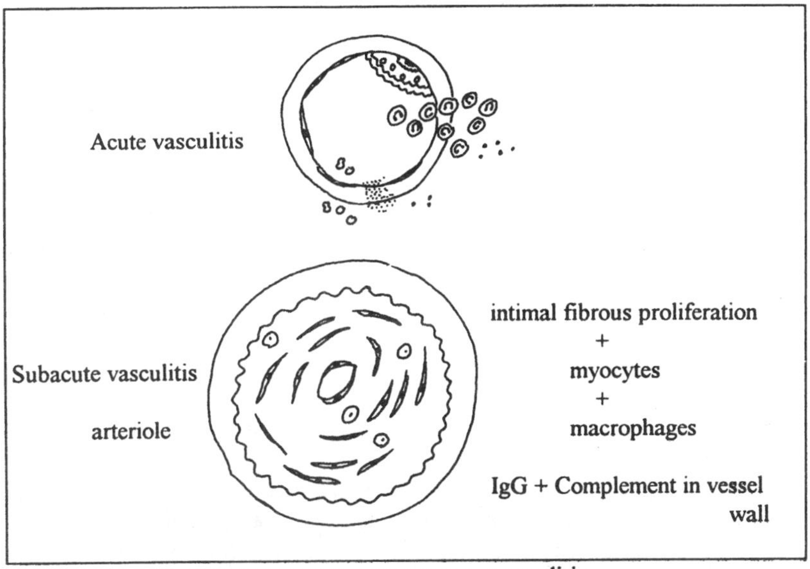

ACUTE HUMORAL REJECTION - vasculitis

Chronic rejection This may be seen where a patient has suffered from a number of previous attacks of acute rejection controlled by immunosuppressive agents. There is a slow progressive diminution in renal function over months. The cortical arteries show very marked intimal fibrosis which leads to ischaemic changes with hyalinisation of some glomeruli, together with tubular atrophy and interstitial fibrosis.

There may also be interstitial mononuclear cell infiltration with plasma cells and eosinophils, perhaps representing chronic cell-mediated rejection.

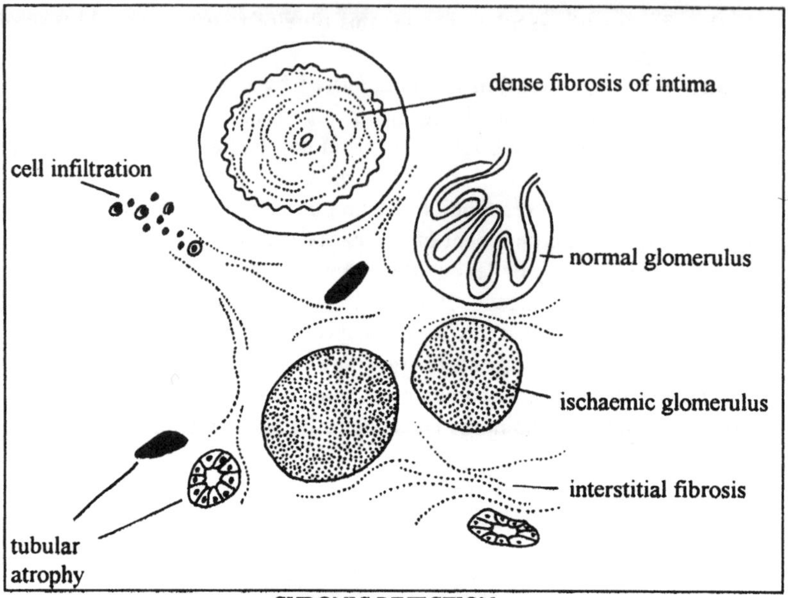

CHRONIC REJECTION

Immunosuppression is necessary in organ transplantation from all donors except an identical twin. The closer the match with regard to HLA antigens, the better the graft survival, but factors such as the number of antigens matched, the population group of the donor and recipient and the nature of immunosuppressive therapy all have to be taken into consideration.

TRANSPLANT OF OTHER ORGANS

Liver transplants are also performed with increasing frequency. Close HLA matching seems to be less important.

Rejection in the liver may be acute, in which case there is infiltration of the portal tracts, the bile ducts and the central veins by inflammatory cells.

Chronic rejection shows continuation of portal inflammation with pronounced bile duct destruction as well as arteriolar thickening and hyalinisation.

Bone marrow transplants are used in cases of leukaemia where large doses of radiation need to be given to destroy malignant cells, destroying all the haemopoietic marrow in the process. There are two problems:

- There may be rejection of the transplant or
- There may be a graft-versus-host reaction with the lymphocytes of the transplant reacting against the recipient's tissues.

The reaction is characterised by necrosis of the liver cells leading to **jaundice,** skin leading to **rash**, and gut mucosa leading to **diarrhoea.** T cells may bring about this reaction, but if they are removed from the transplant it is less likely to take.

AUTOIMMUNE DISEASES

These are diseases in which the immune system of the body reacts against its own tissues. It is not enough merely to demonstrate the existence of autoantibodies, as these can be found in the absence of disease following injury to tissues. There *must* be autoantibodies which appear to be the primary cause of the disease in the absence of any other cause.

Using these criteria, a number of disease states have been classified as being autoimmune. These range from diseases affecting one organ or one cell type to diseases affecting many systems of the body. New diseases keep being added to the list (and others subtracted) as our knowledge about the basis of disease increases.

The immune system does not normally react against the body's own tissues. That is, a state of self-tolerance exists. It seems as if immunologically competent cells learn to recognise the body's own antigens at an early stage.

Three mechanisms are proposed for the development of this tolerance:

- **Clonal deletion:** T lymphocytes with receptors for the body's own antigens are removed in the thymus during the maturation of T cells. Some B cells may also be lost during their development process.
- **Clonal anergy** occurs when cells do not react to the presentation of antigen, because of the lack of co-stimulants (T cells) or because of immaturity (B cells).
- **Suppression by T cells:** Suppressor T cells which are CD8+ may act to prevent autoimmune reactions if potentially self-reactive cells are not eliminated by the previous two mechanisms.

There are a number of mechanisms which have been postulated as leading to loss of self-tolerance and the possible development of autoimmune disease.

- **Release of sequestered antigens** such as spermatozoa, myelin protein and lens material which are normally maintained in isolation from the immune mechanisms. This probably only applies in a few special situations such as those quoted.
- **Loss of T-suppressor cell function**
- **Breakdown of T-helper cell tolerance.** The need for tolerant helper-T cells to assist specific B cells in the production of antibody can be overcome if there is:
  - Alteration of the body's own molecule for example in viral infection, or by combination with a drug (for example, haemolytic anaemia caused by methyl dopa).
  - Cross reaction in which an external antigen can stimulate the formation of an antibody which will cross react with the body's tissues, for example, streptococcal antibodies can be formed which act against the heart in rheumatic fever. Molecular mimicry
  - Stimulation of anergic self specific immune cells by endotoxins or following infection with Epstein-Barr virus

- **Genetic factors** Autoimmune diseases have a familial distribution in some instances. Some autoimmune disease are associated with specific HLA antigens, for example, HLA-B27 with ankylosing spondylitis.
- **Infection** with bacteria or viruses could potentially act in a variety of ways to bring about autoimmunity, but specific evidence of this is lacking at present.

There are two major groups of autoimmune diseases:

- Those which are organ-specific such as thyroiditis and haemolytic anaemia
- Those which are generalised such as systemic lupus erythematosus and the collagen vascular diseases.

It is important to remember that not everyone who has circulating antibodies has overt autoimmune disease. Quite often family members of a patient with undisputed disease may possess autoantibodies of similar type without evidence of there being anything wrong with them.

The following table lists some of the organ specific autoimmune diseases and the nature of the antibodies which may be found in these diseases.

SOME ORGAN SPECIFIC AUTO-IMMUNE DISEASES	
Disease	**Antibodies to**
Hashimotos' thyroiditis	TSH receptors; thyroid cell microsomes etc.
Graves' disease (thyrotoxicosis)	TSH receptors
Autoimmune atrophic gastritis of pernicious anaemia (Vit B_{12} deficiency)	Intrinsic factor; parietal cells; intrinsic factor/B_{12} complex
Idiopathic adrenal atrophy	Adrenal cell microsomes; ACTH receptor
Insulin dependent diabetes mellitus	Pancreatic islet B cells
Goodpasture's syndrome	Basement membrane of renal glomerulus and alveolar septa of lung
Myasthenia gravis	Acetyl choline receptors of striated muscle neuromuscular junction
Pemphigus vulgaris	Epidermal intercelluar bridges
Autoimmune haemolytic anaemia thrombocyopenic purpura leucopenia	Red cells Platelets White cells

HASHIMOTO'S DISEASE

This was the first organ-specific autoimmune disease to be described. The disease is not related to iodine deficiency. Goitre or thyroid swelling is the presenting symptom. There is a marked female preponderance (10:1) and patients are usually middle aged around 30-50 years, although the disease can occur in children.

There is a familial incidence and there seems to be an association with the HLA-DR5 class II major histocompatibility complex.

There is also an association with other autoimmune diseases such as systemic lupus erythematosus, rheumatoid arthritis, Sjögren's syndrome, autoimmune gastric atrophy, insulin-dependent diabetes and also primary thyrotoxicosis or Graves' disease

There is a variety of thyroid autoantibodies, including TSH-receptor antibodies and thyroid microsomal antibodies, as well as antibodies to thyroglobulin and thyroid hormones. These may act to kill cells by antibody-dependent cell-mediated toxicity and complement-mediated cytotoxicity (type II hypersensitivity), immune complex deposition (localised type III hypersensitivity) and T cell-mediated toxicity (type IV hypersensitivity). TSH receptor antibodies may stimulate hormone production (Thyroid-Stimulating Immunoglobulin - TSI) or thyroid growth (Thyroid-Growth Immunoglobulin - TGI). The latter predominate in Hashimoto's disease, whilst the opposite holds good in Graves' disease.

Thyroid function is normal or impaired.

Macroscopically, the gland is symmetrically enlarged, firm and rubbery.

Microscopically, there is massive infiltration of the thyroid gland by lymphocytes. Germinal follicles are present and some plasma cells may be seen. Thyroid cells have either been destroyed or altered into Hürthle cells or oncocytes. These are large cells with prominent pink cytoplasm packed with mitochondria. Hürthle cells are inactive. In some forms there may be fibrosis.

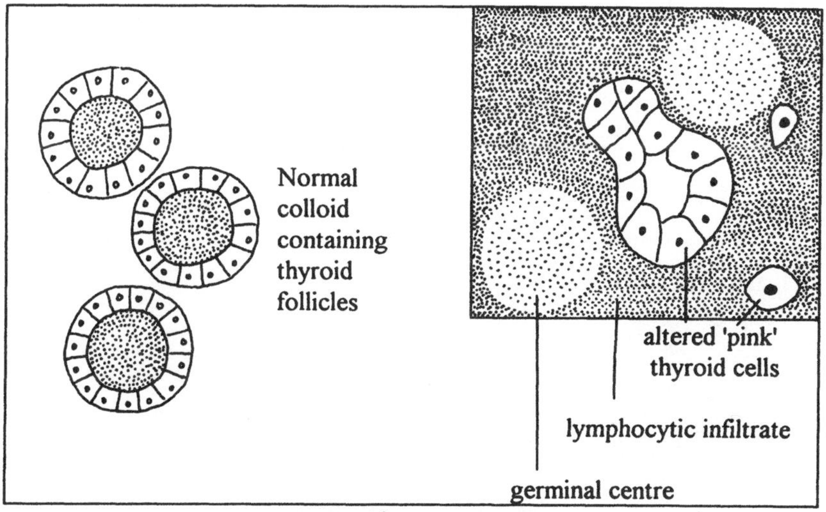

HASHIMOTO'S THYROIDITIS

Clinically, the end result of Hashimoto's disease may be hypofunction of the thyroid, which may get worse with the passage of time.

INSULIN-DEPENDENT DIABETES MELLITUS

This disease has already been described in a previous chapter. There is a genetic susceptibility to the disease with most patients having either HLA-DR3 or HLA-DR4 alleles.

Evidence for the autoimmune nature of the disease is present in that there are islet cell antibodies and there is also a lymphocyte infiltrate of the the islets in the early stages of the disease.

There is also an increased incidence of other autoimmune diseases, for example, thyroiditis, Graves' disease or pernicious anaemia. This seems to be a common feature, particularly of the organ-specific auto-immune diseases, that one of these diseases may predispose to another. In this type of diabetes it is thought that some environmental agent, possibly a virus, triggers off the autoimmune reaction in the first place.

Other single organ autoimmune diseases will be described in detail in systematic pathology. This sort of disease can be suspected when there is:

- A familial incidence
- The presence of autoantibodies
- Lymphocytic or mononuclear infiltrate in diseased tissues
- There are other conditions which are known or suspected to be autoimmune in nature.

GENERAL PATHOLOGY - CHAPTER THIRTY EIGHT

GENERALISED AUTOMIMMUNE DISEASES

There are a group of diseases in which the major disturbances seem to be concentrated in the connective tissue and the vessels throughout the body. These were grouped together as the 'collagen' diseases or 'collagen-vascular' diseases fifty years ago, on account of the morphological features which they showed, including fibrinoid necrosis in collagen and in vessel walls.

Since then many of the original diseases falling under this heading (rheumatic fever, rheumatoid arthritis, systemic lupus erythematosus, polyarteritis nodosa and scleroderma) have been shown to be associated with the presence of autoantibodies which react to the components of most cells, for example, antinuclear antibodies. A few other diseases in which there are circulating autoantibodies and which have somewhat similar features have been added to the list (Sjögren's syndrome, dermatomyositis and mixed connective tissue disease). These are now regarded as generalised, non-organ-specific autoimmune diseases. The disease process seems to be of the type III hypersensitivity variety, with immune complex deposition in tissues.

SOME GENERALISED AUTOIMMUNE DISEASES **The collagen-vascular diseases**	
Original collagen diseases (Klemperer, 1942)	**to which can be added**
(Rheumatic fever)*	Sjögren's syndrome
Rheumatoid arthritis (RA)	Polymyositis/dermatomyositis
Systemic lupus erythematosus (SLE)	Mixed connective tissue disease
(Polyarteritis nodosa)‡	
Scleroderma	

* Rheumatic fever is probably due to the formation of streptococcal antibodies that cross react with human tissues in heart, joints and less commonly at other sites. This condition tends to be excluded from modern classifications of autoimmune disease.

‡ Polyarteritis nodosa is one of the types of vasculitis; immune complexes, not necessarily autoimmune, cannot be demonstrated in most cases. This disease is also excluded from the more modern classifications.

PATHOLOGICAL FEATURES OF COLLAGEN-VASCULAR DISORDERS

These diseases have a number of features in common:

- There is widespread degeneration of collagen with fibrinoid necrosis. Fibrinoid necrosis is not true necrosis in terms of the definition of the changes following on cell death in the living body, although there may be destruction of endothelial cells and fibrocytes. It is a change that occurs in blood vessel walls and in the extracellular matrix of collagen.

Fibres become less well defined, fuzzy in outline and more deeply eosinophilic and take on the staining properties of fibrin - hence the name fibrinoid or fibrin-like. In vessel walls this change seems to be associated with immune complex deposition and activation of complement and immunofluorescent staining techniques will demonstrate immunoglobulin, complement and fibrinogen.

Although a characteristic of the collagen vascular diseases fibrinoid necrosis is not specific for these diseases. It may also be found in the walls of arterioles in malignant hypertension and in vasculitis as well as in tissues following irradiation.

- There is often associated inflammatory change with cellular infiltration and oedema.
- There may be reactive fibrosis.

In summary, **fibrinoid necrosis** (most marked in SLE), **inflammation** (most marked in rheumatoid arthritis) and **fibrosis** (most marked in scleroderma) are the pathological changes which may be found in the collagen vascular diseases in many tissues of the body.

There may be a certain amount of overlap both clinically and pathologically between different conditions, for example, SLE and rheumatoid arthritis, and also in mixed connective tissue disease in which some features of SLE, dermatomyositis and scleroderma are found together.

Some general features of collagen-vascular diseases:

- These diseases tend to affect young or middle aged people. There is an increased incidence in some congenital immunological deficiency syndromes. More females are affected than males.
- The effects are widespread throughout the body. There is a marked increase in the erythrocyte sedimentation rate, often anaemia and usually hyperglobulinaemia. There is frequently fever and also weight loss.
- A number of types of circulating autoantibodies, usually antinuclear antibodies, are present.
- Corticosteroid therapy usually provides some relief from the effects of the disease, probably by virtue of the anti-inflammatory effect (phospholipase inhibitor). Immunosuppressive drugs may also be used. None the less these diseases are important causes of long term chronic illness.

SYSTEMIC LUPUS ERYTHEMATOSUS

This disease has inherited (from syphilis, which is now quite uncommon in its later forms) the title of the 'great mimic'. The reason is that it can present with a multitude of symptoms. The disease is fairly common and occurs chiefly in young persons. There is a female:male ratio of 10:1.

It can affect a variety of sites although the skin, kidneys, serous membranes, joints and heart are the most likely. Pathologically there are vascular lesions with fibrinoid necrosis and inflammation. Immunologically there is a large variety of antibodies mostly directed against cell nuclei, or various nuclear components.

The American Rheumatism Association has published a list of clinical symptoms which may be found. The diagnosis of SLE be only be made if four of the eleven criteria are present

Clinical findings in systemic lupus erythematosus are:

Clinical findings in Systemic Lupus Erythematosus

1.	Skin rash in butterfly area of face	7.	Renal disorder with proteinuria or casts
2.	Discoid lupus of skin	8.	Neurologic disorder: seizures or psychosis
3.	Photosensitivity	9.	Autoimmune anaemia, leucopenia, or thrombocytopenia
4.	Oral ulcers	10.	Specific antibodies to DNA or ribonucleoprotein (anti-Sm).
5.	Arthritis		Positive LE cell test. False positive serologic test for syphilis
6.	Inflammation of serous membranes: pleurisy, pericarditis	11.	Antinuclear antibodies in high titre

As has been mentioned, the basic pathological lesion is that of fibrinoid necrosis of small arteries and arterioles, often with a surrounding infiltrate of mononuclear cells. These vessels may later show fibrous thickening with stenosis.

Changes occur in specific organs:

The skin: There is a red rash in the butterfly area on either side of the nose, which may be worsened by exposure to sunlight. Microscopically there is degeneration of the basal layer of the epidermis and a round cell infiltrate. On immunofluorescence, granular deposits of immunoglobulin are present at the dermal epidermal junction in both involved and uninvolved skin.

Localised round discoid lesions can occur anywhere and may be present without systemic involvement.

Serous membranes: There may be inflammation of pleura or pericardium with or without effusion.

Kidney: This is almost always involved if examined carefully enough. There may be:

- Normal appearance
- Mesangial deposits of immune complex
- Focal endothelial and mesangial proliferation and inflammation
- Diffuse proliferative glomerulonephritis
- Membranous glomerulonephritis with severe proteinuria and the nephrotic syndrome.

Deposits of DNA-anti-DNA complexes may be seen within the glomeruli on either side of the basement membrane. Localised thickening of the basement membrane by deposits can give rise to a 'wire loop' appearance.

The heart: There may be pericarditis or a rather non-specific myocarditis. Libman-Sacks endocarditis consists of non-infected warty deposits on either side of the valve surface. These are composed of fibrinoid material and inflammatory cells.
Joint involvement: Although a common cause of clinical symptoms, there may only be minor pathological changes of synovial swelling and inflammation.
The spleen: This may be slightly enlarged and show perivascular fibrosis with 'onion skin' lesions of the penicillar arteries.
The brain: No constant lesions may be seen, despite clinical neurological symptoms.

In fact, any organ may show changes with vasculitis, fibrinoid deposits and mononuclear cell infiltrate. Haematoxyphil bodies, which are purplish staining degenerative nuclei that have been damaged by antibodies, may be seen in necrotic areas.

There are a large number of antinuclear autoantibodies, to DNA, histones, non-histone proteins bound to DNA and to nucleolar proteins, which may be detected by an indirect immunofluorescent technique. Antibodies to double-stranded DNA and to a ribonucleoprotein called the Smith antigen are characteristic of SLE.

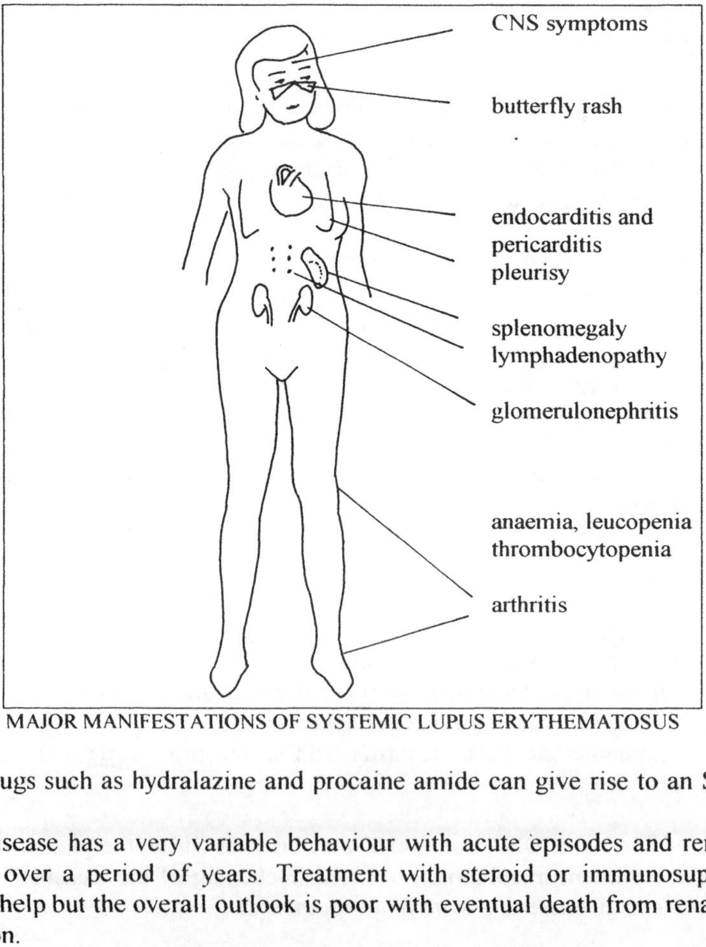

MAJOR MANIFESTATIONS OF SYSTEMIC LUPUS ERYTHEMATOSUS

Certain drugs such as hydralazine and procaine amide can give rise to an SLE-like state.

The disease has a very variable behaviour with acute episodes and remissions occurring over a period of years. Treatment with steroid or immunosuppressive drugs can help but the overall outlook is poor with eventual death from renal failure or infection.

RHEUMATOID ARTHRITIS

This is a chronic inflammatory connective tissue disease affecting mainly the joints, but also many other sites. It is a common disease affecting some one per cent of the population. The disease usually commences between 20 and 40 years and the female:male ratio is 3-5:1.

The disease typically commences in the small joints of the hands and feet with inflammation of the synovium. There is proliferation of synovial cells and villus formation, a plasma cell infiltrate, lymphocytic aggregates and proliferation of capillaries in the subsynovial tissue. Proliferating synovium covers the surface of the articular cartilage as a sheet or pannus. There is periarticular inflammation and joint swelling.

The joints are painful and stiff, particularly in the morning. Movement is restricted and may be lost completely in the late stages if there is fibrosis and immobilisation of the joint. Joints are swollen and deformed.

Rheumatoid nodules may occur on the arms or legs. These are areas of fibrinoid necrosis surrounded by palisaded macrophages in fibrous tissue. They may measure 1-2cm in diameter. They may also be present in viscera.

As in other generalised autoimmune diseases lesions may occur in pleura, pericardium and lungs. Vasculitis can result in ischaemic changes in, for example, the skin, gut, brain and heart. There may be lymph node enlargement with marked follicular hyperplasia. Pulmonary lesions can occur and usually result in fibrosis. In patients who have pneumoconiosis, particularly large rheumatoid nodules can develop in the lungs.

Reactive amyloidosis (AA type) may occur as a late complication.

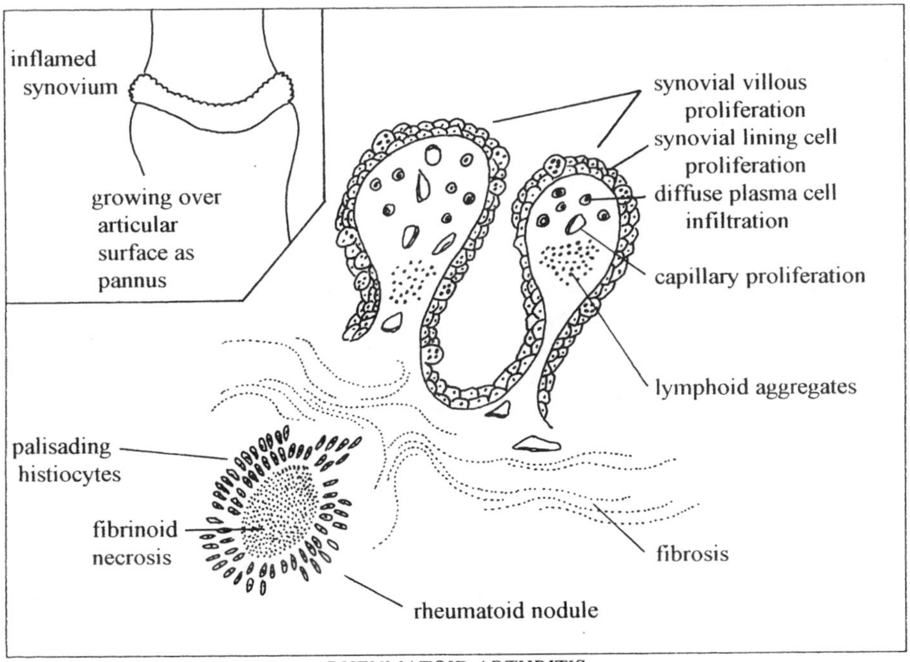

RHEUMATOID ARTHRITIS

Some people may be more at risk because they carry susceptible HLA-DR4 or HLA-DR1 alleles. The disease seems to be produced by some event that triggers helper-T cells to produce a wide variety of cytokines which bring about inflammation in the joint space. The B cells are activated to produce antibodies. Rheumatoid factor is an autoantibody directed against the Fc portion of IgG. In the joints IgG/anti-IgG complexes are deposited, activating the complement system and causing or perpetuating inflammation.

Variants of rheumatoid arthritis may be found in children.

There is another group of diseases characterised by inflammation of the vertebral and sacro-iliac joints which have some similarities with rheumatoid arthritis but which lack rheumatoid factor and which are associated with the HLA-B27 allele. Ankylosing spondylitis, Reiter's syndrome and psoriatic arthropathy may be included in this group.

SJÖGREN'S SYNDROME

Primary Sjögren's syndrome consists of dry eyes (keratoconjunctivitis sicca) and dry mouth (xerostomia) resulting from the destruction of the lachrymal and salivary glands due to an autoimmune process. It may also occur in secondary form in association with other autoimmune diseases such as rheumatoid arthritis and SLE.

The affected glands are heavily infiltrated with lymphocytes and plasma cells. There may be germinal follicle formation. Dried mucous membranes may become inflamed and ulcerate. Other secretory glands, mainly in the upper parts of the respiratory tract may also be involved. There may be a mild interstitial nephritis in the kidney with tubular defects and loss of amino acids in the urine. Vasculitis may occur in the skin and also other organs in extraglandular sites.

There is a wide variety of antibodies which may be found including rheumatoid factor and antinuclear antibodies, even in the absence of overt RA or SLE.

Over half the cases are secondary. In a small proportion of cases (less than one per cent) a B cell lymphoma may develop.

SCLERODERMA

Scleroderma is now better known as systemic sclerosis, because it is not confined to the skin. It is ultimately the systemic lesions that result in disability and death. There is excess production of collagen (fibrosis or sclerosis) in affected tissues, and the skin is involved in 90 per cent of cases. The disease can occur at any age and like most autoimmune disease is more frequent in females (female:male ratio is 3:1). There is:

- A diffuse form which is more rapidly progressive, and
- A more localised form in which skin involvement is limited to the hands and face and which is less likely to progress to internal organs. This is called the **CREST** syndrome because of **C**alcification, **R**aynaud's phenomenon, o**E**sophogeal fibrosis, **S**clerodactyly and **T**elangiectasia.

The changes are most marked in the skin which is thickened and stiffened, to such an extent that facial expression may be lost and movement of fingers restricted.

Microscopically there is an increase in dermal collagen and loss of skin adnexal structures. Small vessels are thick walled and hyalinised. There may be dermal or subcutaneous calcification, especially in the localised form.

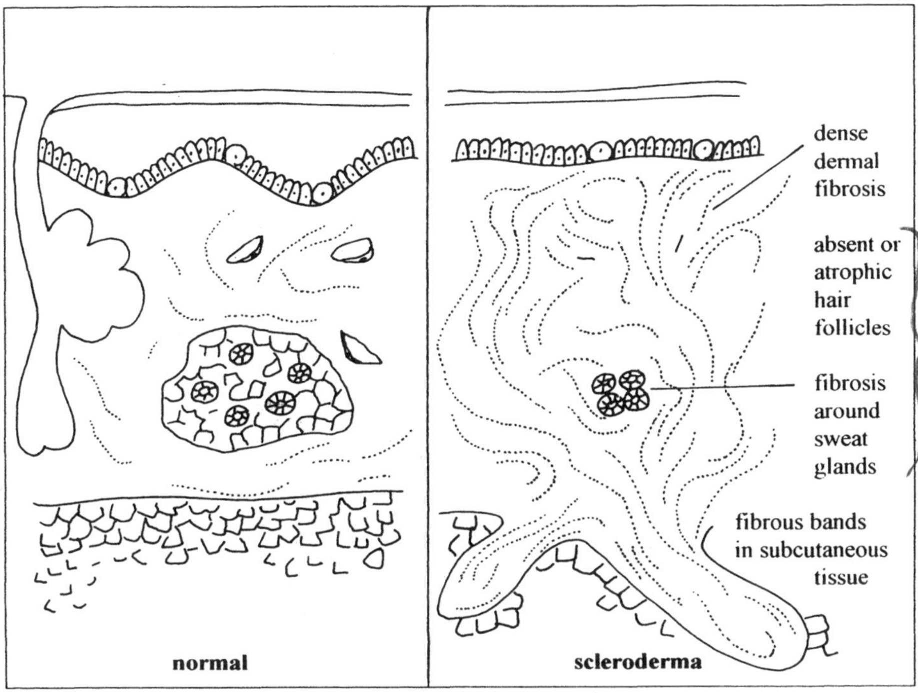

SCLERODERMA (SYSTEMIC SCLEROSIS) IN THE SKIN

Visceral lesions can occur in:

- The gut, where oesophogeal fibrosis may lead to dysphagia (difficulty in swallowing).
- The lungs which may show interstitial fibrosis
- The kidneys which may show the vascular lesions of hypertension, regardless of the level of the blood pressure. In fact about one half of those with renal involvement may have hypertension and in about one third of these this may be the severe or malignant form.
- Other organs including joints, muscles and heart and nerves may show fibrotic changes.

The actual cause of the disease is unclear. Cytokine release giving rise to growth factors attracting and stimulating fibroblasts to grow and synthesise collagen could play a part. Endothelial injury to small vessels may also be involved. There is a wide range of antibodies present, with many cases having antinuclear antibodies and rheumatoid factor.

Even the diffuse form of the disease is relatively slowly progressive and increasing disability may take many years to develop.

DERMATOMYOSITIS (POLYMYOSITIS)

This is a generalised disease affecting the skin and the muscles. If the skin is not involved it is called **polymyositis**.

In the skin a rash affecting the upper eyelids, knuckles elbows and knees may be seen. Histologically there may be oedema and mononuclear cell infiltrate around vessels.

In the muscles there is inflammation with patchy muscle fibre degeneration. There is proximal muscle weakness which spreads distally..

General features include fever, malaise and muscle tenderness as well as weakness. The incidence of visceral cancer appears to be increased in older patients with some ten per cent of patients having underlying carcinomas.

A number of autoantibodies may be found and there is often overlap with other connective tissue diseases such as SLE, rheumatoid arthritis and systemic sclerosis.

In the absence of underlying malignancy the patients can respond to therapy with immunosuppressive agents.

MIXED CONNECTIVE TISSUE DISEASE

This is a term given to a disease which has features of SLE, polymyositis and scleroderma together with antibodies to nuclear ribonucleoprotein. Renal involvement is rare and the disease responds well to steroid therapy. It is still uncertain whether this is a separate entity or whether it represents a mixed group of disorders in which there is overlap of signs and symptoms of various autoimmune diseases.

AUTOANTIBODIES THAT MAY BE FOUND IN VARIOUS AUTOIMMUNE DISEASES

Antibody	Nature of antigen	SLE	Drug induced SLE	Rheumatoid arthritis	Sjögren's syndrome	Systemic sclerosis diffuse	Systemic sclerosis CREST	Polymyositis/ dermato-myositis	Mixed connective tissue disease
Generic ANA (indirect IF)**	Many nuclear antigens	>95% H S M	>95% H	?	50-80% S N	70-90% S N	70-90% S N	40-60% H S	>95% S
Anti- double-stranded DNA	native DNA	40-60%	-	-	-	-	-	-	-
Anti-histone	Histones	70%	>95%	-	-	-	-	-	-
Anti-Sm	Ribonucleo-protein (Smith ag)	20-30%	-	-	-	-	-	-	-
Anti-Nuclear RNP	Ribonucleo-protein	40%	-	-	-	-	10%	-	>95%
Anti-SS-A (Ro)	Ribonucleo-protein	30-40%	-	-	70-95%	-	-	10%	-
Anti-SS-B (La)	Ribonucleo-protein	15%	-	-	60-90%	-	-	-	-
Anti-Scl-70	DNA topo-isomerase 1	-	-	-	-	40-70%	10%	-	-
Anti-centromere	Centromeric proteins	-	-	-	-	<10%	90%	-	-
Anti-Jo-1	Histidyl - tRNA synthetase	-	-	-	-	-	-	25%	-
Rheumatoid factor	IgG (Fc piece)	20-30%	<25%	85-95%	80%	25-40%	25-30%	30%	50%

**Antinuclear antibodies are demonstrated by indirect immunofluorescence using patients' serum on rat liver sections and staining for IgG classes.

Nuclear positivity may be:
- Homogeneous (H) due to antibody to DNA-histone complex
- Speckled (S) due to antibody to saline extractable antigens (e.g. Sm, RNP, Ro, La, Scl 70 and Jo-1)which are ribonucleoproteins
- Membranous (M) around the rim of the nucleus due to antibody to double stranded DNA
- Nucleolar (N) due to antibodies to nucleolar RNA.

INDEX

Abscess, 143
Acanthosis nigricans, 292
Acid base balance, 72,73
Acidosis, 72,73
    diabetic, 72
    lactic, 72
Actinomycosis, 190-191
Adenoma, 210
Adrenal neuroblastoma, 283-284
Aetiology, definition, 4
Agenesis, 195, 281
Alanine transaminase (ALT), 64
Albinism, 53
Aldosterone, 66
Alkalosis, 71-74
Alpha particles, 295
Amniotic fluid embolism, 99, 114
Amyloid, 37 *et seq.*
    definition, 37
    diagnosis, 43
    electron microscopic appearance, 38
    localised, 41
    mechanisms of disease, 43
    pancreatic, 307
    stains, 38
    structure, 38-39
Amyloidosis, 40-41, 139
    familial, 41
    haemodialysis, 41
    in chronic inflammation, 40
    in immunocytic dyscrasias, 40
    in lepromatous leprosy, 187
    in rheumatoid arthritis, 341
    in tumours, 292
Anaemia,
    auto-immune haemolytic, 334
    deficiency, 326
    haemolytic, 60
Anaplasia, 203
Anasarca, 75
Angioma, 211, 282
Anoxia, 19
Antidiuretic hormone (ADH), 65
Aplasia, 196
Apoptosis, 31, 195
Arachidonic acid metabolism, 104, 134
Arginine vasopressin (AVP), 65
Arterial obstruction,
    causes, 116
    effects, 118
Arterial thrombosis, 109
Arteriolar hyalinisation, 307
Arthus reaction, 330
Aschoff giant cell, 16, 153
Ascites, 75
Aspartate transaminase (AST), 27, 64, 121
Atheroma, 116, 121, 307
Atrial natriuretic peptide (ANP), 66
Atrophy, 195-197
    brown, 53-54
    definition, 195
    radiation, 299
Auto-immune disease, 333-336, 337 *et seq.*
    definition, 333
    general features, 338
    localised, 333-336
    mechanisms, 333-334
    organ specific, 334
    pathologic features, 337-338
Auto-immune haemolytic anaemia, 334
Autolysis, 31
Autophagy, 195
Autopsy, 6

Bacteraemia, 108
Bacterial disease, 170-173
Bacteriology, 1
Barr body, 10
Basophils, 136
Bacille-Calmette-Guérin (BCG), 182
Bence Jones protein, 276
Beri beri, 327
Beta particles, 295
Bilirubin, 57 *et seq.*
    conjugated, 58
    unconjugated, 58
Bladder hypertrophy, 201
Blood, clotting 103
    post mortem clot, 33

Boil, 171
Bone,
  changes in hyperparathyroidism, 48
  chronic osteomyelitis, 149
  osteoma, 210
  tuberculosis, 182
Bone marrow,
  hyperplasia, .198-199
  leukaemias, 275
  multiple myeloma, 211, 277
  radionecrosis, 299
  tumours,
  tumours in children, 288
Bowen's disease, 250
Brain,
  in shock, 97
  infarct, 122
  medulloblastoma, 287
  oedema, 79
  tumours in children, 288
Breast,
  carcinoma, 215
  fibroadenoma, 212
  hyperplasia, 198
Bronchitis, chronic, 150
Bronchopneumonia, 156
Brown induration, 86
Burkitt's lymphoma, 274

Cachexia, 290
Calcification, 45 *et seq.*
  definition, 45
  dystrophic, 45
  mechanisms, 50
  metastatic, 46
  causes of, 45-49
Calcinosis, 49
Calcium,
  metabolism, 47
  plasma level, 47
Calculi, 49
Calmodulin, 12
Cancer, causes of, see carcinogenesis
Cancerous change in cells, 203,217-218,
Carbohydrate metabolism, 303 *et seq.*
Carcinogenesis, 235 *et seq.*
  causes, 236-238
  cell growth, 238
  chemical agents, 237
  oncogenes, 240-241
  physical agents, 236-237
  radiation, 236-237
  tumour growth, 239
  tumour suppressor genes, 242-243
  viruses, 238, 241
Carcinoid syndrome, 291
Carcinoma, 210
  basal cell carcinoma, 253
  carcinoma-in-situ, 203, 248, 250
  squamous cell carcinoma, 251
Cardiac 'cirrhosis', 86-87
  failure, 83-87, 100
  infarction, 121-122
  oedema, 78
Cardinal signs of inflammation, 127
Cartilage, tumours of, 210
Cells, 7 *et seq.*
  connective tissue, 8
  cytoplasm, 13
  electron microscopic appearance 18
  endothelial, 8
  epithelial, 7
  growth cycle, 238
  injury, 19 *et seq.*
  junctions, 15-16
  labile cells, 155
  membrane, 10
  membrane function, 11-13, 19
  mesothelial, 8
  multinucleate giant cell, 16-17, 153
  nucleus, 9
  permanent, 155
  stable, 155
Centriole, 14
Chemical mediators of inflammation, 133
Chemotaxis, 131
Childhood tumours, 281 *et seq.*
Chlamydial diseases, 169-170
Cholestasis, 61-62
Chondroma, 210
Choristoma, 209, 281
Cicatrisation, 163
Cilium, 14
Clathrin, 12
Clot, post mortem, 33
Clotting of blood, 103

Cloudy swelling, 21
Clubbing of fingers, 292
Coagulation of blood, 103
Coagulative necrosis, 29, 121-122, 123, 124
Collagen diseases, 337
Commensal organisms, 165
Complement, 133
Congestion, 83 *et seq.*
  causes, 83-84
  definition, 83
Connective tissue
  cells, 8
  diseases, 337 *et seq.*
  tumours, *et seq.*
Creatine kinase (CK), 27, 121
Criggler-Najjar disease, 64
Cushing's syndrome, 291
Cystic fibrosis, 316
Cytokines, 135
Cytology, 1
Cytological diagnosis of malignancy, 217-219
Cytoplasmic granules, 15
Cytoskeleton, 14
  microtubules, 14
  actin filaments, 14
  intermediate filaments, 15

Death,
  point of, 27
  post mortem changes, 31 *et seq.*
  somatic, 31
Decompression sickness, 113
Deficiency disease, 5, 325-328
Degeneration, cellular, 20-22
Dermatofibroma (fibrous histiocytoma), 261
Dermatofibrosarcoma protruberans, 261
Dermatomyositis, 292, 344
Desmoid tumour, 260
Diabetes mellitus, 303 *et seq.*
  causes of death, 310
  clinical course, 310
  complications,
    nervous, 309
    ocular, 309
    renal, 308
    vascular, 307
  diagnosis, 303
  insulin dependent, 305-306, 336
  non-insulin dependent form, 305
Diabetic acidosis, 72
Diagnosis, definition, 4
  steps in, 4
  of malignancy, 218-220
Diapedesis, 131,
Differentiation of tumours, 207, 217
Disease,
  acquired, 5-6
  classification by cause, 5-6
  congenital, 5
  definition, 3
  genetic, 5, 313
  infectious, 165 *et seq.*, 175 *et seq.*, 185 *et seq.*
Diseases, connective tissue, 337
Disseminated intravascular coagulation, 99
D.N.A., 9, 236-242
Dubin-Johnson syndrome, 64
Dysplasia, 195, 202-203
Dysplastic naevi, 255

Ecchymosis, 89
Edema, see Oedema, 75 *et seq.*
Electrolytes, 65 *et seq*
Electron microscope, 6
Embolism, 103 *et seq.*
  amniotic fluid, 111, 114
  arterial, 113
  fat, 111, 114
  gas, 111, 113
  parasitic, 111, 114
  pulmonary, 111
  thrombotic, 111-113
  tumour, 111, 114
  types, 111
Embolus, definition, 107, 111
Emigration of white blood cells, 131
Endocarditis,
  Libman-Sacks, 340
  non-bacterial thrombotic, 292
Endocrine diseases, 5
Endocytosis, 12
Endoplasmic reticulum, 13, 19

Endocytosis, 12
Eosin, 6
Eosinophilic granuloma, 280
Eosinophils, 136
Epithelial cells, 7
Epstein Barr virus, 241
Exocytosis, 11
Extracellular fluid, 66 *et seq.*
Exudate, 142, 147
  formation, 139
Eye, retinoblastoma, 286

Fasciitis, nodular, 260
Fat embolism, 111, 114
Fat necrosis, 30
Fatty change, 22 *et seq.*
  causes, 23
Fernandez reaction, 187
Ferruginisation, 85
Fever, 139, 290
Fibrinoid necrosis, 30, 298, 337-338
Fibroma, 211
Fibromatosis, palmar/plantar, 260
Fibrosarcoma, 260
Fibrosis, radiation, 299
Fibrous tissue tumours, 260-261
Finger clubbing, 292
Fluid balance, 65 *et seq.*
Free radicals, 21
Foreign body giant cell, 21, 153
Foreign body granuloma, 152
Fungus diseases, 190-191

Galactosaemia, 310
Gall stones, 49, 322
Gamma rays, 295
Gangrene, 33-35
  definition, 33
  dry, 34
  gas, 35
Gas embolism, 111, 113
Gastro-intestinal tract,
  amyloid, 42-43
  gangrene, 24
  infarcts, 124
  radiation change, 299
Gaucher's disease, 318
Ghon complex, focus, 178
Giant cells, 16-17, 153
Gilbert's disease, 64
Glomerular filtration rate, 66
Glomerulosclerosis, diabetic, 308
Glucagon, 303, 304
Glucose, levels, metabolism, 303
Glycogen,
  infiltration, 26, 308
  storage diseases, 311
Golgi apparatus, 13
Gonads, radiation change, 299
Gout, 319-320
Graft rejection, 329-333
Graft-versus-host reaction, 332
Granulation tissue, 158-159
Granuloma, 150-152
Granuloma pyogenicum, 267
Growth disorders, 195 *et seq.*
Gumma, 188
Gummatous necrosis, 30

Haemangioma, benign, 268, 282
Haemangioendothelioma, 268
Haemangiopericytoma, 268
Haemangiosarcoma, 268, 282
Haematin, 55
Haematoidin, 54, 56
Haematology, 1
Haematoxylin, 6
Haemochromatosis, 53, 55
Haemoglobin, 54
Haemopoietic tissue, radiation change, 299
Haemorrhage, 89 *et seq.*
  causes, 89, 90
  definition, 89
  effects, 90
Haemosiderin, 54
Haemosiderosis, 54
Haemostasis, 103
Hamartoma, 209, 281
  definition, 209
Hand-Schuller-Christian disease, 279-280
Hashimoto's' disease, 334-335
Healing, 155 *et seq.*
  complications, 163-164

Healing, 155 *et seq.*
definition, 155
factors influencing, 162-163
mechanism, 158-159
of surgical wound, 160
of wound with separated edges, 161
Heart,
brown atrophy, 53-54
failure, 84, 85, 86, 87
fatty change, 23-24
healing, 158
hypertrophy, 200
infarction, 121-122
in Marfan's syndrome, 315
muscle fibrosis, 118, 158
thrombosis, 109-110
Hepatoblastoma, 287
Hepatocellular damage, 63
Hernia, strangulated, 34
Heterophagy, 195
Histamine response, 129
Histiocytoma, fibrous
(dermatofibroma), 261
malignant fibrous, 261
Histiocytoses, 211,271, 279
Histiocytosis X, 279-280
Hodgkin's disease, 211, 271-273
staging, 272
classification, 272-273
Host, resistance to infection, 165
Hyaline degeneration, 25
droplets, 25
Mallory's, 25-26
Hydropic degeneration, 21\
Hydropericardium, 75
Hydrothorax, 75
Hyperaemia, 83 *et seq.*
Hypercalcaemia, 47,48
Hypercholesterolaemia, 313, 315
Hyperlipidaemia, 313-315
Hyperparathyroidism, 47-49
Hyperplasia, 105, 197-198
Hypertrophy, 195, 199-201
Hypertriglyceridaemia, 314, 315
Hypoglycaemia, 291
Hypoplasia, 195, 281
Hypoxia, 20, 115
types of, 115

Immune disease, 5
Immunoelectrophoresis, 93,94
Immunofluorescence, 6
Immunology, 1
Inclusion bodies, 168
Infarct,
classification, 120
definition, 29
Infarction, 115 *et seq.*
Infection, 5,
spread of, 165-168
Infectious diseases, 165 *et seq.*, 175 *et seq.*, 185 *et seq.*
Infiltration, cellular, 20 *et seq.*
Inflammation, acute, 125, *et seq.*, 144-145
cardinal signs, 127, 130
causes, 126
microscopic features of, 127-133
outcome, 145
reaction to, 139
Inflammation, chronic, 144, 147 *et seq.*
effects, 153
granulomatous, 150-152,
microscopic features, 147-148
non-specific, 149-150
Inflammation, classification, 141 *et seq.*
Inflammatory exudate, 142-143
Injury, 5
Insulin, 303
Intestine, infarct, 124
Intracellular fluid, 65,
Iron,
deficiency, 326
excess, 54-55
Perl's stain for, 54
Irradiation, 295 *et seq.*
causes of death, 300
effects, 296, 298-299
Ischaemia, 20, 115 *et seq.*
causes, 116-117
definition, 115

Jaundice, 57 *et seq.*
classification, 59
definition, 59
genetic causes, 64
haemolytic, 61

Jaundice, 57 *et seq.*
hepatocellular, 63
obstructive, 62

Kaposi's sarcoma, 268-269
Karyolysis, 28
Karyorrhexis, 28
Keloid, 163
Keratosis, 248, 250
Kernicterus, 61
Kidney,
amyloid, 42
calcification, 46
diabetic, 308
glycogen infiltration, 26
hyperplasia,
infarct, 123
lupus erythematosus, 339
multiple myeloma, 278
nephroblastoma, 284-285
radiation change, 299
shock (acute tubular necrosis), 97-98
stones, 49
tuberculosis, 182
Kinin system, 134
Koch's postulates, 167
Kwashiorkor, 323-324

Langhans giant cell, 16, 153
Lactic dehydrogenase (LDH), 27, 121
Leiomyoma, 210, 264
Leiomyosarcoma, 265
Leprosy, 185-187
lepromin test, 187
Letterer-Siwe disease, 273
Leucocytosis, 139
Leukaemia, 211, 271, 275, 288
Ligands, 12
Lipofuscin, 53, 195
Lipoma, 210, 263
Liposarcoma, 263
Lithopaedion, 45
Liver,
amyloid, 42
cirrhosis, 157-158
congestion (nutmeg), 86
cystic fibrosis, 316
fatty change, 22-23
glycogen infiltration, 26
healing, 157
hepatoblastoma, 287
regeneration, 157
Low density lipoproteins, (LDL), 12, 313-314
Lungs,
brown induration, 86
congestion, 85
cystic fibrosis, 316
gangrene, 34
haemosiderin deposition 85
healing, 156
infarct, 123
oedema, 79
radiation change, 299
repair, 156
resolution, 156
shock, 97
Lupus erythematosus, 338-340
Lymphangiosarcoma, 282
Lymphatic,
emboli, tumour, 227-28
permeation, 228-229
Lymphoid tissue, radiation change, 299
Lymph nodes,
tumour spread, 227-229
Lymphocytes, 138
Lymphoma, 271 *et seq.*, 288
Burkitt's, 274
Lysosome, 13
Lysosomal storage diseases,
Gaucher's disease, 318
Niemann-Pick disease, 317
Tay-Sachs disease, 317
Mucopolysaccharidoses, 319

Macrophages, 137, 147,
Malabsorption in cystic fibrosis, 316
Malignancy,
causes of, 235 *et seq.*
characteristics, 215 *et seq.*
spread of, 225 *et seq.*
Malnutrition, 196, 321, 323
Mantoux test, 178, 182
Marasmus, 324-325

Mast cells, 136
Measles, 168-169
Medulloblastoma, 287
Melanin, 51
Melanoma, malignant, 256-259, 282
  acral lentiginous, 258
  lentigo maligna, 256-257
  nodular, 258
  superficial spreading, 257
Melanosis coli, 53
Mesothelial cells, 7
Mesothelioma, malignant, 267
Metabolic diseases, 5, 313 *et seq.*
Metaplasia, 195, 201-202
Metastasis, 226-234
Microbiology, 1
Mitochondria, 13, 19
Mitsuda reaction, 187
Monocytes, 137
Mononuclear phagocytes, 137
Morbid anatomy, 1
Mucopolysaccharidoses, 319
Multiple myeloma, 211, 277-278
Muscle,
  glycogen infiltration, 26
  tumours, 210, 264-266
Mycobacterium,
  atypical, 183
  leprae, 185
  tuberculosis, 175-176
Mycology, 1
Myeloma, multiple, 211, 277-278
Myocardial infarction, 121-122
Myopathy, 291
Myxomatous tumours, 211, 263

Naevi, 254-255, 282
Necrosis, 20, 27 *et seq.*
  caseous, 29-30, 151, 177-178
  coagulative, 29, 119, 121, 123, 124
  colliquative, 29, 122
  fat, 30
  fibrinoid, 30, 298, 337-338
  gangrenous, 29, 33-35
  gummatous, 30, 152, 188-189
  suppurative, 29, 143
  types of, 29-30
Neoplasia, 195, 198, 203
Neoplasm, definition, 205
Neoplastic disease, 5
Nephroblastoma, 284-285
Nephrotic syndrome, 308, 339
Nerve sheath tumours, 262-263
Neurilemmoma, 211, 262
Neuroblastoma, 283-284
Neurofibroma, 262
Neurofibrosarcoma, 263
Neuropathy, diabetic, 309
  malignant 291
Neutrophils, 135-136
Neutrophil granules, 132
Nicotinamide, 328
Niemann-Pick disease, 317-318
Normal cell, 9, 18
Nucleus, 9
Nucleolar organising region (NOR), 10
Nutmeg liver, 86
Nutritional disorders, 5, 321 *et seq.*

Obesity, 321-322
  complications, 322
Ochronosis, 53
Oedema, 75 *et seq.*
  mechanisms, 76
  cerebral, 79
  pulmonary, 79-80, 82
Opsonins, 132
Organisation, 158
Organisms,
  commensal, 165
  parasitic, 165
  pathogenic, 165
  saprophytic, 165
  symbiotic, 165
Osteoma, 210
Osteomalacia, 327
Osteomyelitis, 149
Ovary, teratoma, 213

Pancreas,
  in diabetes, 306-307
  fibrocystic disease of, 316
Papilloma, 210
  basal cell, 252
  simple squamous, 249

Paraneoplastic syndromes, 291-292
Parasitology, 1
Parathyroid,
adenoma, 47
hyperplasia, 47
Pathogenesis, definition 4
Pathology, definition, 1
Pasteur, 6
Patient, 1,3
Pellagra, 328
Perl's stain for iron, 54
Petechiae, 89
Phagocytosis, 12, 132
Phlebolith, 45
Phlebothrombosis, 106
Pigmentation, 51 *et seq.*
Pigments,
endogenous, 51
exogenous, 51
Pinocytosis, 12
Plasma cells, 138
dyscrasias, 271, 276-279
Plasmacytoma, 278
Platelet activating factor, 135
Platelets, in thrombus formation, 103-105
Pneumonia,
bronchopneumonia, 156
lobar pneumonia, 156, 171-172
tuberculous, 179
Polycythaemia, 291
Polymorphonuclear leucocytes, 135-136
Polymyositis, 344
Post mortem examination, 1
changes, 31-33
Potassium,
deficiency, 71
excess, 71
Prognosis, definition, 4
of tumours, 220-222
Prostate, hyperplasia, 198
Psychiatric disease, 6
Pulmonary embolism, 111
Pulmonary infarction, 123
Purpura, 89
Pus, definition, 143
Putrefaction, 33
Pyaemia, 108
Pyknosis, 28

Radiation, 295 *et seq.*
cell sensitivity, 297-298
effects on tissues, 298-299
ionising, effects, 296
measurement, 297
whole body irradiation, 299-300
Radiotherapy, 300-301
Reed-Sternberg cell, 17, 271-272
Regeneration, 153, 157, 198
Rejection, transplant, 329-332
acute, 330
chronic, 331
hyperacute, 329
Renin-angiotensin-aldosterone mechanism, 66
Renal failure, 49
Renal tubular acidosis, 49
Repair, 155 *et seq.*
Resolution, 155
Retinoblastoma, 286
Rhabdomyoma, 210, 265
Rhabdomyosarcoma, 265-266
alveolar, 266
embryonal, 265, 287
pleomorphic, 266
Rheumatic fever, 337
Rheumatoid arthritis, 341-342
Riboflavine, 328
Rickets, 327
Rickettsial disease, 170
Rotor syndrome, 64
Russell bodies, 25

Saprophytic organisms, 165
Sarcoidosis, 193-194
Sarcoma, 211, 259
Scar formation, 159
Schaumann body, 193
Scleroderma, 342-343
Scurvy, 328
Septicaemia, 108
Sequestrum, 149
Shock, 93 *et seq.*
causes of death in, 100
classification, 94
definition, 93
irreversible, 96

mediators, 96
Shock, 93 *et seq.*
pathogenesis, 97, 101
Siderosis, African, 68
transfusion and haemolytic, 68
Sjögren's syndrome, 342
Sinus, 149
Skin,
basal cell carcinoma, 253
basal cell papilloma, 252
malignant melanoma, 256-258
naevi, 254-255
premalignant lesions, 250
radiation changes, 299
simple squamous papilloma, 249
staphylococcal infection, 171
tumours, 247 *et seq.*
viral wart, 249
Smooth muscle tumours, 264-265
Sodium and water deficiency, 68, 74
Sodium and water excess, 70, 74
Spleen,
amyloid, 41-42
congestion, 87
infarct, 124
Starvation, 290, 323
Steatorrhoea, 316
Strangulated hernia, 124
Striated muscle tumours, 294
Stromal fatty infiltration, 25
Sulphur granules, 190
Surgical pathology, 1
Symbiotic organisms, 165
Synovial sarcoma, 267
Synovioma, benign, 266
Syphilis, 187-190
Systemic lupus erythematosus, 338-340
Systemic sclerosis, 342-343

Tay-Sachs disease, 317
Teratoma, 213-214, 282-283
Thiamine, deficiency, 327
Thrombophlebitis, 106, 108
Thrombosis, 103 *et seq.*
Thrombus,
ante-mortem, 33
definition, 103
formation 104-107
pale, and red, 105
Thyroid hyperplasia, 198
Tick-bite fever, 170
Tolerance, 333
loss of, 333-334
Touton giant cell, 16, 153
Transforming growth factor (TGF), 201
Transplant rejection, 329-332
Transudate, 76-77
Tubercle, 151-152
Tuberculosis, 29-30, 175 *et seq.*
adult/secondary form, 180-181
childhood/primary form, 178-180
HIV and, 182
isolated organ tuberculosis, 181-182
spread, 179-181
treatment and prevention, 182
Tumour,
definition, 205
emboli, 114
suppressor genes, 242-243
Tumours,
benign, 215-216
causes, 235 *et seq.*
classification, 205 *et seq.*
connective tissue, 259 *et seq.*
effects, 289 *et seq.*
embryonal and childhood, 281 *et seq.*
epithelial, 247 *et seq.*
malignant, characteristics, 215 *et seq.*
mixed, 212
spread, 225 *et seq.*
Typhoid 172-173
Tyrosine, 52

Vascular permeability, 127
Vasculitis, 331
Vasoactive amines, 133
Venous thrombosis, 106
Virchow, 6
Virchow's triad, 105
Virology, 1
Viral disease, 168-169
oncogenic viruses, 241
Vital reaction, 31

Vitamin deficiencies, 326-328
A deficiency, 326
B complex deficiency, 327-328
B_{12} deficiency, 326
C deficiency, 328
D deficiency, 327
folic acid deficiency, 326
K deficiency, 327
Volvulus, 34, 124

Waldenstrom's macroglobulinaemia, 278
Warthin-Finkeldey giant cell, 17, 168-169
Water, body, 65
deficiency, 67, 74
excess, 69, 74
Wound healing, 160-161
complications, 163

Xanthomas, 309, 315
X-rays, 295

Zebra bodies 317, 319
Ziehl-Neelsen stain, 175
Zonal necrosis of liver, 63